# Topological Defects and the Non-Equilibrium Dynamics of Symmetry Breaking Phase Transitions

# NATO Science Series

*A Series presenting the results of activities sponsored by the NATO Science Committee. The Series is published by IOS Press and Kluwer Academic Publishers, in conjunction with the NATO Scientific Affairs Division.*

| | |
|---|---|
| **A. Life Sciences** | IOS Press |
| **B. Physics** | Kluwer Academic Publishers |
| **C. Mathematical and Physical Sciences** | Kluwer Academic Publishers |
| **D. Behavioural and Social Sciences** | Kluwer Academic Publishers |
| **E. Applied Sciences** | Kluwer Academic Publishers |
| **F. Computer and Systems Sciences** | IOS Press |
| **1. Disarmament Technologies** | Kluwer Academic Publishers |
| **2. Environmental Security** | Kluwer Academic Publishers |
| **3. High Technology** | Kluwer Academic Publishers |
| **4. Science and Technology Policy** | IOS Press |
| **5. Computer Networking** | IOS Press |

**NATO-PCO-DATA BASE**

The NATO Science Series continues the series of books published formerly in the NATO ASI Series. An electronic index to the NATO ASI Series provides full bibliographical references (with keywords and/or abstracts) to more than 50000 contributions from international scientists published in all sections of the NATO ASI Series.
Access to the NATO-PCO-DATA BASE is possible via CD-ROM "NATO-PCO-DATA BASE" with user-friendly retrieval software in English, French and German (WTV GmbH and DATAWARE Technologies Inc. 1989).

The CD-ROM of the NATO ASI Series can be ordered from: PCO, Overijse, Belgium

**Series C: Mathematical and Physical Sciences – Vol. 549**

# Topological Defects and the Non-Equilibrium Dynamics of Symmetry Breaking Phase Transitions

edited by

Yuriy M. Bunkov

and

Henri Godfrin

Centre de Recherches sur les
Très Basses Température,
C.N.R.S. Grenoble, France

**Kluwer Academic Publishers**

Dordrecht / Boston / London

Published in cooperation with NATO Scientific Affairs Division

Proceedings of the NATO Advanced Study Institute on
Topological Defects and the Non-Equilibrium Dynamics of Symmetry Breaking Phase Transitions
Les Houches, France
16-26 February 1999

A C.I.P. Catalogue record for this book is available from the Library of Congress.

ISBN 0-7923-6204-7

---

Published by Kluwer Academic Publishers,
P.O. Box 17, 3300 AA Dordrecht, The Netherlands.

Sold and distributed in North, Central and South America
by Kluwer Academic Publishers,
101 Philip Drive, Norwell, MA 02061, U.S.A.

In all other countries, sold and distributed
by Kluwer Academic Publishers,
P.O. Box 322, 3300 AH Dordrecht, The Netherlands.

*Printed on acid-free paper*

---



Printed in the Netherlands.

Table of contents

## PREFACE

Topological defects formed at symmetry-breaking phase transitions play an important role in many different fields of physics. They appear in many condensed-matter systems at low temperature; examples include vortices in superfluid helium-4, a rich variety of defects in helium-3, quantized magnetic flux tubes in type-II superconductors, and disclination lines and other defects in liquid crystals. In cosmology, unified gauge theories of particle interactions suggest a sequence of phase transitions in the very early universe some of which may lead to defect formation. In astrophysics, defects play an important role in the dynamics of neutron stars.

In 1997 the European Science Foundation started the scientific network "Topological defects" headed by Tom Kibble. This network has provided us with a unique opportunity of establishing a collaboration between the representatives of these very different branches of modern physics.

The NATO-ASI (Advanced Study Institute), held in Les Houches in February 1999 thanks to the support of the Scientific Division of NATO, the European Science Foundation and the CNRS, represents a key event of this ESF network. It brought together participants from widely different fields, with diverse expertise and vocabulary, fostering the exchange of ideas. The lectures given by particle physicists, cosmologists and condensed matter physicists are the result of the fruitful collaborations established since 1997 between groups in several European countries and in the U.S.A.

The scientific goals of the network, to improve the transfer of information and expertise within the community and to enable efficient dissemination of results, are reflected in the contents of this book. It opens with a Condensed Matter – High Energy & Cosmology dictionary prepared by Grigory Volovik, which places on a common basis ideas pertaining to these branches of modern physics. The following lectures, given by experts in these fields, provide the fundamental concepts of symmetry-breaking phase transitions discussed during the ASI. We hope that the reader will share the excitement that we, the theorists and experimentalists participating in the ESF Network, felt during this journey through this interdisciplinary advanced topic of modern physics.

Yuriy M. Bunkov<br>Henri Godfrin

# DICTIONARY

*Condensed matter - High energy physics & cosmology*

G.E. VOLOVIK
*Low Temperature Laboratory, Helsinki University of Technology, Box 2200, FIN-02015 HUT, Espoo, Finland*
*and*
*Landau Institute for Theoretical Physics, Moscow, Russia*

There are two main levels of analogies: (i) analogy in terms of the symmetry breaking scheme and (ii) analogy in terms of the fermionic spectrum and effective gauge and gravitational fields arising in the low-energy corner. These two levels are indicated, when necessary.

## 1. Fermionic vacuum state, fermionic excitations

| | | |
|---|---|---|
| - Superfluid ground state (no excitations) | | Quantum physical vacuum (no elementary particles) |
| - Normal component of superfluid liquid (collection of excitations) | | Matter (collection of elementary particles) |
| - Bogoliubov quasiparticles (fermionic excitations) in $^3$He-A | level (ii) | Elementary chiral fermions (quarks and leptons) in Standard Model |
| - Bogoliubov-Nambu Hamiltonian for fermions in $^3$He-B | | Dirac Hamiltonian for massive fermions |
| Bogoliubov-Nambu Hamiltonian for fermions in $^3$He-A | level (ii) | Weyl Hamiltonian for massless chiral fermions |
| - Fermi point (point gap node) in the energy spectrum of $^3$He-A at $\mathbf{p} = p_F\hat{\mathbf{l}}$ | level (ii) | Point $\mathbf{p} = 0$ in momentum space, where the energy of the chiral fermion is zero |
| - Fermi velocity $v_F$ | level (ii) | Speed of light along $\hat{\mathbf{l}}$ |
| - Pair breaking velocity $\Delta_0/p_F$ | level (ii) | Speed of light |
| - Amplitude of the gap $\Delta_0$ | level (ii) | Planck energy $(\hbar c^5/G)^{1/2}$ |



*Y.M. Bunkov and H. Godfrin (eds.),*
*Topological Defects and the Non-Equilibrium Dynamics of Symmetry Breaking Phase Transitions,* 1–6.

| | | |
|---|---|---|
| - Broken symmetry of ground state | | Broken symmetry of the vacuum |
| - Broken symmetry $SO(3)_L \times U(1)_N \rightarrow U(1)_{L_3-N/2}$ in $^3$He-A | level (i) | Broken symmetry $SU(2)_T \times U(1)_Y \rightarrow U(1)_{T_3+Y/2}$ in electroweak interactions |
| - Number of particles $N$ | level (i) | Hypercharge $Y$ |
| - Group of orbital rotations $SO(3)_L$ | level (i) | Group of isotopic rotations $SU(2)_T$ |
| - Generalized angular momentum $L_3 - N/2$ | level (i) | Electric charge $Q = T_3 + Y/2$ |
| - Broken relative spin-orbit symmetry in $^3$He-B $SO(3)_L \times SO(3)_S \times U(1) \rightarrow SO(3)_{S+L}$ | level (i) | Broken relative color-flavour symmetry $SU(3)_f \times SU(3)_c \rightarrow SU(3)$ in color superconductivity of quark matter; broken chiral symmetry in strong interactions |
| - $\hat{\mathbf{l}}$-vector | level (i) | Direction of isospin in electroweak interactions |

## 2. Collective modes and fields

| | | |
|---|---|---|
| - Bosonic collective modes (sound, spin waves, orbital waves, other modes of the order parameter, etc.) | | Gauge fields, Higgs field, gravitational field |
| - Quanta of collective modes (phonons, etc.) | | Photons, gravitons, Higgs bosons, etc |
| - Orbital momentum unit vector $\hat{\mathbf{l}}$ | level (ii) | Vector potential $\mathbf{A}$ of electromagnetic field |
| - $\hat{\mathbf{l}}$-texture with $\vec{\nabla} \times \hat{\mathbf{l}} \neq 0$ | level (ii) | Magnetic field $\mathbf{B} = \vec{\nabla} \times \hat{\mathbf{A}}$ |
| - $\partial_t \hat{\mathbf{l}}$ | level (ii) | Electric field $\mathbf{E}$ |
| - Orbital dynamics | level (ii) | Maxwell equations |
| - Orbital waves | level (ii) | Electromagnetic waves |
| - Quanta of orbital waves | level (ii) | Photons |
| - Spin space anisotropy axis $\hat{\mathbf{d}}$-vector in $^3$He-A | level (ii) | Direction of isospin in electroweak interactions |
| - Spin waves in $^3$He-A | | Goldstone bosons |
| - Spin-orbital waves in $^3$He-A | level (ii) | W-bosons |

| | | |
|---|---|---|
| - Collective clapping modes in $^3$He-A | level (ii) | Gravitons |
| - $N$-fold degeneracy of Fermi point | level (ii) | $SU(N)$ gauge field |
| - Orbital part of the order parameter $\mathbf{e}_1 + i\mathbf{e}_2$ | level (ii) | Tetrad (vierbein) |
| - $\hat{\mathbf{l}}$-texture and superflow | level (ii) | Gravitational field |
| - Superfluid velocity $\mathbf{v}_s$ | level (ii) | Spin connection |
| - Superfluid velocity $\mathbf{v}_s$ | level (ii) | Gravimagnetic potential $g_{0i} = v_{si}$ |
| - Vorticity $\vec{\nabla} \times \mathbf{v}_s$ | level (ii) | Gravimagnetic field |
| - Kinetic energy of superflow $v_s^2/2$ | level (ii) | Gravitational potential $\Phi = -v_s^2/2$, $g_{00} = c^2 + 2\Phi$ |
| - Counterflow of superfluid and normal components $\mathbf{v}_n - \mathbf{v}_s$ | level (ii) | Excess of the righthanded fermions $\mu_R - \mu_L = p_F \hat{\mathbf{l}} \cdot (\mathbf{v}_n - \mathbf{v}_s)$ |
| - Relation between energies of fermions and bosons | | Supersymmetry |

## 3. Topological defects

| | | |
|---|---|---|
| - Abrikosov vortex in superconductors | level (i) | Local cosmic string |
| - Vortex in superfluids | level (i) | Global cosmic string |
| - $N = 1$ vortex in $^3$He-A | level (ii) | Torsion string or tetrad string |
| - $N = 1/2$ vortex in $^3$He-A | level (i) | Alice string |
| - $N = 2$ singular vortex in $^3$He-A | level (i) | Electroweak $Z$-string |
| - $N = 2$ continuous vortex (Anderson-Toulouse-Chechetkin vortex) in $^3$He-A | level (i) | Stringy texture |
| - $N = 2$ continuous vortex with broken parity | level (i) | Stringy texture with persistent current |
| - $N = 1$ continuous Mermin-Ho vortex in $^3$He-A | level (i) | Meron |
| - Mass-flow vortex in $^3$He-A | level (ii) | Spinning cosmic string |

| | | |
|---|---|---|
| - Mass-flow vortex in $^3$He-B | level (i) | $\eta$'-string in chiral symmetry breaking in strong interactions |
| - Spin-flow vortex in $^3$He-B | level (i) | Pion string |
| - Disgyration in $^3$He-A | level (i) | Electroweak $W$-string |
| - Radial disgyration in $^3$He-A | level (ii) | Antigravitating local cosmic string |
| - $\hat{\mathbf{l}}$-hedgehogue as termination point of $N=2$ singular vortex in $^3$He-A | level (i) | Dirac monopole as terminating point of physical $Z$-string |
| - $\hat{\mathbf{l}}$-hedgehogue as termination point of $N=2$ singular vortex in $^3$He-A | level (ii) | Gravimagnetic monopole |
| - $\hat{\mathbf{d}}$-vector hedgehog in $^3$He-A | level (ii) | t'Hooft-Polyakov monopole |
| - $N=2$ singular vortex in supercoductor with the $^3$He-A order parameter | level (i) | Semilocal string |
| - Mass-flow vortex in $^3$He-B with nonaxisymmetric core | level (i) | Witten's superconducting string |
| - Vortex ring | level (i) | Vorton |
| Soliton terminating on $N=1/2$ vortex | level (i) | Domain wall terminating on string |
| - Soliton terminating on spin-flow vortex in $^3$He-B | level (i) | Domain wall terminating on string |
| - $N=1$ $^3$He-B vortex terminating on the phase boundary between B phase and A-phase | level (i) | Dirichlet topological defect |
| - Parity violation in the core of $^3$He-B vortex | level (i) | Broken discrete symmetry in the string core |
| - Crossing point of $N=2$ continuous vortex and soliton in $^3$He-A, which has a nonzero Hopf invariant; loop of $N=2$ continuous vortex with nonzero Hopf invariant | level (i) | $\pi_3$ texture, skyrmion, hopfion |
| - Vortex sheet: vortices as kinks living within the soliton | | ??? |
| - Boojums | | ??? |

| | | |
|---|---|---|
| - Dynamical process of changing of the topological invariant | level (i) | Instanton |
| - Intermediate saddle point configuration | level (i) | Sphaleron |
| - Homogeneously precessing domain in $^3$He-B | level (i) | Q-ball |
| - Iordanskii force acting on vortex | level (ii) | Gravitational Aharonov-Bohm effect on spinning string |
| - Low-energy fermions localized in the vortex core | | Fermions zero modes on strings |
| - Kopnin force acting on a vortex due to spectral flow of fermion zero modes in the vortex core | | Baryogenesis by cosmic string |
| - Vortex core instability induced by fermion zero modes | | Instability of electroweak $Z$-string induced by fermion zero modes |
| - Interface between $^3$He-A and $^3$He-B | level (i) | Interface between $SU(3)_C \times SU(2)_T \times U(1)_Y$ and $SU(4) \times U(1)$ vacua in $SU(5)$ theory |
| - Moving interface | level (ii) | Perfect mirror moving in quantum vacuum |

## 4. Polarization of vacuum, vacuum instability, axial anomaly

| | | |
|---|---|---|
| - Logarithmically divergent textural energy in $^3$He-A $\ln\left(\frac{\Delta_0^2}{T^2}\right)\frac{p_F^2 v_F}{24\pi^2\hbar}\left((\hat{\mathbf{l}}\cdot\nabla)\hat{\mathbf{l}}\right)^2$ | level (ii) | Energy of magnetic field in curved space $\frac{\sqrt{-g}}{2\gamma^2}g^{ij}g^{kl}F_{ik}F_{jl}$ |
| - Logarithmically growing prefactor $\frac{1}{12\pi^2\hbar}\ln\left(\frac{\Delta_0^2}{T^2}\right)$ | level (ii) | Running coupling constant $\gamma^{-2}$ (zero charge effect) |
| - Nucleation of the linear momentum of quasiparticles by $\hat{\mathbf{l}}$-textures in $^3$He-A | level (ii) | Axial anomaly |
| - Anomalous current in $^3$He-A | level (ii) | Chern-Simons topological charge; baryonic charge of skyrmion |

| | | |
|---|---|---|
| - Linear momentum carried by texture | level (ii) | Fermionic charge of the vacuum |
| - Landau critical velocity $v_L$ | level (ii) | Speed of light $c$ |
| - Quasiparticles nucleation by supercritical superflow $v_s > c$ | level (ii) | Hawking radiation |
| - Superfluid in rotating container | level (ii) | Rotating vacuum |
| - Region with negative quasiparticle energy in the moving superfluids, $E(\mathbf{p}) + \mathbf{p} \cdot \mathbf{v}_s < 0$ | level (ii) | Ergoregion |
| - Radiation of quasiparticles by cylinder rotating in superfluids | level (ii) | Radiation by rotating black hole |
| - Instability of counterflow of superfluid and normal components towards $\hat{\mathbf{l}}$-texture | level (ii) | Instability of the excess of righthanded fermions towards production of the hypermagnetic field (magnetogenesis) |
| - Quasiparticle nucleation by time-dependent $\hat{\mathbf{l}}$-vector | level (ii) | Schwinger pair creation in electric field |
| - Quasiparticle nucleation by nonuniformly moving object in superfluids | | Unruh effect |
| - Gap in the spectrum of high frequency orbital waves | level (ii) | Photon mass in relativistic plasma |
| - Action for orbital dynamics | level (ii) | Wess-Zumino action |

# CLASSIFICATION OF TOPOLOGICAL DEFECTS AND THEIR RELEVANCE TO COSMOLOGY AND ELSEWHERE

T.W.B. KIBBLE
*Blackett Laboratory, Imperial College*
*London SW7 2BZ, United Kingdom*

These lectures will review symmetry-breaking phase transitions and the formation of topological defects, primarily in the context of cosmology but also with reference to condensed matter. The idea that early in its history the Universe went through a series of phase transitions will be discussed. Following a discussion of the basic ideas of spontaneous symmetry breaking, the classification of defects in terms of homotopy groups of the vacuum manifold will be reviewed, covering domain walls, cosmic strings or vortices, monopoles and textures and also composite objects of various kinds. The importance of the central problem of estimating the density of defects formed at a phase transition will be emphasized, with reference both to cosmology and to recent low-temperature experiments.

## 1. Introduction

The study of topological defects in the early Universe has long been a very interdisciplinary field, on the interface between particle physics and cosmology. But in the last few years, a new dimension has emerged: an interaction with the apparently very different field of condensed-matter physics.

One of the obvious problems of early-Universe physics is the impossibility of making experimental tests of our ideas. It is not possible to test directly the idea that the Universe underwent a series of high-temperature symmetry-breaking phase transitions at some of which topological defects may have been formed. But what we can do is to look for analogous processes in experimentally accessible condensed-matter systems. This enterprise has not only helped our understanding of the cosmology of the early Universe, but at the same time has thrown up some novel and exciting ideas about condensed-matter physics.

I shall start by recalling briefly our reasons for thinking that there were phase transitions in the immediate aftermath of the Big Bang. In Section 2,



*Y.M. Bunkov and H. Godfrin (eds.),*
*Topological Defects and the Non-Equilibrium Dynamics of Symmetry Breaking Phase Transitions*, 7–31.

I describe the basic ideas of symmetry breaking and defect formation. I then review the homotopy classification of individual defects — domain walls, strings or vortices, monopoles and textures — and in Section 3 I discuss the extension to the case of composite defects. Section 4 is devoted to a resume of the application of these ideas to helium-3, by far the most interesting from a topological point of view of the condensed matter systems. Then in Section 5 I consider what we can hope to learn about cosmological defects by looking at analogous condensed-matter systems, and briefly review the current status of the experiments and theory. Section 6 is devoted to brief comments on some open problems.

### 1.1. BIG BANG PHASE TRANSITIONS

The presently very successful standard model of particle physics is a gauge theory, based on the symmetry group SU(3)×SU(2)×U(1). The SU(3) factor is the colour symmetry group describing the strong interactions, while SU(2)×U(1) is the electro-weak symmetry, which is visible at energies well above 100 GeV (around the masses of the W and Z bosons), but hidden at lower energies by the process usually described as spontaneous symmetry breaking: only the U(1) of electromagnetism survives.

From the behaviour of the three gauge couplings, $g_3, g_2$ and $g_1$, as functions of energy [1, 2, 3], there is good reason to believe that even more symmetry, described by a Grand Unified Theory (GUT), emerges at extreme energies, above $10^{16}$ GeV or so. At the Planck scale, around $10^{19}$ GeV, even gravity becomes strong, and perhaps superstring theory or $M$-theory takes over. However these ideas are as yet untestable.

Symmetry breaking often signals a phase transition, and there may indeed be an electro-weak phase transition at a temperature of the order of 100 GeV. There may also be a transition at lower energy, around 100 MeV, the 'quark-hadron transition' at which the initial dense soup of quarks and gluons condenses into individual hadrons, which may be associated with the breaking of the approximate chiral symmetry (which would be exact if the quarks were massless).

But even these 'low-energy' transitions are still very poorly understood. Indeed, it is far from certain that either of these is actually a true phase transition. Evidence from lattice simulations suggests that for realistic values of the Higgs mass ( $\gtrsim$ 80 GeV), the electro-weak theory does not exhibit a true phase transition, only a smooth 'crossover' [4]. Nor is it yet sure whether the quark-hadron transition really exists (though in that case experimental evidence may help to resolve the issue in the fairly near future).

At first sight the possibility that there are no true phase transitions may be surprising. As usually envisaged, the breaking of a symmetry im-

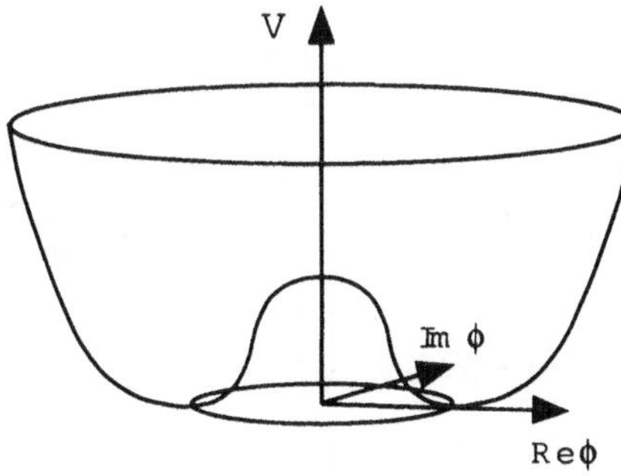


*Figure 1.* The 'Mexican hat' potential.

plies a sudden change from a symmetric state to an ordered state, in which some symmetry-breaking order parameter acquires a non-zero value. (I discuss this process in more detail in the following section.) However, strictly speaking a *local gauge* symmetry can *never* be broken because we would want an observable to serve as an order parameter, but all such are necessarily gauge-invariant; this is the content of *Elitzur's theorem* [5, 6, 7]. It is nevertheless often useful to talk in terms of a symmetry-breaking order parameter. This is done routinely in superconductivity, and is quite legitimate if one introduces a gauge-fixing term that breaks the symmetry explicitly, for example by choosing the Coulomb gauge. However, this possibility does not always guarantee that a phase transition must exist. The precise conditions for the existence of a phase transition in these circumstances are not fully known. This is an important issue, but one I shall not pursue further in these lectures.

## 2. Spontaneous Symmetry Breaking and Topological Defects

### 2.1. U(1) COSMIC STRINGS

To be specific let us discuss linear defects — cosmic strings or vortices (for reviews, see [8, 9]). The simplest model in which such defects appear involves a single complex field $\phi$, with potential

$$\mathcal{V} = \tfrac{1}{2}\lambda(\phi^*\phi - \tfrac{1}{2}\eta^2)^2 \qquad (1)$$

(see Fig. 1). The symmetry here is a U(1) phase symmetry. The minima of $\mathcal{V}$ lie on a circle, and the degenerate vacua are labelled by a phase angle $\alpha$:

$$\langle\phi\rangle = e^{i\alpha}\eta/\sqrt{2}. \qquad (2)$$

There is a phase transition in this model at a critical temperature $T_c \sim \eta$ (in units with $c = \hbar = k_B = 1$). For $T > T_c$, there are large fluctutations in

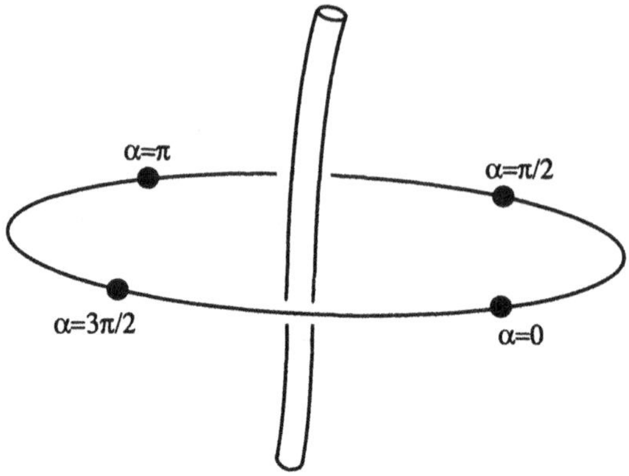


*Figure 2.* Cosmic string, around which the phase $\alpha$ varies by $2\pi$, with $\phi = 0$ in the core.

the field, but on average $\langle\phi\rangle = 0$; below $T_c$, $\phi$ settles down towards some point in the valley of the potential, and in doing so has to choose a phase. In a large system such as the Universe, the phase will be chosen differently in different regions of space. (In absolute terms the Universe at the relevant time is quite small, but the relevant thing is that the time scale for the transition is fast compared to the expansion time.)

It may happen (Fig. 2) that around some large loop in space, the chosen phase $\alpha$ varies from 0 to $2\pi$. Then, to satisfy the requirement of continuity, somewhere within the loop the field $\phi$ must vanish. In fact, it must do so all along a curve that threads its way through the loop — the core of the cosmic string.

In the core, the field has to go over the central hump in the potential. When the string is formed, the energy density everywhere is large, of order ${T_c}^4$. In most regions it decreases rapidly as the Universe expands, but on the string the energy is trapped and cannot dissipate. If the U(1) is a *global* symmetry, the energy per unit length of an isolated straight string is actually infinite, because the angular part of the gradient energy falls off only like $1/r^2$ with distance from the string. Nevertheless the total energy per unit volume of a system of strings will be finite, because the typical inter-string distance $\xi$ provides a natural cutoff. The energy per unit length is of order $\eta^2 \ln(\eta\xi)$.

If the U(1) is a *local* symmetry, then by introducing a non-trivial azimuthal gauge field $A_\theta$ one can make the covariant derivative

$$D_\mu\phi = \partial_\mu\phi + iqA_\mu\phi \tag{3}$$

fall off exponentially. In that case the energy per unit length is of order $\eta^2$, and the strings carry a magnetic flux, quantized in units of $2\pi/q$.

It is important to note that the string is stable for topological reasons. It can of course move, but it cannot break. Strings cannot have ends; they are either infinite or form closed loops. A closed loop may shrink and eventually vanish, but a small piece of string cannot just disappear.

Actually this statement requires a caveat: it is theoretically possible for a string to terminate on a black hole. In principle, a string can break with the formation of a pair of black holes [10, 11, 12]. However, this is an extremely improbable tunnelling process.

The gravitational effects of a string are governed by the dimensionless number $G_N\eta^2 \sim (T_c/M_{Pl})^2$, with $M_{Pl} = 1.22 \times 10^{19}$ GeV. For GUT scale strings, this number is of order $10^{-6}$ or $10^{-7}$, a cosmologically interesting range: such strings could have significant effects on the large-scale density distribution and the anisotropy of the cosmic microwave background. They have provided an alternative to the popular inflationary mechanism for formation of density perturbations. In a critical-density Universe this theory is in some difficulty [13], but in a low-density Universe or one dominated by a cosmological constant, the idea seems to work better [14]. Theories with both inflation and defects are in some ways particularly attractive [15].

## 2.2. HOMOTOPY CLASSIFICATION OF STRINGS — THE FUNDAMENTAL GROUP

It is useful to have more general criteria for deciding which theories exhibit defects. This is provided by the homotopy groups of the vacuum manifold [16]. In particular strings are classified by the *fundamental group* $\pi_1(\mathcal{M})$. The fact that there are strings in the theory of a complex scalar field depends on the existence of non-trivial loops in the vacuum manifold (a circle).

Suppose we have a theory with symmetry group $G$, and that $\phi_0$ is a point on the surface $\mathcal{M}$ of minima of $\mathcal{V}$. Then all the points $\phi$ obtained from $\phi_0$ by the action of elements of $G$, $\phi = g\phi_0$, with $g \in G$, also lie on $\mathcal{M}$. But not all $g\phi_0$ are distinct, because there is a subgroup of elements that leave $\phi_0$ unchanged, the isotropy subgroup $H \subset G$:

$$H = \{g \in G : g\phi_0 = \phi_0\}. \tag{4}$$

Then the points of the vacuum manifold $\mathcal{M}$ are in one-to-one correspondence with the left cosets $gH$ of $H$ in $G$. In other words, $\mathcal{M}$ is homeomorphic to the quotient space $G/H$: I shall write this as a simple equality: $\mathcal{M} = G/H$.

The *fundamental group* $\pi_1(\mathcal{M})$ is defined in terms of closed loops in $\mathcal{M}$, continuous curves starting and ending at $\phi_0$. Two such curves are *homotopic* if one can be continuously deformed into the other. The elements of $\pi_1(\mathcal{M})$ are homotopy classes of such loops. The product $ba$ of two loops $a$ and $b$

is defined as the loop obtained by following $a$ and then $b$. This defines a product of homotopy classes, so that $\pi_1(\mathcal{M})$ becomes a group. The identity is the class containing the trivial loop $t \mapsto \phi(t) = \phi_0$.

For example, for $G = \mathrm{U}(1)$, the isotropy group is $H = \mathbf{1}$, the group consisting of the identity element alone, so the vacuum manifold is a circle: $\mathcal{M} = G/H = S^1$. Here the different homotopy classes of loops are distinguished by the number of times they wind around the circle. The fundamental group is simply the additive group of integers, $\pi_1(S^1) = \mathbf{Z}$.

The condition that the theory supports topologically stable cosmic strings is that the fundamental group be non-trivial: $\pi_1(\mathcal{M}) \neq \mathbf{1}$. Topologically distinct strings are labelled by the elements of this group, for example by the winding number in the U(1) case. For a typical string of winding number $n$, the scalar field asymptotically takes the form

$$\lim_{r\to\infty} \phi(r, \theta, z) = \phi_0 e^{ni\theta}, \qquad \phi_0 = \eta/\sqrt{2}. \tag{5}$$

For $|n| > 1$, these strings may or not be stable. For the Type-I case, where the scalar self-coupling is large, $\lambda > e^2$, they are stable for all $n$, but for Type-II, where $\lambda < e^2$, it is energetically favourable for strings with $|n| > 1$ to break up into $|n|$ strings of winding number $\pm 1$.

As another example, consider an SO(3) theory with two scalar fields in the vector representation, with a potential of the form

$$\mathcal{V} = \tfrac{1}{8}\alpha(\vec{\phi}^2 - \eta^2)^2 + \tfrac{1}{8}\beta(\vec{\psi}^2 - \sigma^2)^2 + \tfrac{1}{4}\gamma(\vec{\phi}\cdot\vec{\psi})^2, \tag{6}$$

where $\alpha, \beta$ and $\gamma$ are positive constants, as are $\eta$ and $\sigma$. The manifold on which $\mathcal{V}$ takes its minimum is then given by

$$\vec{\phi}^2 = \eta^2, \qquad \vec{\psi}^2 = \sigma^2, \qquad \vec{\phi}\cdot\vec{\psi} = 0. \tag{7}$$

Here $H = \mathbf{1}$, so the vacuum manifold is a *projective* three-sphere: $\mathcal{M} = \mathrm{SO}(3) = \mathbf{RP}^3$. In this case, $\pi_1(\mathcal{M}) = \mathbf{Z}_2$, the group with two elements. The homotopically non-trivial loops are those starting at the identity and ending at a rotation through $2\pi$.

In this example, there is only one type of string. Strings of winding number 2 are not topologically stable and can disappear, because loops in SO(3) extending to a $4\pi$-rotation are continuously deformable to the trivial loop.

## 2.3. OTHER DEFECTS

Strings are one-dimensional, but there are also defects of other dimensions, two-dimensional domain walls, zero-dimensional monopoles and (in a sense!) (−1)-dimensional textures.

To determine whether or not the theory supports topologically stable monopoles, we have to look at the second homotopy group, $\pi_2(\mathcal{M})$, whose elements correspond to homotopy classes of closed two-dimensional surfaces in $\mathcal{M}$.

A typical example is an SO(3) model with a single scalar field $\vec{\phi}$ in the vector representation, where we take

$$\mathcal{V} = \tfrac{1}{8}\lambda(\vec{\phi}^2 - \eta^2)^2. \tag{8}$$

The degenerate vacua here are labelled by two angles; the vacuum manifold is a two-sphere: $\mathcal{M} = \mathrm{SO}(3)/\mathrm{SO}(2) = S^2$. Here again, homotopy classes of closed surfaces in $\mathcal{M}$ are labelled by a winding number. In a typical monopole corresponding to winding number $n$ the scalar field has the asymptotic form

$$\lim_{r\to\infty} \phi(r, \theta, \varphi) = \eta \begin{pmatrix} \sin\theta\cos n\phi \\ \sin\theta\sin n\phi \\ \cos\theta \end{pmatrix}. \tag{9}$$

In general, topologically stable monopoles exist if and only if $\pi_2(\mathcal{M}) \neq \mathbf{1}$.

Domain walls are also classified by what is often called the 'zeroth homotopy group', though in fact it is not in general a group, and would be more accurately called the *zeroth homotopy set*, $\pi_0(\mathcal{M})$, whose elements are in one-to-one correspondence with distinct connected components of $\mathcal{M}$. For example, in the model of a real scalar field with a degenerate double minimum in $\mathcal{V}$, the symmetry group $G = \mathbf{Z}_2$ is broken to $H = \mathbf{1}$, and $\mathcal{M} = \mathbf{Z}_2$ consists of two points only, so $\pi_0(\mathcal{M})$ also has two elements.

In general, however, unless $\mathcal{M}$ itself is a group, there is usually no way to define a multiplication law on $\pi_0(\mathcal{M})$.

There are also topological defects classified by the *third* homotopy group $\pi_3(\mathcal{M})$, which in the cosmology literature are called *textures*, though in condensed matter that term is used with a different meaning. In space-time, domain walls trace out three-dimensional hypersurfaces; cosmic strings, two-dimensional world sheets; and monopoles, one-dimensional world lines. Textures correspond to zero-dimensional pointlike defects in space-time. Typically, there is a twist in the field which gradually collapses to a point, unwinds, and then expands once more. Though they have no permanence, textures might play a role in generating density perturbations in the early Universe, and have provided a possible alternative to strings [17, 18].

## 2.4. EXACT SEQUENCES

There are a number of mathematical results that make it easier to compute homotopy groups, particularly for *homogeneous spaces* which are defined as quotient spaces of Lie groups, of the form $\mathcal{M} = G/H$.

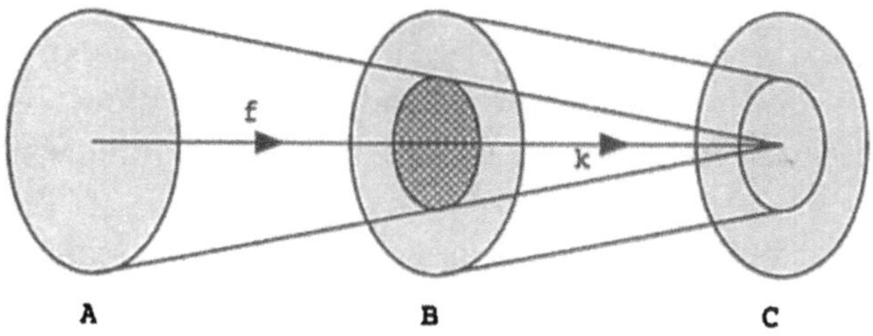


*Figure 3.* An exact sequence of maps; the cross-hatched region denotes the image of $f$ and the kernel of $k$.

The key to these results is the idea of an *exact sequence* [16]. If $f : A \to B$ and $k : B \to C$ are maps, the sequence

$$A \xrightarrow{f} B \xrightarrow{k} C \tag{10}$$

is said to be *exact at $B$* if the image of $f$, $\operatorname{im} f = fA$ coincides with the kernel of $k$, $\ker k = k^{-1}e$ (the cross-hatched region in Fig. 3), where $e$ is the identity element in $C$. A longer sequence of maps is *exact* if it is exact at every intermediate point.

Now since $H$ is a subgroup of $G$, there is a canonical injection $i : H \to G$ that identifies each $h \in H$ with the corresponding $h \in G$. Also there is a canonical projection $p : G \to G/H$ which maps each element $g \in G$ to its left coset $gH$. Moreover, the composite map $pi$ is obviously trivial: it maps every $h \in H$ into the null element of $G/H$, the coset comprising $H$ itself. Because of the existence of these two canonical maps, we can define for each $n$ corresponding maps $i_* : \pi_n(H) \to \pi_n(G)$ and $p_* : \pi_n(G) \to \pi_n(G/H)$. For example, any loop in $H$ can obviously be regarded as a loop in $G$, thus defining a map $i_* : \pi_1(H) \to \pi_1(G)$. Similarly, each loop in $G$ defines a loop in the quotient space $G/H$. A loop in $G$ maps into a trivial loop in $G/H$ if and only if it is homotopic to a loop lying entirely in $H$. Thus the sequence

$$\pi_n(H) \xrightarrow{i_*} \pi_n(G) \xrightarrow{p_*} \pi_n(G/H) \tag{11}$$

is exact.

But this sequence can be continued further. Consider a loop in $G/H$. We can lift it to a *curve* in $G$, a curve whose image under $\pi_*$ is the given loop. Since the loop returns to its starting point, the curve must end at an element of $H$. (See Fig. 4.) If this final element belongs to the same connected component as the identity $e \in H$, we can deform it continuously until the end-point is $e$ and the curve forms a loop in $G$. Thus each homotopy class of loops in $G/H$ defines a connected component of $H$, so there is a map $\partial$ from $\pi_1(G/H)$ to $\pi_0(H)$, and more generally a map $\partial : \pi_n(G/H) \to \pi_{n-1}(H)$ for all positive $n$.

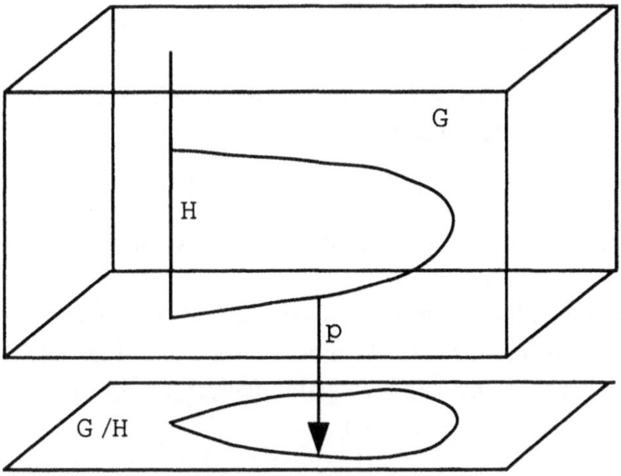


*Figure 4.* Lifting of a loop in $G/H$ to a curve in $G$.

The construction also shows clearly that

$$\pi_n(G) \xrightarrow{p*} \pi_n(G/H) \xrightarrow{\partial} \pi_{n-1}(H) \xrightarrow{i*} \pi_{n-1}(G) \tag{12}$$

is exact. By combining these results we can construct a long exact sequence [16], terminating with $\pi_0(G/H)$:

$$\begin{aligned} \cdots \xrightarrow{i*} \pi_2(G) \xrightarrow{p*} \pi_2(G/H) \xrightarrow{\partial} \pi_1(H) \xrightarrow{i*} \pi_1(G) \xrightarrow{p*} \\ \pi_1(G/H) \xrightarrow{\partial} \pi_0(H) \xrightarrow{i*} \pi_0(G) \xrightarrow{p*} \pi_0(G/H). \end{aligned} \tag{13}$$

This is particularly useful because of a well-known theorem to the effect that the second homotopy group of any Lie group is always trivial, $\pi_2(G) = \mathbf{1}$. Morever, if $\pi_1(G)$ is not trivial, which means that $G$ is not simply connected, we can always replace $G$ by its universal covering group $\tilde{G}$ (for example, replacing U(1) by **R** and SO(3) by SU(2)). If $\tilde{H}$ is the inverse image of $H$ under the projection of $\tilde{G}$ onto $G$, then $\tilde{G}/\tilde{H} = G/H$. Now if $\mathbf{1} \to A \to B \to \mathbf{1}$ is exact, it is easy to see that $A$ and $B$ are isomorphic. Hence we find the important isomorphism

$$\pi_2(G/H) = \pi_1(\tilde{H}). \tag{14}$$

So the condition for topologically stable monopoles to exist is that $\tilde{H}$ be non-simply connected. This is what makes monopoles generic in GUTs. The low-energy symmetry group $H$ must contain a U(1) factor, namely $\mathrm{U}(1)_{\mathrm{em}}$ of electromagnetism. If $G$ is a simple group, or product of simple groups, we then find that $\pi_1(\tilde{H})$ must contain a factor $\mathbf{Z}$; it cannot be trivial. The fact that it is hard to get rid of these monopoles is what leads to the *monopole problem*, one of the original reasons for the invention of the theory of inflation.

If we also restrict attention to the connected part $G_0$ of $G$, and define $H_0 = H \cap G_0$, then we similarly have

$$\pi_1(G/H) = \pi_0(\tilde{H}_0). \tag{15}$$

For example, when $G = \mathrm{U}(1)$ and $H = \mathbf{1}$, we have $\tilde{G}_0 = \tilde{G} = \mathbf{R}$ and $\tilde{H}_0 = \mathbf{Z}$, so $\pi_0(\tilde{H}_0) = \mathbf{Z}$.

Thus topologically stable strings exist if and only if $\tilde{H}_0$ has disconnected pieces. Note that in our $SO(3)$ model, $\tilde{G}_0 = \mathrm{SU}(2)$, and so $\tilde{H}_0 = \mathbf{Z}_2$, the subgroup comprising the two elements $\pm 1 \in \mathrm{SU}(2)$. There are strings because this space is disconnected.

## 3. Composite Defects

The defects so far considered by no means exhaust the possibilities. In particular, especially when there is more than one phase transition, we may have composite defects, domain walls bounded by strings, and strings terminated by monopoles.

### 3.1. EXAMPLES OF COMPOSITE DEFECTS

It is simplest to begin by thinking of a two-stage symmetry breaking — though composite defects are more generic than this. Suppose that at the first transition $\phi$ acquires a non-zero expectation value, breaking the symmetry from $G$ to $H \subset G$, while at the second another scalar field $\psi$ does so, breaking the symmetry further to $K \subset H$.

Let us choose $G$ to be connected and simply connected. But suppose that $H$ contains disconnected pieces, so that $\pi_1(G/H) = \pi_0(H)$ is non-trivial, and strings are formed. Now suppose that $\pi_0(H/K)$ is non-trivial. Then, looking only at what happens at the second transition, we should expect that domain walls will be formed. What happens to the strings at this transition? As we have seen, the strings are classified by the elements of $\pi_0(H)$: there is one type of string for each connected piece of $H$. However, some of these may become unstable. Since the overall symmetry breaking is from $G$ to $K$, the topologically stable strings are classified by $\pi_1(G/K) = \pi_0(K)$, *i.e.*, by the connected components of $K$.

Now for a particular string it may happen that the relevant component of $H$ contains *no* elements of the final symmetry group $K$. In that case, the string can no longer be topologically stable. In fact, what happens is that the string becomes the boundary of a domain wall.

A simple example is provided by a U(1) theory with two scalar fields [19, 20]. At the first symmetry breaking a scalar field $\phi$ with U(1) charge 2

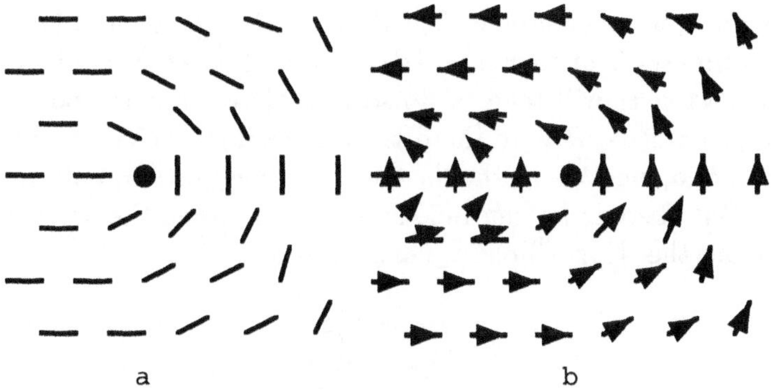


*Figure 5.* Effect of a two-stage symmetry breaking: (*a*) string formed when $\phi \neq 0$; (*b*) domain wall formed when also $\psi \neq 0$.

acquires a non-zero expectation value. This breaks the U(1) symmetry to $\mathbf{Z_2}$. Then at a second, lower-energy transition a second scalar field $\psi$ with unit charge becomes non-zero, breaking the remaining symmetry. If there is a coupling between the two, say $\lambda\mathrm{Re}(\phi^*\psi^2)$, then when $\psi$ becomes non-zero, it will try to match its phase to half that of $\phi$ — but this of course involves a choice between the two possibilities differing by $\pi$. Around a string, however, this choice is frustrated. At large distances we have $\phi = \eta e^{i\theta}$, so we should expect $\psi = \sigma e^{i\theta/2}$, but there is inevitably a discontinuity somewhere, say at $\theta = 0$, a surface across which $\psi$ changes sign. This is the domain wall.

This is illustrated in Fig. 5. In (*a*), half the phase of $\phi$ at each point is represented by the angle of the line; opposite directions here are equivalent. Then (*b*) shows the effect of the phase choice for $\psi$, which amounts to embellishing each line with an arrowhead. Clearly along one axis this is impossible, and we cannot keep to the phase matching between $\phi$ and $\psi$.

Now let us consider a slightly different case. Suppose that $H$ is non-simply connected, so that monopoles are formed at the first transition, labelled by the homotopy classes of closed loops in $H$, the elements of $\pi_1(H)$.

But we also suppose that $\pi_1(H/K)$ is non-trivial, so that at the second transition, we expect strings to be formed. As before, the second transition also affects the stability of the monopoles. In the final phase, the topologically stable monopoles are labelled by the elements of $\pi_1(K)$. If a non-trivial loop in $H$ can be deformed to lie entirely within $K$, then it will remain stable after the second transition. But if there are classes of loops in $H$ that cannot be deformed to lie in $K$, the corresponding monopoles must become unstable. In fact, what happens is that they become joined by strings [21].

As an example, take an SU(2) model with $\vec{\phi}$ in the triplet representation, and $\psi$ a spinor, with a coupling $\lambda\psi^\dagger(\vec{\phi}\cdot\vec{\sigma})\psi$ [22, 23]. When $\vec{\phi}$ becomes non-zero the symmetry is of course broken to U(1). If later $\psi$ acquires a vacuum expectation value it will tend to do so in a direction corresponding to the +1 eigenvector of $\vec{\phi}\cdot\vec{\sigma}$, but there is an arbitrary choice of phase factor. Around a monopole, $\vec{\phi}$ is in the 'hedgehog' configuration, pointing radially outwards. But there is no continuous way of choosing the arbitrary phase. Starting from the 'North Pole' we can choose

$$\psi = \begin{pmatrix} \cos\frac{1}{2}\theta \\ \sin\frac{1}{2}\theta e^{i\varphi} \end{pmatrix}, \tag{16}$$

but then around the South Pole the phase changes by $2\pi$. There must be a string emerging from the monopole at this point.

At first sight it would seem that there is another intriguing theoretical possibility, namely that textures could be formed at the first transition, and monopoles at the second, but with no truly stable monopoles in the final phase. Then the texture events would represent the creation or annihilation of monopoles. Unfortunately, the conditions for this to happen are that $\pi_3(G/H)$ and $\pi_2(H/K)$ are both non-trivial while $\pi_2(G/K) = \mathbf{1}$, and these can be shown to be inconsistent.

### 3.2. CLASSIFICATION OF COMPOSITE DEFECTS

There is a simple extension of the results on exact sequences quoted above that allow us to classify composite defects.

If $K \subset H \subset G$, then clearly there is an injection $i : H/K \to G/K$ carrying each coset of $K$ in $H$ into the corresponding coset in $G$, and there is a projection $p : G/K \to G/H$ mapping each coset of $K$ in $G$ into the coset of $H$ of which it forms a part. Moreover it is clear that $pi$ is trivial. From this it is fairly easy to show that there is an exact sequence very similar to (13), but involving homotopy groups of the three quotient spaces:

$$\begin{aligned}\cdots \xrightarrow{i*} \pi_2(G/K) \xrightarrow{p*} \pi_2(G/H) \xrightarrow{\partial} \pi_1(H/K) \xrightarrow{i*} \pi_1(G/K) \xrightarrow{p*} \\ \pi_1(G/H) \xrightarrow{\partial} \pi_0(H/K) \xrightarrow{i*} \pi_0(G/K) \xrightarrow{p*} \pi_0(G/H).\end{aligned} \tag{17}$$

Each group here classifies defects at one or another transition:

| | |
|---|---|
| $\downarrow$ | |
| $\pi_2(H/K)$ | monopoles created at second transition |
| $\downarrow$ | |
| $\pi_2(G/K)$ | stable monopoles in final phase |
| $\downarrow$ | |
| $\pi_2(G/H)$ | monopoles in intermediate phase |

$$\downarrow$$

| | |
|---|---|
| $\pi_1(H/K)$ | strings created at second transition |
| $\downarrow$ | |
| $\pi_1(G/K)$ | stable strings in final phase |
| $\downarrow$ | |
| $\pi_1(G/H)$ | strings in intermediate phase |
| $\downarrow$ | |
| $\pi_0(H/K)$ | domain walls created at second transition |
| $\downarrow$ | |
| $\pi_0(G/K)$ | stable domain walls in final phase |
| $\downarrow$ | |
| $\pi_0(G/H)$ | domain walls in intermediate phase. |

Thus for example the strings created at the first transition are classified by $\pi_1(G/H)$. If the corresponding element of this group is in the image of the map from $\pi_1(G/K)$, then the string remains stable at the second transition. But if not, it maps to a non-trivial element of $\pi_0(H/K)$ so that domain walls are formed at the second transition with these strings as their boundaries.

## 4. Helium-3

There are many condensed-matter systems that exhibit symmetry-breaking phase transitions with defect formation. But in many ways the most interesting is helium-3, because of its unique eighteen-dimensional order parameter.

There are in fact intriguing analogies between the classification of composite defects and of defects in $^3$He. Only one transition is involved in going from the normal to the superfluid $A$ or $B$ phase, but there are two very different length scales that play a similar role to the two transition energies.

### 4.1. SYMMETRY BREAKING IN HELIUM-3

Since a $^3$He atom is a fermion, the mechanism of superfluidity is very different from that in $^4$He, where there is essentially a Bose condensation. As with electrons in superconductors, the fermions form bound Cooper pairs, but in $^3$He they are in the $^3P$ state rather than $^1S$. We have both $L = 1$ and $S = 1$, so the corresponding order parameter — the macroscopic wave function of the Cooper pairs — is a 3×3 complex matrix $\Phi$. We could use a basis of eigenstates of $L_z$ and $S_z$, but for our purposes it is easier to use the equivalent form $\Phi_{jk}$ where $j$ and $k$ are vector indices. The fact that we have an 18-dimensional order-parameter space yields an amazingly rich symmetry-breaking structure (for reviews, see [24, 25]).

The coupling between orbital and spin angular momenta, via a dipole interaction, is very weak, so to a good approximation the symmetry group is

$$G = \mathrm{SO}(3)_{\mathbf{L}} \times \mathrm{SO}(3)_{\mathbf{S}} \times \mathrm{U}(1)_Y, \tag{18}$$

where the subscripts denote the generators; **L** and **S** are the orbital and spin angular momenta, while $Y$ generates the U(1) phase transformation $\Phi \mapsto \Phi e^{i\alpha}$.

There is however a weak $L$-$S$ coupling whose effect is seen only over very long distances. If we take that into account, the symmetry is reduced to

$$G' = \mathrm{SO}(3)_{\mathbf{J}} \times \mathrm{U}(1)_Y, \tag{19}$$

where $\mathbf{J} = \mathbf{L} + \mathbf{S}$ is the total angular momentum.

There are two distinct superfluid phases, $A$ and $B$, with very different symmetry properties. (In a magnetic field there is third superfluid phase, $A1$.)

## 4.2. THE B PHASE

If normal $^3$He at low pressure is cooled below about 2 mK, it enters the $B$ phase. Here the order parameter takes the form

$$\Phi_{jk} = \eta R_{jk} e^{i\alpha}/\sqrt{3}, \tag{20}$$

where $\eta$ is a fixed constant, $R$ is an arbitrary real orthogonal matrix and $\alpha$ an arbitrary phase factor. By some transformation in $G$ we could transform it for example to a multiple of the unit matrix: $\Phi_{jk} = \eta\delta_{jk}/\sqrt{3}$. The subgroup $H_B$ of $G$ that leaves this $\Phi$ unaltered is clearly

$$H_B = \mathrm{SO}(3)_{\mathbf{J}}, \tag{21}$$

though for some other choice of $R$, the generators would not be $\mathbf{L}+\mathbf{S}$ but $\mathbf{L} + R \cdot \mathbf{S}$.

Clearly in this case the manifold of degenerate ground states is

$$\mathcal{M}_B = G/H_B = \mathrm{SO}(3) \times U(1) = \mathbf{RP}^3 \times S^1. \tag{22}$$

The zeroth and second homotopy groups are both trivial, so there are no domain walls or monopoles. But $\pi_1(\mathcal{M}_B) = \mathbf{Z} \times \mathbf{Z}_2$, so there are two quite different types of string or vortex.

One is very similar to the more familiar vortices in superfluid $^4$He: the phase $\alpha$ runs from 0 to $2\pi$ as we go round the vortex. In the $B$ phase the superfluid velocity is proportional to the gradient of $\alpha$, so, as in $^4$He, there is a quantized circulation around the vortex, quantized in units of

$\kappa = \pi\hbar/m_3$, where $m_3$ is the mass of a $^3$He atom (so that the mass of the Cooper pair is $2m_3$).

The other type of string is created by a rotation through $2\pi$: we have for example $\Phi = \exp(i\theta L_z)\Phi_0$. Here there is no actual circulation around the string, but instead a relative rotation of the orbital and spin angular momenta.

So far we have ignored the $L$-$S$ coupling, but over large distances this becomes important. The true symmetry group is not $G$ but $G'$, so the degeneracy among the states described by (20) is partly lifted, in particular between states where $R$ is a rotation through different angles. The dipole energy is proportional to

$$|\mathrm{tr}\Phi|^2 + \mathrm{tr}(\Phi^*\Phi) - \tfrac{2}{3}\mathrm{tr}(\Phi^\dagger\Phi). \tag{23}$$

When $\Phi$ is of the form (20), with $R$ a rotation about some axis through angle $\theta$, this is equal to

$$2\cos\theta + 4\cos^2\theta - 2, \tag{24}$$

so the energy is minimized if $\theta$ is equal to the so-called *Leggett angle*, $\theta = \arccos(-\frac{1}{4}) = 104°$. In this case, while the expectation values of $\mathbf{L}$ and $\mathbf{S}$ are both zero, the expectation value of $\mathbf{L}\cdot\mathbf{S}$ is non-zero. (Note however that 104° is *not* the angle $\psi$ between the $\mathbf{L}$ and $\mathbf{S}$ vectors. If $\psi$ is defined in the obvious way, it is in fact $\arccos(-\frac{1}{6}) = 100°$, but it is purely accidental that the two are close, since $\psi$ clearly decreases as $\theta$ increases, from $\psi = \pi$ when $\theta = 0$.)

If the axis of rotation is chosen to be the $z$ axis, then the subgroup $H'$ of $G'$ that leaves $\Phi$ invariant is simply

$$H'_B = \mathrm{U}(1)_{J_z}, \tag{25}$$

corresponding to rotations of both $\mathbf{L}$ and $\mathbf{S}$ together about the $z$ axis. Correspondingly, the manifold of completely degenerate ground states is

$$\mathcal{M}'_B = G'/H'_B = S^2 \times S^1, \tag{26}$$

where the $S^2$ represents the arbitrariness in the direction of the rotation axis, and $S^1$ as before the phase.

We can now examine the homotopy groups of $\mathcal{M}'_B$. The zeroth group is trivial as before, but now we find $\pi_1(\mathcal{M}_B) = \mathbf{Z}$ and $\pi_2(\mathcal{M}_B) = \mathbf{Z}$. So at this level there are both strings and monopoles. In the monopole the axis of the rotation $R$ points outwards in a hedgehog configuration.

What is particularly interesting is to compare the short-distance structures described by homotopy groups of $\mathcal{M}_B$ with the long-distance ones described by those of $\mathcal{M}'_B$. The relationship is quite similar to that of a two-stage symmetry breaking, but more complicated, because $\mathcal{M}_B = G/H_B$ and $\mathcal{M}'_B = G'/H'_B$ are not quotients of the *same* group like $G/H$ and $G/K$. Correspondingly, since $\mathcal{M}_B$ and $\mathcal{M}'_B$ are not groups, the quotient $\mathcal{M}_B/\mathcal{M}'_B$ is not defined. (A rather closer analogy is with a *semi-local* theory, with a symmetry group $G$ and a subgroup $G'$ of local, gauge symmetries [26].)

However, there is a substitute for the exact sequence (17), involving the *relative* homotopy groups, $\pi_n(\mathcal{M}_B, \mathcal{M}'_B)$. For example, to construct $\pi_1(\mathcal{M}_B, \mathcal{M}'_B)$ we consider paths in $\mathcal{M}_B$ starting at the base point but ending at *any* point of $\mathcal{M}'_B$. Just as the exact sequence (17) can be used to classify the composite defects, we can do exactly the same thing here:

| | |
|---|---|
| $\downarrow$ | |
| $\pi_3(\mathcal{M}_B, \mathcal{M}'_B)$ | monopoles 'created' at long range |
| $\downarrow$ | |
| $\pi_2(\mathcal{M}'_B)$ | stable monopoles at long range |
| $\downarrow$ | |
| $\pi_2(\mathcal{M}_B)$ | monopoles at short range |
| $\downarrow$ | |
| $\pi_2(\mathcal{M}_B, \mathcal{M}'_B)$ | strings 'created' at long range |
| $\downarrow$ | |
| $\pi_1(\mathcal{M}'_B)$ | stable strings at long range |
| $\downarrow$ | |
| $\pi_1(\mathcal{M}_B)$ | strings at short range |
| $\downarrow$ | |
| $\pi_1(\mathcal{M}_B, \mathcal{M}'_B)$ | domain walls 'created' at long range |
| $\downarrow$ | |
| $\pi_0(\mathcal{M}'_B)$ | stable domain walls at long range |
| $\downarrow$ | |
| $\pi_0(\mathcal{M}_B)$ | domain walls at short range |

Note that the dimensions of the groups are different: for example, strings created at the second transition are described by $\pi_2(\mathcal{M}_B, \mathcal{M}'_B)$, rather than $\pi_1(H/H')$. As a matter of fact, this group can be expressed as the first homotopy group of another space, the space of *curves* in $\mathcal{M}$ starting at the base point and ending at any point of $\mathcal{M}'$. In general, if $x_0 \in Y \subset X$, then

$$\pi_n(X, Y) = \pi_{n-1}([X : Y]), \qquad (n \geq 1) \tag{27}$$

where $[X : Y]$ denotes the space of curves in $X$ starting at the base point $x_0$ and ending at any point in $Y$.

In our particular case, the relative homotopy groups are

$$\pi_3(\mathcal{M}_B, \mathcal{M}'_B) = \mathbf{Z}, \quad \pi_2(\mathcal{M}_B, \mathcal{M}'_B) = \mathbf{1}, \quad \pi_1(\mathcal{M}_B, \mathcal{M}'_B) = \mathbf{Z}_2. \tag{28}$$

What this means is that monopoles and domain walls are created at this transition.

So far as the monopoles are concerned, the implication is that there is no short-range defect structure; the monopole configuration is non-singular in that sense. Indeed the form of the order parameter is

$$\Phi(\mathbf{r}) = \exp[-i\psi(r)\mathbf{r}\cdot\mathbf{L}/r]\Phi_0, \tag{29}$$

where $\psi(0) = 0$ and $\psi(\infty) = \arccos(-\frac{1}{4})$.

The two kinds of strings formed at the 'first transition' behave differently. The simple U(1) phase vortices remain stable; they are classified by the non-trivial elements of $\pi_1(\mathcal{M}_B)$ which are images of elements in $\pi_1(\mathcal{M}'_B)$. The long-distance $L$-$S$ coupling has no effect on their stability.

But the other type of string is a very different case. It corresponds to an element of $\pi_1(\mathcal{M}_B)$ that is not an image of any element in $\pi_1(\mathcal{M}'_B)$; there is no stable configuration of this type that extends out to infinity. Exactly as in the case of a composite defect, what happens is that the string becomes the boundary of a domain wall. The homotopy class of this string maps to the nontrivial element of the relative homotopy group $\pi_1(\mathcal{M}_B, \mathcal{M}'_B) = \mathbf{Z}_2$.

### 4.3. THE A PHASE

If normal $^3$He is cooled at pressures above about 21 bar, it makes a transition to a different superfluid phase, the $A$ phase. The form of $\Phi$ that minimizes the free energy in this region is quite different from (20), namely

$$\Phi_{jk} = \eta(e_j + if_j)d_k/\sqrt{2}. \tag{30}$$

Here $\mathbf{d}$, $\mathbf{e}$ and $\mathbf{f}$ are unit vectors with $\mathbf{e}\cdot\mathbf{f} = 0$; $\mathbf{d}$ is the spin anisotropy axis: the state is one in which the component of $\mathbf{S}$ along the direction $\mathbf{d}$ is zero. There is also an orbital anisotropy axis, $\mathbf{l} = \mathbf{e}\wedge\mathbf{f}$; the component of $\mathbf{L}$ along $\mathbf{l}$ is $+1$. It is important to note, however, that reversing the signs of all three vectors leaves $\Phi$ unchanged.

The form (30) is invariant under spin rotations about the axis $\mathbf{d}$, and also under a combination of orbital rotation about $\mathbf{l}$ with phase change. If we take $\mathbf{d}$, $\mathbf{e}$ and $\mathbf{f}$ in the $z$, $x$ and $y$ directions, then the subgroup of $G$ that leaves (30) invariant is easily seen to be

$$H_A = \mathrm{U}(1)_{L_z+Y} \times \mathrm{U}(1)_{S_z} \times \mathbf{Z}_2. \tag{31}$$

The reason for the $\mathbf{Z}_2$ factor is that there is a combination of spin and orbital rotations through $\pi$ that simply reverses all three vectors $\mathbf{d}$, $\mathbf{e}$ and $\mathbf{f}$, leaving $\Phi$ unchanged. The manifold of degenerate ground states is now

$$\mathcal{M}_A = \mathrm{SO}(3) \times S^2/\mathbf{Z}_2, \tag{32}$$

where the SO(3) represents the orientation of the orthonormal triad $(\mathbf{e}, \mathbf{f}, \mathbf{l})$ and $S^2$ the direction of $\mathbf{d}$. The factor $\mathbf{Z}_2$ is present because the combination of an orbital rotation through $\pi$ about $\mathbf{l}$ and a spin rotation through $\pi$ about an axis perpendicular to $\mathbf{d}$ clearly leaves (30) unchanged.

The homotopy groups of $\mathcal{M}_A$ are:

$$\pi_3(\mathcal{M}_A) = \mathbf{Z} \times \mathbf{Z}, \quad \pi_2(\mathcal{M}_A) = \mathbf{Z}, \quad \pi_1(\mathcal{M}_A) = \mathbf{Z}_4, \quad \pi_0(\mathcal{M}_A) = \mathbf{1}. \tag{33}$$

This means that there are both monopoles and vortices in the $A$ phase, as well as two distinct families of textures, one of which I will discuss later.

The monopoles are labelled by an integer-valued winding number $n$. In the case $n = 1$, the spin axis $\mathbf{d}$ is in a hedgehog configuration, while the orbital axis $\mathbf{l}$ has no singularity and may be uniform.

The vortices are labelled by an index $n$ that is an integer modulo 4, so there are three non-trivial types, with $n = \pm 1$ and $n = 2$ (of course $n = 3$ is equivalent to $n = -1$).

Now, as before, let us see what difference it makes to take into account the small $L$-$S$ coupling. In this case, the degeneracy of different states (30) is broken, and we find that the lowest energy states are those in which $\mathbf{d}$ is parallel (or anti-parallel) to $\mathbf{l}$. The subgroup $H'_A$ of $G'$ that leaves (30) invariant is

$$H'_A = \mathrm{U}(1)_{J_z+Y}. \tag{34}$$

The manifold of precisely degenerate ground states is now simply

$$\mathcal{M}'_A = \mathrm{SO}(3). \tag{35}$$

We do not need to take the quotient by $\mathbf{Z}_2$ because we can always choose between the two possible directions of $\mathbf{d}$ by requiring that it be parallel rather than antiparallel to $\mathbf{l}$. This time we find

$$\pi_3(\mathcal{M}'_A) = \mathbf{Z}, \quad \pi_2(\mathcal{M}'_A) = \mathbf{1}, \quad \pi_1(\mathcal{M}'_A) = \mathbf{Z}_2, \quad \pi_0(\mathcal{M}'_A) = \mathbf{1}. \tag{36}$$

So there are no totally stable monopoles, only one class of stable vortices, and one family of textures.

To find out what happens at large distances to the defects present at short distances, we can again examine the relative homotopy groups $\pi_n(\mathcal{M}_A, \mathcal{M}'_A)$. In this case, one finds:

$$\pi_3(\mathcal{M}_A, \mathcal{M}'_A) = \mathbf{Z}, \quad \pi_2(\mathcal{M}_A, \mathcal{M}'_A) = \mathbf{Z}, \quad \pi_1(\mathcal{M}_A, \mathcal{M}'_A) = \mathbf{Z}_2. \tag{37}$$

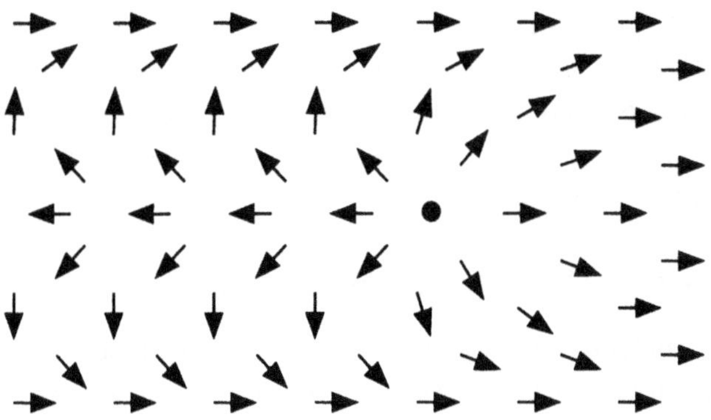

*Figure 6.* Small monopole attached to a fat string.

So large-scale monopoles, strings *and* domain walls are generated by the 'second transition'.

We then have to examine the effect of the maps from one homotopy group to another on the homotopy classes of the defects. Consider first the short-distance monopoles. It is easy to see that they become joined by the new strings. In this case, the **l** vector has no singularity, and may be taken to be constant, say in the $z$ direction. Immediately around a monopole, the spin axis **d** points radially outwards, but at large distances it will bend round because of the dipole interaction to align itself to **l**. However, this cannot be done smoothly everywhere. Along the negative $z$ axis, **d** must still point in the opposite direction (see Fig. 6), bending round away from the axis to its preferred orientation. This gives a fat string attached to the small monopole.

Among the short-distance vortices labelled by the mod-4 index, only those with $n = 2$ remain stable at large distances. For those with $n = \pm 1$, the image of the map is the non-trivial element of $\pi_1(\mathcal{M}_A, \mathcal{M}'_A)$, so these strings become the boundaries of the large-scale domain walls. Around one of these strings, the **d** vector rotates by $\pi$ with a compensating rotation about **l**.

I mentioned earlier the possibility of short-range 'textures' at which long-range monopoles appear or disappear. The inconsistency that prevents composite texture/monopoles from being realised in a two-stage symmetry breaking does not apply here, so one might hope that such configurations could in principle occur in $^3$He-$A$, where the relevant homotopy groups are indeed non-trivial. The configuration of the **d** vector around a texture event would correspond to the well-known homotopically non-trivial map of $S^3$ onto $S^2$. That configuration could only be matched smoothly onto the configuration at large distances, where **d** is parallel to **l**, with the omission of one point in space, the position of the (long-range) monopole. Since the 'monopole' has no singularity, it would perhaps be better described as a skyrmion [27, 28]. Unfortunately, however, in the absence of the higher-derivative terms needed to stabilize skyrmions, the 'monopole' would not be even metastable against radial collapse, so it is unlikely that it would live long enough to be detected.

## 5. Cosmology in the Laboratory

If topological defects are formed in the early Universe, then to predict their observational consequences we have to answer at least four separate questions:

*i.* What theory describes particle interactions at the relevant energies, and what types of defects can exist within this theory?

*ii.* How many defects will be formed when the system undergoes its transition?

*iii.* How do these defects evolve as the Universe expands?

*iv.* How will the defects interact with the matter and radiation in the Universe, to yield observational effects?

For (*iv*) and to a large extent for (*iii*), there is little that comparisons with condensed-matter systems can tell us; evolution in the expanding Universe is very different from evolution in a low-temperature system. The problem with (*i*) is that there is still such a wide range of possibilities; it may be easier to work backwards: to use what we can learn about cosmological defects to constrain the high-energy theory.

It is with regard to (*ii*) and the early stages of (*iii*) that the analogy with low-temperature systems has proved fruitful. Methods have been developed in the cosmological context for estimating the initial number of defects formed at a rapid phase transition. But to test them we have had to turn to condensed-matter analogues.

### 5.1. INITIAL DEFECT NUMBERS

Let us consider specifically question (*ii*). Let us suppose our system as it cools goes through a second-order phase transition at which strings or vortices are formed. In different regions, the system will make different choices of the arbitrary phase. So we expect that a random tangle of strings will be formed. The question then is this: what is the characteristic length scale $\xi_{\rm str}$ of the tangle? One way to define $\xi_{\rm str}$ is to say that in a randomly chosen cube of size $\xi_{\rm str}^3$, we expect on average to find a length $\xi_{\rm str}$ of string, *i.e.*, the length of string per unit volume is $l = 1/\xi_{\rm str}^2$. But what determines $\xi$?

It seems clear that the answer must be related in some way to the correlation length $\xi_\phi$ of the scalar field. But near the transition $\xi_\phi$ is varying rapidly. Indeed, the *equilibrium* correlation length goes to infinity at the transition. So at what time, or equivalently at what temperature, should we compare $\xi_{\rm str}$ and $\xi_\phi$?

An answer to this question was proposed by Zurek [29, 30], based essentially on a causality argument. One way of phrasing it is the following. If we proceed through the transition at a finite rate, the true correlation length can never beome infinite; it cannot grow faster than the speed of light, or, in a non-relativistic case, than some characteristic speed $u$ of the system, such as the speed of second sound or of spin waves. We expect $\xi_\phi$ to behave like the equilibrium correlation length more or less until it reaches the point where $d\xi_\phi/dt = u$, and thereafter to remain approximately constant, at least until the time after the transition when it again meets the falling equilibrium value. The hypothesis is that at this *Zurek time*, $t_Z$, we have $\xi_{\rm str} \approx \xi_\phi$, and so $l(t_Z) \approx 1/\xi_\phi^2(t_Z)$, up to a factor of order unity. This then provides a prediction of the initial defect density.

This is not a very rigorous argument. It needs both to be tested experimentally and to be confirmed or modified by more accurate theoretical work. Both these subjects will be extensively discussed by others at this meeting.

### 5.2. TESTS IN SUPERFLUID HELIUM

The first proposal for experimental tests was made by Zurek [31]. He suggested an experiment involving a pressure quench in superfluid $^4$He. The idea was to start with a sample at high pressure just above the lambda transition, and then by rapid expansion to take it through the lambda line.

The experiment has in fact been performed, by a group at Lancaster University. The technique was to determine the number of vortices present by measuring the attenuation of second sound, which is known to be strongly damped by vortices. Initial results were very encouraging [32]. Unfortunately it was not possible to measure the number of vortices immediately after the transition because the detectors were overloaded. However, extrapolating back from the decaying attenuation measured at longer times, it seemed that the initial defect number could well be in agreement with Zurek's predictions. However, the experiment could not be regarded as conclusive because there were various alternative ways in which vortices might have been formed, for example by hydrodynamic effects at the walls or the capillary filling tube.

More recently, following a number of improvements, the experiment has been repeated, but this time with a null result [33]. Certainly it appears that if any vortices are formed in this transition, their number is much less than Zurek predicted.

Meanwhile, tests have also been performed in superfluid $^3$He, which has a number of very important advantages. Firstly, $^3$He is an excellent neutron absorber, so it is possible to use the reaction

$$n + {}^3\mathrm{He} \to p + {}^3\mathrm{H} + 764\ \mathrm{keV} \tag{38}$$

to heat a small region of the fluid. Two experiments have been done using this technique, one in Helsinki [34] and the other in Grenoble [35]. Both experiments start with the sample in the superfluid $^3$He-$B$ phase and subject it to neutron bombardment to heat a small region above the critical temperature. This region then rapidly cools, going back through the transition in a time of the order of 1 $\mu$s.

Another important difference between the two isotopes is that the correlation length $\xi_\phi$ in $^3$He is much larger, between 10 and 100 nm, compared with less than 1 nm in $^4$He. The important thing is that in $^3$He, $\xi_\phi \gg a$, the interatomic spacing. So a continuum (Ginzburg–Landau) description is a much better approximation. Indeed, phenomenologically, the time-independent Ginzburg–Landau model is a rather good description of $^3$He, but distinctly poor for $^4$He. So in this sense, rather surprisingly perhaps, $^3$He is better understood theoretically than $^4$He.

A further consequence of the large value of $\xi_\phi$ is that, relative to available thermal energies, it takes much more energy to create a vortex in $^3$He, so it is easier to exclude extrinsic vortex formation mechanisms.

Finally, because of the non-zero nuclear spin of $^3$He, it is possible to use nuclear magnetic resonance (NMR) to count vortices. This is done very effectively in the Helsinki experiment [34], which uses a rotating cryostat. Vortices formed following a neutron absorption event are expanded by the transverse Magnus force due to the relative velocity of the normal and superfluid components, and migrate to join a central cluster parallel to the axis, where they can be counted by NMR.

The Grenoble experiment [35] is complementary. It uses $^3$He-$B$ in a non-rotating container, and at a much lower temperature, a long way below $T_c$. Instead of counting individual vortices, the aim is to measure the total energy released. Of the 764 keV input, some 50 keV emerges as ultraviolet radiation, and most of the rest in the form of quasiparticles. But there is a deficit. Energy is clearly going into something else, and the only likely candidate is vortices.

The results of both experiments are in agreement, within a factor of a few, with the predictions of the Zurek model.

## 6. Theoretical Interpretation

We are faced with something of a paradox. The experiments in $^3$He seem to be in good agreement with theory, while the new experiments in $^4$He show no sign of any vortices.[1] How can this be?

There have been several attempts to improve or confirm the theory. In particular, Zurek and collaborators [37, 38] have performed computer simulations of defect formation at a second-order phase transition, using the time-dependent Ginzburg–Landau (TDGL) equation. At least in two dimensions, their results suggest that the expected defect density would be *lower* than the original prediction by a factor of 20 to 150. This certainly makes the $^4$He results seem less anomalous, but on the other hand it appears to destroy the good agreement in $^3$He!

However, it is highly questionable whether the TDGL equation provides a good description of the system near the phase transition, especially in the case of $^4$He. It would be much more satisfactory to use thermal field theory to study the problem. This has been done by Karra and Rivers [39], with very interesting results, though these too are dependent on simplifying assumptions, in particular that Gaussian approximations are adequate, which is certainly questionable. However, they have pointed out that there is a very important difference between $^4$He and $^3$He, concerning the width of the 'Ginzburg region'. The Ginzburg temperature $T_G$ ($< T_c$) is the temperature below which thermal fluctuations of the order parameter on the scale of the correlation length $\xi_\phi$ that take it over the central hump of the potential become improbable. In $^4$He, where $\xi_\phi$ is small, this occurs about half a degree below $T_c$, which means that the experiments have been performed within the Ginzburg region, between $T_G$ and $T_c$. On the other hand, in $^3$He, $\xi_\phi$ is much larger, and the Ginzburg region is very narrow: $T_c - T_G \ll T_c$. The result according to Karra and Rivers is likely to be a strong suppression of defect formation in $^4$He as compared to $^3$He.

One of the central problems to be addressed at this meeting is how to improve the theoretical calculations. The experimental results in $^3$He strongly sugggest that the basic theory is on the right lines, but we do not yet have fully reliable calculational techniques. It is too early to know whether the width of the Ginzburg region is the correct explanation of the observed difference in behaviour between $^3$He and $^4$He, but it certainly seems plausible.

Whatever the final resolution of these questions may be, there is no doubt that the interaction between the apparently very different worlds of cosmology and low-temperature physics, at opposite extremes of the energy range, has been very beneficial to both. Long may this continue!

## Acknowledgements

1Preliminary negative results were also reported at this meeting from experiments with thin superconducting films [36].

These lectures were presented at the NATO Advanced Study Institute on *Topological Defects and the Non-Equilibrium Dynamics of Symmetry Breaking Phase Transitions*, also supported by a European Science Foundation network. I am indebted to numerous participants at the ASI, and especially to Ana Achúcarro, Grigoriy Volovik and Tanmay Vachaspati, for valuable discussions of the material covered in this article.

## References

1. Amaldi, U., de Boer, W. and Fürstenau, H. (1991) Comparison of grand unified theories with electroweak and strong coupling-constants measured at LEP, *Phys. Lett.* **B260**, 447–455.
2. Amaldi, U., de Boer, W., Frampton, P.H., Fürstenau, H. and Liu, J.T. (1992) Consistency checks of grand unified theories, *Phys. Lett.* **B281**, 374–382.
3. Haber, H.E. (1998) The status of the minimal supersymmetric standard model and beyond, *Nuc. Phys. Proc. Supp.* **B62**, 469–484.
4. Kajantie, K., Laine, M., Rummukainen, K. and Shaposhnikov, M. (1996) Is there a hot electroweak phase transition at $m$(H) greater than or similar to $m$(W)?, *Phys. Rev. Lett.* **77**, 2887–2890.
5. Elitzur, S. (1975) Impossibility of spontaneously breaking local symmetries, *Phys. Rev. D* **12**, 3978–3982.
6. Banks, T. and Rabinovici, M. (1979) Finite-temperature behaviour of the lattice Abelian Higgs model, *Nuc. Phys.* **160**, 349–379.
7. Fradkin, E. and Shenker, S.H. (1979) Phase diagrams of lattice gauge theories with Higgs fields, *Phys. Rev. D* **19**, 3682–3697.
8. Kibble, T.W.B. and Hindmarsh, M.B. (1995) Cosmic strings, *Rep. Prog. Phys.* **58**, 477–452.
9. Shellard, E.P.S. and Vilenkin, A. (1994) *Cosmic strings and other topological defects*, Cambridge University Press, Cambridge.
10. Hawking, S.W. and Ross, S.F. (1995) Pair production of black holes on cosmic strings, *Phys. Rev. Lett.* **75**, 3382–3385.
11. Eardley, D.M., Horowitz, G.T., Kastor, D.A. and Traschen, J. (1995) Breaking cosmic strings without monopoles, *Phys. Rev. Lett.* **75**, 3390–3393.
12. Gregory, R. and Hindmarsh, M.B. (1995) Smooth metrics for snapping strings, *Phys. Rev. D* **52**, 5598–5604.
13. Albrecht, A., Battye, R.A. and Robinson, J. (1997) The case against scaling defect models of cosmic structure formation, *Phys. Rev. Lett.* **79**, 4736–4739.
14. Contaldi, C., Hindmarsh, M.B. and Magueijo, J. (1999) Power spectrum of the cosmic microwave background and large-scale structure seeded by local cosmic strings, *Phys. Rev. Lett.* **82**, 679–682.
15. Contaldi, C., Hindmarsh, M.B. and Magueijo, J. (1999) Cosmic microwave background and density fluctuations from strings plus inflation, *Phys. Rev. Lett.* **82**, 2034–2037.
16. Hu, S.-T. (1959) *Homotopy Theory*, Academic Press, New York.
17. Davis, R.L. (1987) Texture — a cosmological topological defect, *Phys. Rev. D* **35**, 3705–3708.
18. Davis, R.L. (1987) Cosmic texture and the microwave background, *Phys. Rev. D* **36**, 997–999.
19. Kibble, T.W.B., Lazarides, G. and Shafi, Q. (1982) Strings in SO(10), *Phys. Lett.* **113B**, 237–239.
20. Kibble, T.W.B., Lazarides, G. and Shafi, Q. (1982) Walls bounded by strings, *Phys. Rev. D* **26**, 435–439.

21. Langacker, P. and Pi, S.-Y. (1980) Magentic monopoles in grand unified theories, *Phys. Rev. Lett.* **45**, 1–4.
22. Everett, A., Vachaspati, T. and Vilenkin, A. (1985) Monopole annihilation and causality, *Phys. Rev. D* **31**, 1925–1930.
23. Copeland, E., Haws, D., Kibble, T.W.B., Mitchell, D. and Turok, N. (1988) Monopoles connected by strings and the monopole problem, *Nuc. Phys.* B298, 445–457.
24. Vollhardt, D. and Wölfle, P. (1990) *The superfluid phases of helium 3*, Taylor and Francis, London.
25. Volovik, G.E. (1992) *Exotic properties of superfluid*$^3$*He*, World Scientific, Singapore.
26. Vachaspati, T. and Achúcarro, A. (1991) Semilocal cosmic strings, *Phys. Rev. D* **44**, 3067–3071.
27. Holzwarth, G. and Schwesinger, B. (1986) Baryons in the Skyrme model, *Rep. Prog. Phys.* **49**, 825–872.
28. Manton, N. (1987) Geometry of Skyrmions, *Comm. Math. Phys.* **111**, 469–478.
29. Zurek, W.H. (1993) Cosmic strings in laboratory superfluids and topological remnants of other phase transitions, *Acta Phys. Polon.* **B24**, 1301–1311.
30. Zurek, W.H. (1996) Cosmological experiments in condensed matter systems, *Phys. Rep.* **276**, 177–221.
31. Zurek, W.H. (1985) Cosmological experiments in superfluid helium, *Nature* **317**, 505–508.
32. Hendry, P.C., Lawson, N.S., Lee, R.A.M., McClintock, P.V.E. and Williams, C.D.H. (1994) Generation of defects in superfluid He–4 as an analog of the formation of cosmic strings, *Nature* **368**, 315–317.
33. Dodd, M.E., Hendry, P.C., Lawson, N.S., McClintock, P.V.E. and Williams, C.D.H. (1998) Nonappearance of vortices in fast mechanical expansions of liquid He–4 through the lambda transition, *Phys. Rev. Lett.* **81**, 3703–3706.
34. Ruutu, V.M.H., Eltsov, V.B., Gill, A.J., Kibble, T.W.B., Krusius, M., Makhlin, Yu.G., Plaçais, B., Volovik, G.E. and Xu, W. (1996) Vortex formation in neutron-irradiated superfluid He–3 as an analogue of cosmological defect formation, *Nature* **382**, 334–336.
35. Bäuerle, C., Bunkov, Yu.M., Fisher, S.N., Godfrin, H. and Pickett, G.R. (1996) Laboratory simulation of cosmic string formation in the early Universe using superfluid He–3, *Nature* **382**, 332–334.
36. Carmi, R. and Polturak, E. (1999) private communication.
37. Laguna, P. and Zurek, W.H. (1997) Density of kinks after a quench: When symmetry breaks, how big are the pieces?, *Phys. Rev. Lett.* **78**, 2519–2522.
38. Yates, A. and Zurek, W.H. (1998) Vortex formation in two dimensions: When symmetry breaks, how big are the pieces?, *Phys. Rev. Lett.* **80**, 5477–5480.
39. Karra, G. and Rivers, R.J. (1997) Initial vortex densities after a temperature quench, *Phys. Lett.* **414B**, 28–33.

# THE NORMAL AND SUPERFLUID STATES OF LIQUID $^3$HE

A. J. LEGGETT
*University of Illinois, Urbana, Il., 61801, USA*

## 1. Introduction

As you all know, the system which has been most extensively used for tests of theoretical ideas about the formation of topological defects in quenching is superfluid liquid $^3$He; it had also, earlier, been the subject of a controlled series of experiments on a closely related problem, namely the nucleation by high-energy radiation of a first-order phase transition. In these two lectures I shall give an introduction to the physics of superfluid $^3$He in which I shall consciously try to focus on those aspects which seem most relevant to the quenching problem. (For a more extended discussion of most of the topics I shall discuss, and much else, see ref.[1]).

To see what we will be talking about, consider the phase diagram of liquid $^3$He below 4 mK (in zero magnetic field)(see e.g. ref. 1, Fig. 4.2). All the experiments discussed at this workshop have been done in the region where the B phase is thermodynamically stable. Note that the transition ("N-A") between the normal and A phases is second-order, as in the N-B transition, while the A-B transition is first-order. A further feature which is not apparent from the figure is that existing experimental and theoretical considerations seem consistent with the belief that if we could superheat the B phase far into the region of thermodynamic stability of the A phase, it would become unstable with respect to the N phase precisely at the actual A-N transition; i.e. at this point the normal phase becomes unstable simultaneously against formation of both A and B phases.

## 2. The Normal Phase

The electronic groundstate of the $^3$He atom (like that of $^4$He) is to a very high degree of approximation $(1s)^2$; the ionization energy is 24.6 eV and the Lyman-$\alpha$ line lies around 20 eV. The nucleus has spin 1/2 (and mag-



*Y.M. Bunkov and H. Godfrin (eds.),*
*Topological Defects and the Non-Equilibrium Dynamics of Symmetry Breaking Phase Transitions, 33–53.*

netic moment $\mu_n \cong$ -2.13 nuclear magnetons) and the atoms thus obey Fermi statistics. The groundstate of the $^3He_2$ dimer, $X^1\Sigma_g^+$ in the language of chemistry, has to a very good approximation the simple structure $(1\sigma_g)^2(1\sigma_u)^2$, and the associated interatomic potential has the characteristic inert-gas form, with an essentially infinite repulsive hard core at around 2Å, a minimum (~10K in depth) at ~3Å and a van der Waals tail $\sim r^{-6}$. This potential has no bound state, as a result of which the groundstate of the $^3He_2$ dimer is not bound, irrespective of the nuclear spin configuration; the typical range of interatomic spacings in the liquid, however, is around 3.5-4Å, as a result of which the interatomic potential can by no means be treated as a small perturbation. In addition to the "gross" interatomic potential just discussed, which is of course invariant under separate rotations of the spatial coordinates and nuclear spins, there is also a very small term in the Hamiltonian arising from the electromagnetic interaction of the nuclear magnetic moments:

$$\hat{H}_D = \mu_n^2 \left\{ \frac{\boldsymbol{\sigma}_1 \cdot \boldsymbol{\sigma}_2 - 3\boldsymbol{\sigma}_1 \cdot \hat{\mathbf{r}}_{12}\, \boldsymbol{\sigma}_2 \cdot \hat{\mathbf{r}}_{12}}{r_{12}^3} \right\}; \quad \boldsymbol{\sigma} = \text{(Pauli matrix)} \quad (1)$$

whose effect, crudely speaking, is to make it energetically slightly more favorable for two atoms with parallel nuclear spins to have a spatial separation parallel to the spins rather than perpendicular to them. Since the associated energy gain is only $\sim \mu_n^2/r_{12}^3 \lesssim 10^{-7}K$, one might at first sight think that when discussing phenomena taking place at a few millidegrees one could afford to neglect this term entirely; however, we shall see that this is not so.

If one were to regard normal liquid $^3He$ as simply a gas of noninteracting fermions, the associated Fermi temperature would be around 5K and thus for $T \lesssim$ 100mK the system should certainly be strongly degenerate. A straightforward textbook analysis would then predict that the specific heat $c_v$ should be linear in the temperature, while the Pauli spin susceptibility and the compressibility should be temperature-independent. This model as it stands gives infinite values for the transport coefficients, so let us modify it phenomenologically by putting in "by hand" weak atom-atom scatterings. In a degenerate Fermi gas, because of the effect of the Pauli exclusion principle, such collisions give rise to a mean free path proportional to $E^{-2}$ where $E$ is the energy measured with respect to the T=0 Fermi energy; thus typical thermally excited particles ($E$ ~T) have a mean free path proportional to $T^{-2}$. As a result, the viscosity $\eta$ and spin diffusion coefficient $D_s$ are both proportional to $T^{-2}$, while the thermal conductivity $\kappa$ varies as $T^{-1}$. All these predicted temperature-dependences are in fact seen in normal liquid $^3He$ below ~100 mK, although the numerical coefficients are different from those expected in the free-gas model; this strongly suggests that liquid $^3$He is behaving at least qualitatively like a degenerate Fermi gas.

In fact, the theory which is today almost universally believed to describe normal liquid $^3$He below ~100 mK, Landau's celebrated "Fermi-liquid" theory, can be formulated as saying that the groundstate and low excited states of the real (interacting) system can be obtained by adiabatic evolution starting from the corresponding states of the noninteracting gas (for an extended discussion, see ref.[2]). For our purposes, it is adequate to represent such states in terms of a quantity $n(\mathbf{p}\sigma)$ which is the number of "quasiparticles" occupying the one-particle state with momentum $\mathbf{p}$ and spin $\sigma$; physically, one can think of a quasiparticle as an atom plus a "dressing cloud" of other atoms. In the groundstate one can take $n(\mathbf{p}\sigma)$ to have the form $\theta(p_F - |\mathbf{p}|)$, where $p_F \equiv (3\pi^2 n)^{1/3}\hbar$ is just the Fermi momentum of the noninteracting system (it is approximately $10^{-24}$ SI units for real liquid $^3$He). The energy of the low-lying states in then expressed as a functional of the deviation $\delta n(\mathbf{p}\sigma)$ of the occupation factor from its value in the groundstate

$$E\{\delta n(\mathbf{p}\sigma)\} = E_o + \sum_{\mathbf{p}\sigma} \left(p^2/2m^*\right) \delta n(\mathbf{p}\sigma) +$$

$$\frac{1}{2}\sum_{\substack{\mathbf{pp}' \\ \sigma\sigma'}} f(\mathbf{pp}', \sigma\sigma')\delta n(\mathbf{p}\sigma)\delta n(\mathbf{p}'\sigma') + \text{(scattering terms)} \tag{2}$$

The scattering terms need to be included if we wish to get finite transport coefficients, but because of the Pauli exclusion principle (see above) their effect is small at low temperatures. If we ignore them, the spectrum is parametrized by a single "one-particle" number, the quasiparticle "effective mass" $m^*$, and a function, the so-called Landau interaction function $f(\mathbf{pp}' : \sigma\sigma')$, which characterizes the inter-quasiparticle interactions; in writing $f$ I have neglected, for the sake of notational simplicity, a slight complication associated with the vector nature of the spins (see e.g. ref.[3]).

An illuminating way of looking at the Landau Fermi-liquid Hamiltonian (2) is as describing a weakly interacting Fermi gas of particles of mass $m^*$, subject to a set of molecular fields. For example, we can pick out from the Landau interaction function a term $\frac{1}{2}(4f_o^a \mathbf{S}^2)$ proportional to the square of the total spin $\mathbf{S}$, where $f_o^a$ is the average over $\mathbf{p}$ and $\mathbf{p}'$ of the spin-dependent part of f, and this then gives rise to a "Weiss" field of magnitude - $4f_o^a\mathbf{S}$. Similarly, the corresponding spin-independent term is of the form $\frac{1}{2}f_o^s(\delta N)^2$, where $\delta N$ is the change in the total number of quasiparticles, and this then is equivalent to a "Hartree" field $f_o^s\delta N$ (the sign change and the factor of 4 in the Weiss case arising simply from the standard conventions in the definition of the Weiss and Hartree fields). A very important general point is that these and other "molecular fields" associated with the Landau interaction function are generated only by macroscopic polarizations of one kind or another; consequently, when no such polarizations occur the "Fermi-gas" formulae are valid with the simple replacement $m \to m^*$. This is the case, for example, for the quantities $c_v, \eta, \kappa$ and $D_s$.

As an example of a quantity which is affected by the molecular fields, consider the Pauli spin susceptibility $\chi_s$, defined as the ratio of the induced spin $\mathbf{S}$ to the external field $\mathbf{H}_{ext}$. To calculate this quantity within the Landau picture, we write

$$\mathbf{S} = \chi_{so}\,\mathbf{H}_{tot}, \tag{3}$$

$$\mathbf{H}_{tot} = \mathbf{H}_{ext} + \mathbf{H}_{mol} \tag{4}$$

where $\chi_{so}$ is the susceptibility of a Fermi gas with mass $m^*$, and $\mathbf{H}_{mol}$ is the "Weiss" molecular field; introducing the conventional dimensionless Landau parameter $F_o^a \equiv f_o^a(dn/d\epsilon)$, where $(dn/d\epsilon)$ is the density of states (of both spins) at the Fermi surface of the free gas (mass $m^*$), we have

$$\mathbf{H}_{mol} = -\,(dn/d\epsilon)^{-1}(4F_o^a)\mathbf{S}. \tag{5}$$

We substitute (5) into (4) and (3) and solve for $\mathbf{S}$. Using the fact that the free-gas Pauli spin susceptibility is just $1/4(dn/d\epsilon)$, we get in this way the simple result

$$\chi_s = \frac{\chi_{so}}{1 + F_o^a} \tag{6}$$

A parallel argument leads to a similar renormalization of the compressibility ("density susceptibility") $\chi_d$:

$$\chi_d = \frac{\chi_{do}}{1 + F_o^s} \tag{7}$$

where $\chi_{do}(= dn/d\epsilon)$ is the free Fermi-gas value.

In real-life liquid $^3$He the value of $F_o^a$ ranges from $\sim -0.67$ at saturated vapor pressure (s.v.p.) to $\sim -0.74$ at the melting pressure (m.p.), so that the molecular-field effects enhance the susceptibility over the value for the corresponding Fermi gas by a factor $\sim 3-4$. By contrast, the spin-independent Landau parameter $F_o^s$ has values ranging from $\sim +10$ at s.v.p. to $\sim +95$ at m.p.; this leads to a strong suppression of the compressibility, and a corresponding speed of hydrodynamic sound c which is much greater than the Fermi velocity $v_F \sim p_F/m^*$.

While the effect of the molecular fields on the static quantities $\chi_s$ and $\chi_d$ is, according to eqns. (6)-(7), merely to multiply them each by an overall factor, the effects on the corresponding dynamic susceptibility may change their structure qualitatively. The most instructive thing to plot is the imaginary part of the various response functions, Im$\chi(q\omega)$, as a function of frequency $\omega$ at constant q($\ll p_F/\hbar$). For the free gas (mass $m^*$) we have $\chi_s(q\omega) = \chi_d(q\omega) \equiv \chi_o(q\omega)$, where Im$\chi_o(q\omega)$ has a triangular shape with a sharp energy cutoff at $\omega = v_F q$. By contrast, in a strongly interacting Fermi liquid with Landau parameters appropriate to real $^3$He, the spin and "density" susceptibilities are modified in opposite directions; in the spin case, the weight of the triangular spectrum is shifted to energies appreciably smaller than $v_F q$, while in the density case the whole spectrum is strongly suppressed at $\omega \leq v_F q$ and the lost weight is shifted into the "zero-sound" peak at $\omega = c_o q$, where $c_o \cong c_s \gg v_F$.

At this point it is convenient to tabulate some parameters of normal liquid $^3$He which are relevant to the quenching problem. Some of these are temperature-dependent, and for each pressure (saturated vapor and melting pressure) I give the values at the temperature $T_c$(P) of transition to the superfluid state at that pressure; note that $T_c$ increases from ~0.9 mK at s.v.p. to ~ 2.6mK at m.p. The parameter values interpolate fairly smoothly between these two pressures, with, however, most of the changes tending to occur in the range 0-10 atm.

| | s.v.p. | m.p. |
|---|---|---|
| Fermi velocity $v_F \equiv p_F/m^*$ | 55 m/sec | 30m/sec |
| "Condensation energy" $c_v T_c$ | 9 eV/$\mu^3$ | 100 eV/$\mu^3$ |
| Quasiparticle mean free path $\ell$ | 78$\mu$ | 3$\mu$ |
| Quasiparticle lifetime $\tau \equiv \ell/v_F$ | 1.5 $\mu$sec. | 0.1 $\mu$sec |

Note the extraordinarily large value of $\ell$ at s.v.p. (almost 0.1 mm!). The reason the (order of magnitude of the) condensation energy is given in the unusual units of electron volts per cubic micron is for application to problems involving high-energy radiation.

## 3. The mechanism of attraction

It is generally believed that when superfluidity (or its "charged" version, superconductivity) occurs in a degenerate Fermi system such as liquid $^3$He, the relevant mechanism is the formation of Cooper pairs (a sort of quasi-diatomic molecules, see below) and that such formation requires that the effective interaction felt by two colliding atoms in states near the Fermi surface should be attractive for at least one value of the relative angular momentum $\ell$; the pairs will then form with the "most attractive" value of $\ell$. The most obvious source of such an attraction is simply the original van der Waals potential. Since this has a very repulsive core extending to ~2.5Å, it was recognized long before the experimental discovery of superfluidity in $^3$He that the interaction was not likely to be attractive in a relative s-state ($\ell = 0$). For $\ell \neq 0$, however, the centrifugal barrier automatically keeps the atoms away from one another's hard cores, and one would expect the interaction to be attractive, very crudely speaking, when $\ell$ is of the order of $k_F r_{min}$ where $r_{min}$ is the value of $r(\sim 3\text{Å})$ where the potential is most attractive. This allows $\ell = 1, 2$ or possibly 3 or higher.

A more sophisticated possible source of attraction, which is usually believed to play a major if not dominant role in $^3$He, is the "spin-fluctuation-mediated" interaction. Suppose we create a spin polarization density $\boldsymbol{\sigma}(\mathbf{r}t)$ at point $\mathbf{r}$ at time t. Then, according to the general prescriptions of Landau theory outlined above, this will produce a molecular field $\sim -f_o^a \boldsymbol{\sigma}(\mathbf{r}t)$. Since the liquid cannot "tell the difference" between a molecular field and an external one, the effect will be to produce a (collective) spin polarization $\boldsymbol{\sigma}(\mathbf{r}'t') \sim -f_o^a \chi_s(\mathbf{r}' - \mathbf{r}, t' - t)$ at neighboring points $\mathbf{r}', t'$, where $\chi_s(\mathbf{r}t)$ is the Fourier transform of the quantity $\chi_s(q\omega)$ discussed above. (There are some subtle points related to the avoidance of double-counting here which I will not go into. cf e.g. ref.[3]). This spin polarization in turn produces its own molecular field $\sim - f_o^a \boldsymbol{\sigma}(\mathbf{r}'t')$ which will be felt by an atom which passes the point $\mathbf{r}'$ at time $t'$. The net effect is to produce an effective interaction between spin-polarized atoms at space-time positions $\mathbf{r}t$ and $\mathbf{r}'t'$ of the form

$$V_{eff}(\mathbf{r} - \mathbf{r}', t - t') \sim -(f_o^a)^2 \, \chi_s(\mathbf{r} - \mathbf{r}', \, t - t')\boldsymbol{\sigma} \cdot \boldsymbol{\sigma}' \tag{8}$$

Note that this effective interaction is (a)retarded in time and (b) independently of the sign of $f_o^a$, attractive for parallel spins (and by the same token repulsive for antiparallel ones). It thus tends to disfavor pairing in even-$\ell$ states (which by the Pauli principle must have total spin $S = 0$) and favor states with $S = 1$ and odd $\ell$.

An illuminating physical picture of the working of the spin-fluctuation-mediated attraction may be obtained as follows: As (say) a spin-up atom moves through the liquid, it tends (for $f_o^a << 0$) to attract towards its path other up-spin atoms and repel down-spin ones, thereby creating in the neighborhood of this path a net collective up-spin polarization, which will then in turn tend to attract a second up-spin atom at a later time. This is somewhat analogous to the standard BCS "phonon-exchange" interaction in a superconducting metal, in which the first electron creates in the neighborhood of its path a positive (ionic) charge density which can then at a later time attract a second electron.

It is generally believed, primarily on experimental grounds, that the Cooper pairs in liquid $^3$He indeed form in a S=1 state, and moreover in a state with relative orbital angular momentum 1 (rather than say 3).

Some theoretical arguments for this state can be given, but they are more complicated that those above and I shall not attempt to give them here.

## 4. Basics of Cooper pair formation

The theory of BCS-type pairing in a degenerate Fermi system is by now very standard, both for the $\ell = 0$ and $\ell \neq 0$ cases, and there seems little point in giving the details of the derivations here; they can be found e.g. in ref.[1] or, more compactly and in the notation of the present lecture, in section V of ref.[3]. Thus, I shall confine myself here to sketching the principal concepts and results. Moreover, for pedagogical simplicity I shall in this lecture treat only the case $\ell = S = 0$ (which is directly relevant to the classical metallic superconductors but not to $^3$He); in the next section I will sketch the generalizations necessary for the $\ell = S = 1$ case which is generally believed to describe superfluid $^3$He.

It will be assumed throughout that the transition temperature $T_c$ of our system to the superfluid (paired) phase is small compared to the Fermi temperature $T_F$; this is certainly true for $^3$He, where the experimental ratio $T_c/T_F$ is $\lesssim 10^{-3}$. Further, it will be assumed for convenience that the effective inter-particle interaction can be taken to be independent of energy (measured relative to the Fermi energy $\epsilon_F$) up to some cutoff $\epsilon_c$ which, while small compared to $k_B T_F$, is large compared to $k_B T_c$; this assumption (made in the original work of BCS) is not at all essential, and the reasons for making it are discussed in ref.[3], Section V.E.

Let's first briefly review the result of the classic calculation of Cooper on the groundstate of two interacting fermions in a relative state $\ell = S = 0$, and excluded by fiat from the Fermi sea. Because the density of states near the Fermi surface is approximately constant, this problem is analogous to the problem of two particles in free space in 2D, for which we know that for a uniformly attractive potential a bound state always exists. Indeed, the result of Cooper's calculation is that for a constant attraction interaction $V_o$ within the "shell" $|\epsilon| << \epsilon_c$, the two fermions can form a state whose energy $E$ relative to $2\epsilon_F$ is negative:

$$E = -2\epsilon_c \, exp(-2/N(0)V_o) \tag{9}$$

where $N(0) \equiv 1/2(dn/d\epsilon)$ is the density of states for one spin at the Fermi surface (it is conventional in BCS theory to use this quantity rather than $dn/d\epsilon$). Moreover, the state associated with this energy is bound in coordinate space, that is, the relative wave function of the two fermions falls off as a function of relative coordinate r roughly as exp - $r/\xi_c$, where the "pair radius" $\xi_c$ is of order $\hbar v_F / |E|$ and hence exponentially large in the limit of small $V_o$. Note that the expression (9) for the binding energy is nonanalytic in $V_o$ and hence cannot be obtained by perturbation theory in the strength of the interaction.

Guided by Cooper's result, BCS constructed an ansatz for the ground-state of the many-body wave function which reflects its qualitative features. While the original formulation relaxed particle number conservation (cf. below) it is helpful to write the "particle-conserving" form: apart from normalization this is

$$\Psi_N(\mathbf{r}_1\sigma_1, \mathbf{r}_2\sigma_2 \ldots r_N\sigma_N) = A\phi(\mathbf{r}_1\mathbf{r}_2\sigma_1\sigma_2)\phi(\mathbf{r}_3\mathbf{r}_4\sigma_3\sigma_4)\ldots\phi(\mathbf{r}_{N-1}\sigma_{N-1}\mathbf{r}_N\sigma_N) \tag{10}$$

where the operator A antisymmetrizes both within and between the brackets. The essential point to note is that the function $\phi$ is the same for each of the N/2 pairs; thus, intuitively speaking, the wave function $\Psi_N$ describes a "Bose condensate" of pseudo-diatomic molecules. Eqn. (10) is rather generic (cf. 1.2), but for the case of Cooper pairs formed with $\ell = S = 0$ and the center of mass at rest the function $\phi(\mathbf{r}_1\mathbf{r}_2\sigma_1\sigma_2)$ reduces to the simple form

$$\phi(\mathbf{r}_1\mathbf{r}_2\sigma_1\sigma_2) = \frac{1}{\sqrt{2}}\left(|\uparrow>_1|\downarrow>_2 - |\downarrow>_1|\uparrow>_2\right)\phi(|\,\mathbf{r}_1 - \mathbf{r}_2\,|) \tag{11}$$

A more convenient procedure, following BCS, is to relax the constraint of conservation of the total particle number and write the wave function in the occupation space of pairs of single-particle states ($\mathbf{k}\uparrow, -\mathbf{k}\downarrow$). In this representation the wave function corresponding to (10) is the original BCS ansatz

$$\Psi_{BCS} = \prod_k \Phi_k \tag{12a}$$

$$\Phi_k = u_k\ |\,00> + v_k\,|\,11> \tag{12b}$$

$$|\,u_k\,|^2 + |\,v_k\,|^2 = 1 \tag{12c}$$

where the ratio $v_k/u_k$ is the coefficient $c_k$ of the Fourier transform of $\phi(|\,\mathbf{r}_1 - \mathbf{r}_2\,|)$ in eqn. (11). In (12b) the notation $|\,00>$ indicates the state in which neither single-particle state $\underline{k}\uparrow$ nor - $\underline{k}\downarrow$ is occupied, and $|\,11>$ indicates the state in which both are occupied. It turns out to be possible to choose all the $u_k$ to be real and positive without loss of generality, so in view of the normalization condition (12c) the state is completely specified by the set of complex numbers $\{v_k\}$.

The actual form of the $\{v_k\}$ which corresponds to the groundstate must be obtained by minimizing the sum of the kinetic and potential energies. For consistency one must subtract from the kinetic energy $\hat{K}$ a Lagrange-multiplier term$-\mu\hat{N}$ where $\mu$ is the chemical potential (which may be shown to coincide in the weak-coupling limit with the Fermi energy $\epsilon_F$ of the normal phase). Then we have, with all single-particle energies measured relative to $\epsilon_F$

$$< \hat{K} - \mu\hat{N} > = \sum_k 2\epsilon_k \mid v_k \mid^2 \tag{13}$$

and this expression by itself is minimized by the choice $\mid v_k \mid = \theta(\epsilon - \epsilon_k)$ which is the special case of (12) representing the normal groundstate.

As regards the potential energy, this may be regarded as a special case of a more general two-particle quantity of the form

$$Q \equiv \frac{1}{2} \sum_{ij} Q(\mathbf{r}_i - \mathbf{r}_j) \tag{14}$$

For the simple BCS ansatz (12), it may be shown that there are only three nonvanishing contributions to $< Q >$. The first two, the Hartree and Fock terms, are already present in the normal phase, and unless the Fourier transform $Q(\mathbf{k})$ of $Q(\mathbf{r})$ is anomalously sharply peaked at small $\underline{k}$ these terms are essentially unchanged by the onset of pairing. The third term is peculiar to the paired phase: it has the form

$$< Q >_{pair} = \frac{1}{2} \int \int d\mathbf{r} d\mathbf{r}' Q(\mathbf{r} - \mathbf{r}') \mid F(\mathbf{r} - \mathbf{r}') \mid^2 \tag{15}$$

where the quantity $F(\mathbf{r} - \mathbf{r}')$, which may be shown to be equal to the so-called "anomalous average" $< \psi_\uparrow^+ (\mathbf{r}) \psi_\downarrow^+(\mathbf{r}') >$, is given in terms of the $u_k$ and $v_k$ by the formula

$$F(\mathbf{r} - \mathbf{r}') = \sum_k u_k v_k exp \left[ i \left( \mathbf{k} \cdot \mathbf{r} - \mathbf{r}' \right) \right] \tag{16}$$

Comparing (15) with the standard expression for $< Q >$ for a simple diatomic molecule, we see that $F(\mathbf{r} - \mathbf{r}')$ plays the role of the effective relative wave function of the Cooper pairs. It is a more physical quantity than the original two-particle function $\phi(\mathbf{r} - \mathbf{r}')$ occurring in (11) (to which it is closely related but not identical) in that it automatically takes into account the (substantial) effects of antisymmetrization in a degenerate Fermi system. We will investigate the general structure of $F$ below.

The equilibrium form of the $u_k$'s and $v_k$'s is obtained by minimizing the sum expectation values of the $KE$, eqn. (13), and the potential energy, which is a special case of (15). It is conventional, and useful, to parametrize the $u$'s and $v$'s by a complex function $\Delta_k$ which is related to the Fourier transform $F_k \equiv u_k v_k$ of $F(\mathbf{r} - \mathbf{r}')$ by

$$\Delta_k \equiv \frac{2F_k \mid \epsilon_k \mid}{(1 - 4 \mid F_k \mid^2)^{1/2}} \tag{17}$$

When expressed in terms of $\Delta_k$ the equation minimizing the groundstate energy with respect to the $u$'s and $v$'s is the celebrated BCS gap equation:

$$\Delta_k = \sum_{k'} V_{kk'} \, \Delta_{k'}/2E_{k'} \tag{18}$$

where

$$E_k \equiv +(\mid \epsilon_k \mid^2 + \mid \Delta_k \mid^2)^{1/2} \tag{19}$$

so that from (12) we have $F_k = \Delta_k/2E_k$. We shall shortly see the physical significance of the quantities $E_k$ and $\Delta_k$.

To generalize the BCS ansatz to finite temperatures, we express the total many-body density matrix in the form of a product over sub-density matrices $\hat{\rho}_k$ describing the state of occupation of the pair of single-particle states $(\mathbf{k}\uparrow, -\mathbf{k}\downarrow)$:

$$\hat{\rho}_{BCS} = \prod_k \hat{\rho}_k \tag{20}$$

However, in $\hat{\rho}_k$ we must now include, as well as the "ground pair" states $u_k \mid 00 > + \, v_k \mid 11 >$, two "broken pair" states $\mid 10 >$ and $\mid 01 >$ (where e.g. $\mid 10 >$ is the state containing a particle in $\mathbf{k}\uparrow$ and none in $-\mathbf{k}\downarrow$), and also the "excited pair" state which is the linear combination of $\mid 00 >$ and $\mid 11 >$ orthogonal to the ground pair, namely $v_k \mid 00 > - \, u_k \mid 11 >$. It turns out that the broken-pair states have energy $E_k$ relative to the groundstate, while the "excited pair" state has energy $2E_k$; thus, $E_k$ is the minimum excitation energy of any state associated with the pair $(\mathbf{k}\uparrow, -\mathbf{k}\downarrow)$, and if $\Delta_k \equiv \Delta$ is independent of $\mathbf{k}$ as in the simple BCS model (cf. below), then the minimum (fermionic) excitation energy of the whole system is $\mid \Delta \mid$. For this reason the quantity $\Delta_k$ is often known as the "energy gap", though it should be remembered that it is actually a complex quantity.

The principle which determines the (thermal average) form of $u_k$ and $v_k$ at finite temperature is that the free energy $< K - \mu N > + < V > -TS$ must be a minimum, where $S$ is the entropy (which can be written in the form $\sum_k Tr(\hat{\rho}_k \, ln \hat{\rho}_k)$). The resulting forms of $u_k$ and $v_k$ are temperature-dependent, as is the gap $\Delta_k(T)$: the generalization of the $T = 0$ equation is

$$\Delta_k(T) = -\sum_{k'} V_{kk'} \, (\Delta_{k'} \, (T)/2E_{k'} \, (T)) \;\; tanh \beta E_{k'}(T)/2$$

$$(\beta \equiv 1/kT) \tag{21}$$

For the simple BCS model ($V_{kk'} = const. = -V_0$ for $|\epsilon_k|, |\epsilon_{k'}|$ both $\lesssim \epsilon_c$, zero otherwise) the solution of (16) has the following properties, among others:

1. $\Delta_k$ = independent of $\mathbf{k} \equiv \Delta(T)$ (22a)

2. $\Delta(0) = 2\epsilon_c exp - 1/(N(0)V_0)$ (22b)

3. $\Delta(T) \to 0$ at $T = T_c = \Delta(0)/1.76$ (22c)

4. for $T \to T_c$ from below, $\Delta(T) \cong 3.06kT_c(1 - T/T_c)^{1/2}$ (22d)

For a plot of the real-space behavior of the "pair wave function" $F(r)$ as a function of relative coordinate $r$, see ref. 1, Fig. 3.1 (for $\ell = 1$). The salient point is that at large distances $F(r)$ falls off as $exp - r/\xi$, where the quantity $\xi$ can be regarded as effectively the "radius" of the pair; it is given at $T = 0$ by $\xi_o = \hbar v_F/\pi k T_c$, and for real-life $^3$He varies from about 150 Å at m.p. to ~ 750Å at s.v.p. It is important to note that unlike the Ginzburg-Landau correlation length $\xi(T)$ which we shall meet in the next lecture, the pair radius $\xi$ is not strongly temperature-dependent and remains perfectly finite as $T \to T_c$.

## 5. The superfluid phases of liquid $^3$He

In the previous section I sketched the essentials of BCS theory for a Fermi superfluid in which the pairs are formed in a relative state with $\ell$ = S = 0. The first task is to generalize the description to the $\ell$ = S = 1 states believed to describe the superfluid (A, B and $A_1$) phases of liquid $^3$He. I shall again proceed somewhat intuitively, referring to refs.[1] and [3] for technical details.

In the context of the line of argument developed in Lecture 1, it is plausible (and turns out to be true) that the correct generalization to pairing with $\ell$, S $\neq$ 0 is to keep eqn. (10) but to allow the function $\varphi(\mathbf{r}_1\mathbf{r}_2\sigma_1\sigma_2)$ to have an arbitrary dependence on $\mathbf{r}_1$-$\mathbf{r}_2$, $\sigma_1$ and $\sigma_2$. (We still wish to keep the center of mass at rest, thus exclude a dependence on $(\mathbf{r}_1+\mathbf{r}_2)/2$.) This then leads to a pair wave function $F(\mathbf{r}-\mathbf{r}',\sigma\sigma')$ which is a generalization of the quantity introduced in eqn. (16). For most purposes connected with the orientational degrees of freedom, it is legitimate to treat F just as if it were the relative wave function of a simple diatomic molecule.

Let's assume for simplicity that the spin state is pure triplet, S = 1. (Mixtures of triplet and singlet states are, in principle, possible, but not of much practical interest). There are then three Zeeman substates, which I represent symbolically as $|\uparrow\uparrow\rangle$ ($S_z$ = +1), $|\downarrow\downarrow\rangle$ ($S_z$ = -1) and $2^{-1/2}(|\uparrow\downarrow\rangle$

$+ | \downarrow\uparrow\rangle)$ ($S_z = 0$), and the pair wave function F can thus be expanded in the form

$$F(\mathbf{r}-\mathbf{r}':\sigma\sigma') = F_{\uparrow\uparrow}(\mathbf{r}-\mathbf{r}')|\uparrow\uparrow\rangle + 2^{-1/2}F_{\uparrow\downarrow}(\mathbf{r}-\mathbf{r}')|\uparrow\downarrow + \downarrow\uparrow\rangle + F_{\downarrow\downarrow}(\mathbf{r}-\mathbf{r}')|\downarrow\downarrow\rangle \quad (23a)$$

or equivalently,

$$F(\mathbf{r}-\mathbf{r}':\sigma\sigma') = \sum_{\alpha\beta} F_{\alpha\beta}|\alpha\beta\rangle \qquad (F_{\alpha\beta} \equiv F_{\beta\alpha}) \quad (23b)$$

where for the moment the various $F_{\alpha\beta}$'s are arbitrary functions of **r**-**r**′, except that to satisfy the antisymmetry requirements they must be odd in their argument. Note that in general we cannot write (23) as a product of independent space and spin functions. The form (23) represents a system which has **nontrivial orientational degrees of freedom** in both spin and orbital space.

In the case of liquid $^3$He it is generally believed, principally on the basis of experiments such as ultrasound absorption which probe the shape of the energy gap over the Fermi surface (see below) that all the superfluid phases are described by a form of the $F_{\alpha\beta}$'s which is almost pure p-wave (with a very small admixture of higher odd-$\ell$ harmonics). The configurations of the principal (A and B) phases are believed to be as follows:

**A Phase:** In this case the function $F(\mathbf{r}\text{-}\mathbf{r}':\sigma\sigma')$ can in fact be written as a product $\varphi(\mathbf{r}\text{-}\mathbf{r}')\ \chi(\sigma\sigma')$.

The spin function $\chi$ corresponds to a state which has $\mathbf{S}\bullet\hat{\mathbf{d}} = 0$, i.e. is an eigenstate of $S_z$, corresponding to $S_z = 0$, along some characteristic axis which is conventionally labelled $\hat{\mathbf{d}}$.

The orbital state is one which, if realized in a diatomic molecule in free space, would correspond to angular momentum $\hbar$ along a second characteristic axis, which is conventionally labelled $\hat{\boldsymbol{\ell}}$.

Thus, **time reversal invariance is spontaneously broken** in the A phase. We can define an "energy gap" $\Delta_\mathbf{k}$ in terms of the Fourier transform of $\varphi(\mathbf{r}\text{-}\mathbf{r}')$ just as in eqn. (17); the resulting $\Delta_\mathbf{k}$ is "nontrivially" complex (i.e. complex by more than an overall phase common to all **k**), and its modulus $|\Delta_\mathbf{k}|$ is independent of $\epsilon_k$ but proportional to $|\hat{\mathbf{k}}\times\hat{\boldsymbol{\ell}}|$, thus it varies over the Fermi surface, with nodes at the points $\hat{\mathbf{k}} = \pm\hat{\boldsymbol{\ell}}$. As a result, the minimum energy of excitation of the A phase is actually zero, since it is possible to excite quasiparticles arbitrarily close to the nodes and thus with arbitrarily low energy; in fact, the temperature-dependences of quantities associated with the normal component (such as the specific heat or normal density) are power laws in the A phase, rather than the exponential forms characteristic of the original ($\ell = S = 0$) BCS theory. The A phase is anisotropic in its spin and orbital properties separately, with the characteristic "orientational axis" being $\hat{\mathbf{d}}$ and $\hat{\boldsymbol{\ell}}$ respectively; under most physical conditions these vectors are aligned parallel by the weak nuclear dipole force.

**B Phase:** In this case the pair wave function $F(\mathbf{r}\text{-}\mathbf{r}':\sigma\sigma')$ cannot be factorized into spin and orbital components. Rather, the simplest way of describing it is to start with a $^3P_o$ state ($L = S = J = 0$) and to rotate the spins **relative** to the orbital coordinates through an angle $\theta$ around an axis $\hat{\omega}$. In practice, $\theta$ is usually fixed at a value $\sim 104°$ by the weak nuclear dipole forces, but the axis $\hat{\omega}$ is fixed only by various even weaker and often mutually competing effects. Since right- and left-handed rotations are not equivalent, **spatial inversion invariance is spontaneously broken in the B phase** (but time reversal invariance is not). The theory of the "energy gap" is a bit more complicated then in the A phase, (see ref. [1] or [3]), but at the end of the day one obtains a simple result: the quasiparticle excitation energy is given as before by the BCS-type formula

$$E_k = (\epsilon_k^2 + |\Delta_k|^2)^{1/2} \tag{24}$$

where, moreover, the gap magnitude $|\Delta_k|$ is **constant** over the Fermi surface (unlike in the A phase). The spin-independent thermodynamic properties (specific heat, normal density etc.) of the B phase are therefore identical to those of a simple BCS superconductor. More generally, the B phase is isotropic as regards its orbital and spin properties individually; however, it is strongly **anisotropic** as regards those properties such as nuclear magnetic resonance which couple the spin and orbital degrees of freedom, with a characteristic orientational axis $\hat{\omega}$

For completeness, I should mention that the $A_1$ phase, which occurs in a strong magnetic field as a sliver in the phase diagram separating the A and N phases, is similar to the A phase except that the spin part of the pair wave function corresponds to a pure $S_z = 1$ state along some axis $\hat{\mathbf{e}}$ (usually that of the applied magnetic field).

## 6. Time-independent Ginzburg-Landau theory: isotropic case

As is well known, the phenomenological Ginzburg-Landau (GL) theory of a Fermi superfluid (superconductor) antedates the microscopic BCS description by several years, and has proved extremely useful in predicting the macroscopic-scale behavior of such systems. Following the seminal work of Gor'kov [4] and others, we can now justify the GL theory in terms of the BCS one, at least in certain limits. For pedagogical simplicity I will start with the $\ell = S = 0$ case and again refer for intermediate steps in the argument to Sections V.D and E of ref. [3].

The GL theory works in terms of a complex scalar "order parameter" $\Psi(\mathbf{r})$, and our first task is to identify this with an appropriate quantity occurring in the BCS theory. In the context of the latter, let's relax the assumption that the COM of the Cooper pairs is necessarily at rest and thus define a more general pair wave function F($\mathbf{r}$ , $\mathbf{r}'$) by

$$F(\mathbf{r},\mathbf{r}') \equiv \langle \psi_\uparrow^\dagger(\mathbf{r})\psi_\downarrow^\dagger(\mathbf{r}')\rangle \equiv F(\mathbf{R},\boldsymbol{\rho}) \tag{25}$$

where

$$\mathbf{R} \equiv \frac{1}{2}(\mathbf{r}_1 + \mathbf{r}_2) \quad , \quad \boldsymbol{\rho} \equiv \mathbf{r}_1 - \mathbf{r}'_2$$

For the $\ell$ = S = 0 case under consideration, and in near-equilibrium situations, the dependence of F($\mathbf{R}$,$\boldsymbol{\rho}$) on the relative coordinate $\boldsymbol{\rho}$ is fixed at given P and T by the energetics, and extends roughly over the pair radius $\xi_o$, a quantity which we recall is not strongly temperature-dependent and is of order $\hbar v_F/\Delta(0)$. Let us assume that the variation of F as a function of the centrum of mass (COM) coordinate $\mathbf{R}$ is small over distances $\sim \xi_o$. Under these conditions (only!) we can make the identification, up to a possible normalization factor which is a matter of convention,

$$\Psi(\mathbf{R}) \equiv F(\mathbf{R}:\boldsymbol{\rho}=0) \tag{26}$$

i.e. the GL order parameter is nothing but the wave function of the COM of the Cooper pairs, its argument being simply the COM coordinate $\mathbf{R}$. It should be stressed that implicit in this definition is a sort of coarse-graining over distances $\sim \xi_o$, so that it is not useful, for a superfluid Fermi system, to try to define an order parameter over scales smaller than this.

We next raise the question: For the spatially uniform case, ($\Psi$(R) = const.), how does the free energy behave as a function of $\Psi$? It is important to realize that this question only makes sense (or more accurately only has a unique answer) under the assumption that all degrees of freedom other than the pair wave function, and in particular the quasiparticle distribution, have attained thermal equilibrium subject to the prescribed value of $\Psi$.

The potential-energy term in the free energy is trivial to calculate, provided we make the usual ansatz V($\mathbf{r}$) = -$V_o\delta(\mathbf{r})$ (with an implicit cutoff in the sums over $\mathbf{k}$ at $|\epsilon_k| = \epsilon_c$ as in the usual BCS theory). In that case we have simply

$$\langle V\rangle = -\frac{1}{2}V_o|\Psi|^2. \tag{27}$$

The calculation of the kinetic-energy and entropy terms as a function of $\Psi$ is considerably more complicated, and is most easily done by introducing a Lagrange multiplier $\Delta$ for $\Psi$ (sometimes regarded as an "off-diagonal field") which turns out, after some nontrivial algebra, to be related to $\Psi$ by the implicit equation

$$\Psi(\equiv \sum_k F_k) = \Delta \sum_k (2E_k)^{-1} tanh\beta E_k/2$$
$$(E_k \equiv (\epsilon_k^2 + |\Delta|^2)^{1/2}) \tag{28}$$

It turns out to be convenient to express the free energy (relative to that of the normal state at temperature) in terms of $\Delta$ rather than $\Psi$, but even so the result is messy and I quote the results only in the limits T = 0 and T $\to$ $T_c$. For T = 0 we have

$$E(\Delta) = N(0)|\Delta|^2\left\{ln|\Delta/\Delta_o| - \frac{1}{2}\right\} \tag{29}$$

where $\Delta_o$ corresponds to the equilibrium value of $\Delta$ and is given by the standard BCS formula $\Delta_o = 2\epsilon_c$ exp-1/(N(0)$V_o$). In the limit T $\to$ $T_c$ the quantities $\Delta$ and $\Psi$ are related by a factor which is temperature-independent in this limit (cf. eqn. (28)), so it does not matter which we use. However, the coefficients take a simpler form if we write F in terms of $\Delta$:

$$F(\Delta : T) = -\alpha(1 - T/T_c)|\Delta|^2 + \frac{1}{2}\beta|\Delta|^4 + \ldots \tag{30}$$

with

$$\alpha \equiv N(0), \quad \beta \equiv (7/8)\zeta(3)(\pi k_B T_c)^{-2} \tag{31}$$

Minimization of F with respect to $\Delta$ then gives back eqn. (22d) of L. 1. (Incidentally, the argument [3] also automatically gives the relation (22c) between $T_c$ and $\Delta(0)$).

So far we considered only the case that $\Psi$ is constant in space. In considering the more general case let's first assume that $\Psi$ is (a) slowly varying and (b) **small**. Then we expect, on general grounds of analyticity etc., that the term of lowest order in the spatial gradients will have the form

$$F_{grad} = \gamma(T)|\boldsymbol{\nabla}\Psi|^2 \tag{32}$$

If this is so, the cofficient $\gamma$(T) may be obtained by considering the special case $\Psi(\mathbf{R})$ = $|\Psi|$ exp($2im\mathbf{v}_s \cdot \mathbf{R}/\hbar$), which corresponds to superfluid flow with superfluid velocity $\mathbf{v}_s$ and hence, by definition, to an extra energy $\frac{1}{2}\rho_s v_s^2$ where $\rho_s$ is the superfluid density as conventionally defined within the two-fluid model [5]. Substituting into (32) and assuming that the normalization of $\Psi$(T) is chosen so as to make it equal to $\Delta$(T), we obtain

$$\gamma(T) = (\hbar^2/8m^2)\rho_s(T)/|\Delta|^2 \tag{33}$$

Now the quantity $\rho_s$(T) (or rather its complement $\rho_n$(T)) can be calculated by the standard Landau formula [5]; $\rho_s$ turns out to be proportionaol to $|\Delta|^2$, so we find in the limit T $\rightarrow$ $T_c$

$$\gamma(T) \sim const. = \frac{N\hbar^2}{8m}\frac{m}{m*}\frac{7\zeta(3)}{4\pi^2}\frac{1}{(k_BT_c)^2} \tag{34}$$

If we add the gradient term (32) to the uniform terms (30) in the free energy, we see that this defines a characteristic length $\xi_{GL}$(T) given by

$$\xi_{GL}(T) \equiv (\gamma/\alpha)^{1/2}(1-T/T_c)^{-1/2} \tag{35}$$

which is usually called the "healing length" or, more commonly, the Ginzburg-Landau correlation length; this is the characteristic length over which the order parameter returns to its bulk value when perturbed e.g. by the effects of a wall. If we substitute eqns. (34) and (31) and use the definition $\xi_o = \hbar v_F/\Delta(0)$ of the "pair radius" we find after a little algebra (note that N(0) = $3N/2m^*v_F^2$!)

$$\xi_{GL}(T) = \left[\frac{7\zeta(3)}{48}\right]^{1/2}(\Delta(0)/\pi k_BT_c)\xi_o(1-T/T_c)^{-1/2} \approx 0.3\,\xi_o(1-T/T_c)^{-1/2} \tag{36}$$

Now, we recall that the GL description is valid only if the spatial variation of $\Psi$ is slow on a scale $\sim \xi_o$. Thus, in view of the small numerical prefactor in (36), the GL theory will be applicable to a typical "healing" situation only when T is very close to $T_c$.

Given this result, one might at first sight think there is no point in trying to generalize the GL description to lower temperatures. However, in a neutral superfluid one not infrequently finds situations where the actual scale of variation is much larger than $\xi_{GL}$ (cf. e.g. Section X.D of ref. [3]), and thus there is no *a priori* objection in a GL description for T far from $T_c$. Both the bulk and the gradient terms in the free energy are extremely messy in the intermediate temperature regime, and there is no point in writing them out explicitly, but I would like to call attention to one point which is, I believe, not universally appreciated, namely that in general the lowest gradient term in F is **not** of the form (32). In fact, one can write two different second-order gradient terms, namely

$$(a)|\Psi|^2(\nabla\varphi)^2 \qquad (b)|\nabla|\Psi||^2 \tag{37}$$

and one finds in general that the coefficients of (a) and (b) are different; indeed, in superfluid $^3$He they (or the related more general quantities, cf. below) may differ by a factor of $\sim$ 10 in some temperature regimes.

## 7. Time-independent GL Theory: Anisotropic case

In the case of pairing with $\ell = S = 1$ as in superfluid $^3$He we need, first, to generalize the definition of the "order parameter." The conventional procedure (which is closely related but not quite identical to that pursued above for the $\ell = S = 0$ case) goes as follows: We consider the case of Cooper pairs at rest, and define a quantity $F_{\alpha\beta}(\mathbf{r}\text{-}\mathbf{r}')$ by

$$F_{\alpha\beta}(\mathbf{r} - \mathbf{r}') \equiv \langle \psi_\alpha^+(\mathbf{r})\psi_\beta^+(\mathbf{r}')\rangle \tag{38}$$

and consider its Fourier transform $F_{\alpha\beta}(\mathbf{k})$: this latter quantity depends on the direction but not (strongly) on the magnitude of $\mathbf{k}$, so we take the limit $|\mathbf{k}| \to k_F$ and define $\mathbf{k} \equiv k_F\hat{\mathbf{n}}$ and

$$F_{\alpha\beta}(\hat{\mathbf{n}}) \equiv i\sum_{i=1}^{3}(\sigma_i\sigma_2)_{\alpha\beta}A_i\hat{\mathbf{n}} \tag{39}$$

Now, to an excellent approximation the quantities $F_{\alpha\beta}$ $(\mathbf{r}\text{-}\mathbf{r}')$ and hence $A_i(\hat{\mathbf{n}})$ are composed entirely of p-wave spherical harmonics, so we can write

$$A_i(\hat{\mathbf{n}}) \equiv \sum_{\alpha=1}^{3} A_{i\alpha}\hat{n}_\alpha \tag{40}$$

The set of 9 complex quantities $A_{i\alpha}$($i = 1,2,3, \alpha = 1,2,3$) then plays the role of a generalized order parameter for the system. In a situation where the state of the Cooper pairs is slowly varying in space, the natural generalization is to allow the $A_{i\alpha}$ to be functions of the center-of-mass coordinate $\mathbf{R}$. (Note that such a situation need not necessarily correspond to overall flow of the pairs; it could, for example, correspond to a "spin supercurrent" or something more complicated, e.g. the interface between the A and B phases). So far the discussion is purely descriptive and is not restricted to any particular regime of temperature, etc.

The form of the free energy as a function of the $A_{i\alpha}$ at general temperatures is even messier then in the isotropic case, and I shall here consider only the limit $T \to T_c$, where one can make an expansion of F up to fourth order in $A_{i\alpha}$ (cf. eqn. (30)). The resulting expression is strongly constrained by the fact that F must be invariant not only under gauge transformations but also (to the extent that we neglect the very small nuclear dipole-dipole interaction) under rotations of the spin and orbital coordinates separately. With these constraints the most general allowed form of F is

$$F\{A_{i\alpha} : T\} = -\alpha(1 - T/T_c)\sum_{i\alpha}|A_{i\alpha}|^2$$

$$+\frac{1}{2}\sum_{\substack{ijkl\\ \alpha\beta\gamma\delta}} \beta_{\substack{ijkl\\ \alpha\beta\gamma\delta}} A^*_{i\alpha}A^*_{j\beta}A_{k\gamma}A_{\ell\delta} \tag{41}$$

where the coefficient $\beta$ can consist only of a sum of 5 different terms with definite symmetries (e.g. $\delta_{ij}\delta_{k\ell}\delta_{\alpha\beta}\delta_{\gamma\delta}$ ). In the context of quenching, a crucial point to notice about (41) is that because the coefficient of $|A_{i\alpha}|^2$ in the second-order term is independent of $i$ and $\alpha$, **all possible p-wave states have the same $T_c$**. Below $T_c$, however, different choices of the $A_{i\alpha}$ may induce different values of the fourth-order term, and the most stable state at any given P and T will obviously be the one corresponding to the minimum value of this term, which will then have a unique form of the $A_{i\alpha}$ up to symmetry transformations. In particular the form of the $A_{i\alpha}$ for the A and B phases described above is

$$A\ phase: A_{i\alpha} = const.(P,T) \times (\delta_{\alpha,x} + i\delta_{\alpha,y}) \tag{42a}$$

$$B\ phase: A_{i\alpha} = const.(P,T) \times \delta_{i\alpha} \tag{42b}$$

but it should be remembered that to the extent that the nuclear dipole interaction and other small symmetry-breaking terms can be neglected, we can start from the expressions (42) and rotate the spin and orbital coordinates separately without affecting the expression (41). (Thus, for example, the most general form of the B phase is represented by $A_{k\alpha}$ = const. (P,T) $\times \hat{R}_{i\alpha}$, where $\hat{R} \in SO(3)$ is a 3D orthogonal matrix.)

The considerations concerning the gradient terms are similar to those in the isotropic case, but there are 3 independent invariants near $T_c$ and more at lower temperatures: see e.g. ref.[1], Ch. 7.

One puzzle which was noticed and, at least in large part, resolved very early in the history of research on superfluid $^3$He is: why is the A phase ever thermodynamically stable? In fact, if one uses a standard BCS-type theory with a fixed interparticle interaction V(r) to calculate the five fourth-order invariants $\beta$ in eqn. (41), one finds that their relative values are such that the phase described by (42b), i.e. the B phase, is always the most stable. The elegant solution to this dilemma proposed by Anderson and Brinkman [6] goes schematically as follows: A large part of the effective attraction in liquid $^3$He is actually of an indirect nature and mediated, just as in the case of a classic superconductor, by the "polarizability" (in this case the spin rather than charge polarizability) of the medium. Now, the important difference between the two cases is that in a superconductor the "medium" (ions) is a different system from the electrons undergoing pairing, and thus their polarizability is virtually unchanged by the onset of the latter, whereas in $^3$He the "medium" is nothing but the atoms themselves, and thus the onset of pairing may be expected to change the polarizability, in a way which is in general anisotropic in spin space. In fact, in the very simplest implementation of this idea the effective interaction (8) is generalized to

$$V_{eff}(\mathbf{r}-\mathbf{r}',t-t') \sim -\sigma_i\sigma_j'\chi_{ij}(\mathbf{r}-\mathbf{r}',t-t') \tag{43}$$

where $\chi_{ij}$ is the **superfluid-state** spin susceptibility and in general a function of the $A_{i\alpha}$. On calculating this dependence, one finds that the resulting contribution $\Delta F_{s.f.}$ to the fourth-order terms in (41) has the form

$$\Delta F_{s.f.} \sim -\sum_{ij}(\sum_{\alpha} A_{i\alpha}^* A_{j\alpha})^2 \tag{44}$$

and the magnitude of this term is larger for the form (42a) then for (42b), thus permitting the A phase possibly to be stable in some regions of the phase diagram.

## 8. The time-dependent GL equation

In the context of discussions of quenching in superfluid $^3$He, an essential generalization of the Gl theory is to make it applicable to situations where the order parameter is changing in time, i.e. to derive a "time-dependent Ginzburg-Landau" (TDGL) equation for $\Psi(\mathbf{R},t)$. Unfortunately, a microscopic justification for such a generalization is available only in very restricted regions of the phase diagram. For simplicity I treat here only the isotropic case and the region close to $T_c$ where the free energy has the simple form (30); anisotropy seems unlikely to make the situation any better though it could conceivably make it worse.

Intuitively, one would expect that the general form of the TDGL equation for sufficiently slow variation of $\Psi(\mathbf{R},t)$ in time would be

$$\delta(T)\frac{\partial\Psi(\mathbf{R},t)}{\partial t} = -\frac{\delta F\Psi(\mathbf{R},t)}{\delta\Psi} \tag{45}$$

where $\delta(T)$ is a real positive quantity. If so, then in the limit $T \to T_c$ we should get (using eqns. (30) and (32) above and choosing the normalization of $\Psi$ so as to make it equal to $\Delta$ in this limit)

$$\delta(T)\frac{\partial\Delta}{\partial t} = +\alpha(1-T/T_c)\Delta - \beta|\Delta|^2\Delta + \gamma\nabla^2\Delta \tag{46}$$

So the question is: (a) can eqn. (46) be justified from microscopic theory, and (b) if it can, what is $\delta(T)$?

In addition to the obvious condition, required already for the validity of the time-independent GL description, that the spatial variation of $\Psi$ be slow over the pair radius $\xi_o$, the validity of (46) puts extremely strong constraints on its variation in time. In particular, it is necessary that the normal component should be close to local thermal equilibrium (i.e. $\omega\tau_n \ll 1$, where $\omega^{-1}$ is a typical time scale of variation of $\Psi$ and $\tau_n$ is the normal-state collision time). Early attempts to derive a TDGL equation proceeded by means of an expansion in $\Delta/\omega$, and therefore reached the conclusion that such an equation would be justified only in the limit $\Delta(T)\tau/\hbar \ll 1$; in view of the fact that in $^3$He the ratio $\hbar/(kT_c\tau_n) \sim 0.5\text{-}3 \times 10^{-2}$, we see from eqn. (22d) that we are in this limit only when T is within $\sim 0.1\%$ of $T_c$. In this regime it can be shown that the quantity $\delta(T)$ is temperature-independent and equal to $(8N(0)kT_c/\pi\hbar)^{-1}$ so that the relaxation time $\tau_\psi = \delta/\alpha$ of the condensate to its equilibrium value is given by $(\pi\hbar/8k_BT_c)(1\text{-}T/T_c)^{-1}$. More recently, it has been shown (see e.g. ref.[7]) that a TDGL equation can be obtained under rather more general conditions, without the restriction $\Delta\tau_n \ll 1$; the result for the relaxation time $\tau_\psi$ of the condensate is

$$\tau_\psi = \frac{\pi}{8kT_c/\hbar} \frac{(1+4\Delta^2(T)\tau_n^2)^{1/2}}{(1-T/T_c)} \tag{47}$$

For $\Delta\tau_n \ll 1$ we recover the result quoted above, while for $\Delta\tau_n \gg 1$ we obtain, using (22d), $\tau_\psi \sim \tau_n(1\text{-}T/T_c)^{-1/2}$. In fact, we see that for **arbitrary** $\Delta\tau_n$ we have the inequality

$$\tau_\psi \geq (4/(3.06\pi))\tau_n(1-T/T_c)^{-1/2} \tag{48}$$

The RHS of eqn. (48) is thus the **minimum** time necessary for the condensate to relax to its equilibrium value under conditions (e.g.) of a "temperature quench" (cf. below). It should be noted that for fast quenches this time, evaluated at the "freeze-in" value $\epsilon_c$ of $1\text{-}T/T_c$, may be orders of magnitude larger then the estimate $\tau \sim \xi(T)/c_s(T)$ ($c_s$ = speed of second sound) (an estimate which, while originally made in ref.[8] for $^4$He, has been subsequently widely used in discussions of quenching in $^3$He).

Although the above is the established wisdom in the existing literature on the TDGL equation, it is almost certainly not the whole story as regards the quenching problem. In particular, the derivation [7] of (45) implicitly assumes, inter alia, (a) that the deviation of physical quantities (distribution functions etc.) from their equilibrium values is at all times small, and (b) that the parameters of the Hamiltonian, in particular the pairing interaction $V_o$, are constant in time. Assumption (a) is certainly not true for a typical quench (if it were, the process would not deserve the name "quench"!) and assumption (b) may in effect fail for a "pressure" quench. I believe therefore that it may be necessary to reconsider this problem from scratch, and hope to do so elsewhere in the future; for the present, all one can say is that the considerations of this section should, at the least, make us cautious in applying simple hydrodynamic considerations to a very degenerate Fermi system such as liquid $^3$He where the kinetics of collisions may turn out to play a dominating role.

## References


1. Vollhardt, D. and Wölfle, P. (1990) *The Superfluid Phases of $^3$He*, Taylor and Francis, London.
2. Nozières, P. (1964) *Theory of Interacting Fermi Systems*, W. A. Benjamin, Inc., New York.
3. Leggett, A. J. (1975) A theoretical description of the new phases of liquid $^3$He, *Rev. Mod. Phys.* **47**, pp.331-420.
4. Gor'kov, L. P. (1958) *Zh. Eksp. Teor. Fiz.* **36**, pp.1918-1925 ; translation, (1959)*Soviet Physics JETP* **36**, pp.1364-1375.
5. Khalatnikov, I. M. (1965) *An Introduction to the Theory of Superfluidity*, Benjamin, New York.
6. Anderson, P. W. and Brinkman, W. F. (1973) Anisotropic Superfluidity in $^3$He *Phys. Rev. Lett.* **30**, pp.1108-1111.
7. Ivlev, B. N. and Kopnin, N. B. (1984) Electric current and resistive states in thin superconductors, *Adv. Phys.* **33**, pp.47-114.
8. Zurek, W. H. (1985) Cosmological experiments in superfluid helium ? *Nature* **317**, pp.505-508.

# FORMATION, INTERACTION AND OBSERVATION OF TOPOLOGICAL DEFECTS

TANMAY VACHASPATI
*Physics Department, Case Western Reserve University*
*Cleveland, OH 44106-7079, USA.*

In these lectures, I will describe the formation of defect distributions in first-order phase transitions, then briefly discuss the relevance of defect interactions after a phase transition and the observational signatures of cosmic strings. Some open questions are also discussed.

## 1. Formation of Defects

A number of talks at this school have been dedicated to the density of topological defects formed at a phase transition. Here, I will mostly be concerned with the distribution of defects formed at a phase transition.

I will start with a description of the usual procedure to study the formation of defects (focussing on the case of strings), point out some shortcomings, and then move on to describe the connection of the defect formation problem with froths and percolation. An enormous range of problems is still left untouched and I will end up by describing some of these.

### 1.1. $U(1)$ STRING NETWORK: CONVENTIONAL ANALYSES

The formation of strings can be studied numerically by assigning the $U(1)$ phase, $\alpha$, randomly on lattice sites - say of a cubic lattice. Then one can evaluate differences in $\alpha$ along the edges of each plaquette of the lattice. To do this, it is necessary to interpolate between the values of $\alpha$ at two neighbouring sites. Then one finds the integral

$$\Delta\alpha \equiv \oint d\alpha \tag{1}$$

around a plaquette. If this is non-zero, it indicates that there is a string or anti-string passing through the plaquette. In this way, all the strings are found. Then they are connected and information about the distribution of string is stored.



*Y.M. Bunkov and H. Godfrin (eds.),*
***Topological Defects and the Non-Equilibrium Dynamics of Symmetry Breaking Phase Transitions*, 55–75.**

The surprising result that emerges from numerical simulations [1] is that most of the energy in the string network is in infinite strings. Furthermore, the strings are Brownian on large scales, and the loop distribution is scale invariant. Let us explain these results in more detail:

- *Brownian strings:* This means that the length $l$ of a string is related to the end-to-end distance $d$ by

$$l = \frac{d^2}{\xi} \tag{2}$$

  where $\xi$ is a length scale also called the step length which would be roughly given by the lattice spacing. This result is valid for large $l$ to a good approximation. (The numerical results give an exponent of about 1.92 rather than 2.00.) For smaller lengths, the walk is not Brownian as lattice effects are present.
- *Loop distribution:* Scale invariance means that there is no preferred length scale in the problem apart from the lattice cut-off. Then, scale invariance would imply that the number density of loops having size between $R$ and $R + dR$ is given by dimensional analysis:

$$dn(R) = c\frac{dR}{R^4}$$

  where $c \sim 6$. Using eq. (2), this may be written as:

$$dn(l) = \frac{c}{2\xi^{3/2}}\frac{dl}{l^{5/2}} \ .$$

  Note that the scale invariance is in the *size* of the loops and not in their *length.* (This is because, if we were to examine the network with a magnifying glass, the sizes of the loops would be rescaled by the magnification factor but not their lengths.)
- *Infinite strings:* With the implementation of the geodesic rule, the density in infinite strings was estimated to be about 80% of the total density in strings. The way this estimate was made [1] was to perform the simulation on bigger and bigger lattices and to keep track of the length in the strings that were longer than a large critical length (compared to the lattice size). As the lattice was made bigger, the fraction of string in long strings tended to stabilize around 80%. Simulations on other lattices and with periodic boundary conditions also yield infinite strings but the estimated fraction can vary upward from about 74%. Analytic estimates of the fraction which assume that the strings are random walks on a lattice, are consistent with these estimates [2].

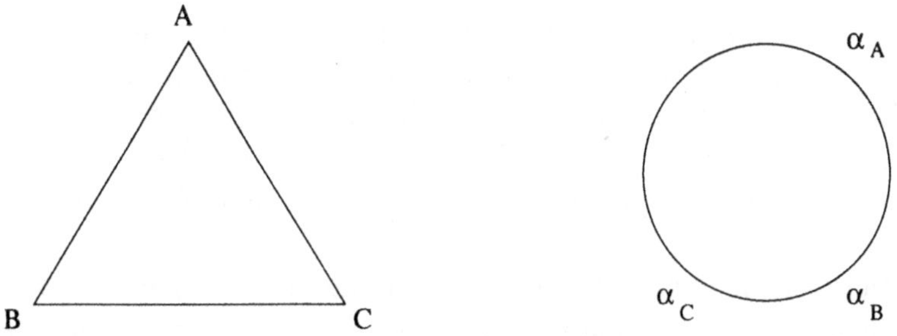


*Figure 1.* The triangle is a plaquette in space and the circle denotes the vacuum manifold. At each vertex of the plaquette a phase is assigned at random. In traversing from $A$ to $B$ on the triangle, the phase must change from $\alpha_A$ to $\alpha_B$. However, there is an infinite degeneracy in the path from $\alpha_A$ to $\alpha_B$ on the vacuum manifold since the path can wrap around the entire circle any number of times.

Can one analytically see the presence of infinite strings? This is an open question. Some progress can be made if one assumes that strings perform a Brownian walk [2]. It is known that random walks do not close in 3 dimensions and this tells us that infinite strings will be present. Furthermore, estimates can be obtained for the fraction of length in infinite strings and the result is similar (though not identical) to the one obtained in simulations.

For cosmological applications such as the formation of large-scale structure, the existence of infinite strings can be vital. The reason is that the small closed loops can decay by emitting gravitational and other forms of radiation but the infinite strings are destined to live forever because of their topological character[1]. So only the infinite strings (and their off-spring loops) could live to influence late time cosmology and also to tell us the story of the cosmological phase transition[2].

A subtle issue in calculating the integral in eq. (1) is the interpolation as we go from one lattice site to another. Consider a $U(1)$ string simulation as shown in Fig. 1. As we traverse the triangle ABC in space, the phase varies from $\alpha_A$ to $\alpha_B$ to $\alpha_C$ and then back to $\alpha_A$. These are simply points on a circle and we know that there are infinitely many paths joining any two points on a circle. (The paths can go around the circle infinitely many times.) So at every stage of the construction, we need to interpolate between the phases and there is an infinite-fold ambiguity in this interpolation. How do we resolve this ambiguity?

1The two ways in which they could decay are: a) a string meets an antistring and annihilates, and, b) a string snaps leading to a gravitational singularity. Neither process is expected to occur at a rate that would be cosmologically interesting.

2If the loop density is high enough, the loops could reconnect and lead to the formation of infinite strings [3].

In the case of global strings, it is assumed that the shortest of the infinitely many paths is the correct one. The rationale for this choice is that the free energy density gets contributions from a term $|\nabla\alpha|^2$ and this is least for the shortest path. The rule of choosing the shortest path to interpolate between two points on the vacuum manifold is known as the "geodesic rule".

In the case of gauge strings, the rationale for the geodesic rule breaks down since the contribution to the free energy involves the covariant derivative of $\alpha$ and not the ordinary derivative. Now which path should be chosen? Following the logic of the global case, it should be the path that minimizes $|\nabla\alpha - eA|^2$, but this would mean keeping track of the gauge field $A$ as well, which would make the simulation much more difficult.

I will now discuss how one might do away with the assumptions of the geodesic rule and the regular lattice in the simulations of string formation.

1.1.1. *Relaxing the Geodesic Rule*

A possible cure for the ambiguity in choosing the path on the vacuum manifold (discussed above) is to relax the geodesic rule and assume that the phase difference between two lattice sites is given by a probability distribution [4]. If the values of the phases at lattice sites 1 and 2 are $\alpha_1$ and $\alpha_2$, the phase difference will be

$$\Delta\alpha = \alpha_2 - \alpha_1 + 2\pi n \equiv \delta\alpha + 2\pi n$$

where $n$ is a random integer drawn from some distribution. A convenient choice for the distribution is

$$P_n = \int_{n-0.5}^{n+0.5} dm\ 2\sqrt{\pi\beta} e^{-\beta(\delta\alpha+2\pi m)^2} \ . \tag{3}$$

with $\beta \geq 0$ being a parameter. This probability distribution is consistent with the idea that longer paths on the circle should be suppressed but the amount of suppression depends on the choice of $\beta$. Note that $\beta$ plays the role of inverse temperature since lower values of $\beta$ (that is, higher temperatures) allow for larger values of $n$ while larger values of $\beta$ reduce the algorithm to the geodesic rule.

In simulations that relax the geodesic rule [4], it is found that the fraction of infinite strings gets larger with smaller values of the parameter $\beta$ (see Fig. 2). To understand this result, note that the smaller the value of $\beta$, the higher is the total amount of string per plaquette because the chance of going around the vacuum manifold increases. Now the higher the string density, the more difficult is it for a string to close since there are more ways for it to connect with other strings and wander away. Hence, the smaller the

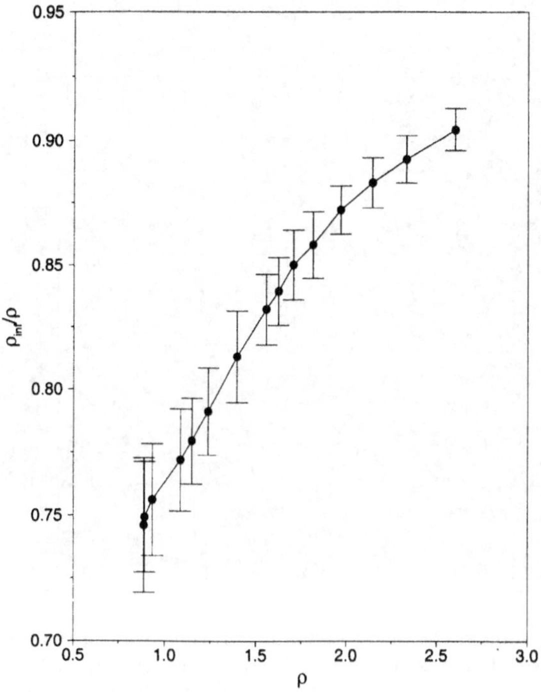


*Figure 2.* A plot of the infinite string density fraction versus the total string density. The total string density increases as the parameter $\beta$ is lowered. The geodesic rule is recovered in the limit that $\beta$ becomes very large.

value of $\beta$, the higher is the fraction of infinite strings. Stated differently, relaxing the geodesic rule leads to a greater fraction of infinite string.

### 1.1.2. *Problems with Lattice Based Simulations*

The simplest way to see that lattice based simulations might be suspect is to realize that the critical percolation probability depends on the lattice that is used.

Consider the case of domain walls in which one throws down one of two phases (+ and −) on the sites of a lattice. Let us denote the probability of laying down a + by $p$. When the critical percolation probability, $p_c$, is less than 0.5, there are three possible phases:

- $p < p_c$: the + domains are islands in a sea of −.
- $p_c < p < 1 - p_c$: the + and − both form seas.
- $1 - p_c < p$: the − domains form islands in a sea of +.

In the unbiased case, $p = 0.5$, and we get seas of + and −. Then the boundary between the + and − regions are also infinite. That is, the domain walls are infinite in size.

If $p_c > 0.5$, the picture is quite different. Now we have:

*Figure 3.* The black squares denote + domains and the white squares denote − domains in a simulation with $p = 0.5$ on a two-dimensional square lattice. Neither the + nor the − domains percolate in this case and the domain walls, which are the boundaries between black and white squares are all finite.

- $p < 1 - p_c$: the + domains are islands in a sea of −.
- $1 - p_c < p < p_c$: the + and − both form islands.
- $p_c < p$: the − form islands in a sea of +.

Again, in the unbiased case, $p = 0.5$ and so both the + and the − form islands. The interfaces between the islands are finite in extent and so there are no infinite domain walls.

What is quite interesting is that, in two spatial dimensions, $p_c = 0.5$ for a triangular lattice and $p_c = 0.59$ for a square lattice. Hence the domain walls in two dimensions with $p = 0.5$ are (marginally) infinite on a triangular lattice and are all finite on a square lattice (see Fig. 3). Which lattice is the correct one to use to study phase transitions?

One expects the same problems to arise in the lattice based study of strings and monopoles. In fact, the study of domain walls is fundamental to understanding strings and monopoles since strings, for example, may be viewed as the intersection of two types of domain walls - one on which the real part of a complex scalar field vanishes and the other on which the imaginary part vanishes [5]. If the two types of domain walls are all finite, the strings will also be finite. Hence it is suitable to first understand the percolation of domain walls.

## 1.2. LATTICE-FREE SIMULATIONS

First-order phase transitions proceed by the nucleation of bubbles of the low temperature phase in a background of the high temperature phase. The bubbles then grow, collide, and coalesce, eventually filling space with the low temperature phase. In a variety of circumstances, the low temperature phase is not unique. Here we will mainly consider the case where there are two low temperature phases, which we call plus (+) and minus (−). Our goal is to determine the percolation probability, $p_c$. If $p_c$ is found to be less than 0.5, then a range of $p$ exists for which both the + and − phases will percolate and, in this case, infinite domain walls will be formed [6].

### 1.2.1. *Random bubble lattice*

Let us begin by studying the structure of the random bubble lattice that is produced during a first order phase transition and later discuss percolation on this lattice. We write the bubble nucleation rate per unit volume as $\Gamma$, and we assume that the bubble walls expand at constant speed $v$. From these quantities we can define a length scale $\xi$ and a time scale $\tau$ by:

$$\xi = \left(\frac{v}{\Gamma}\right)^{1/4}, \quad \tau = \frac{1}{(v^3\Gamma)^{1/4}}, \tag{4}$$

where the exponents have been shown for bubbles in three dimensions. By rescaling all lengths (such as bubble radii) and all times by $\xi$ and $\tau$ respectively, the dependence of the problem on $\Gamma$ and $v$ is eliminated. Therefore dimensionful quantities such as the number density of bubbles of a given size can be rescaled to a universal distribution, and dimensionless quantities, such as the the critical percolation probability, will be independent of $\Gamma$ and $v$.

The scaling argument given above relies on the absence of any other length or time scales in the problem. Potentially such a scale is provided by $R_0$, the size of bubbles at nucleation, and our assumption is that $R_0 << \xi$. Also, note that we have taken all bubbles to expand at the same velocity $v$. This is justified if the low temperature phases within the bubbles are degenerate. If this degeneracy is lifted, different bubbles can expand at different velocities and this may result in lattices with varying properties. We are primarily interested in the exactly degenerate case which is relevant to the formation of topological defects.

In [7], the nucleation and growth of bubbles leading to the completion of the phase transition was simulated according to the scheme described in Ref. [8]. There are two ways to view this scheme. The first is a dynamic view where, as time proceeds, the number of nucleation sites are chosen from a Poisson distribution, bubbles keep growing and colliding until they fill space. The second equivalent viewpoint is static and more convenient for simulations. A certain number of spheres whose centers and radii are drawn from uniform distributions are placed in the simulation box. This corresponds to a snapshot of the bubble distribution. If the number of spheres that are laid down is large, they will fill space and the snapshot would be at a time after the phase transition has completed.

It is worth comparing the present model with currently existing models of froth. The main distinguishing feature is that the bubbles continue to grow even after they collide. This is in sharp distinction with the models used in crystal growth such as the Voronoi and the Johnson-Mehl models. In these models, crystals nucleate randomly inside a volume, grow and then, once they meet a neighboring crystal, stop growing in the direction of that neighbor. (In the Voronoi model, all crystals are nucleated at one instant while in the Johnson-Mehl model, they can nucleate at different times.) This difference between the phase transition model and the Voronoi type models is significant and the resulting lattices have different properties. Another model considered in the literature is called a "Laguerre froth". Here the snapshot of the domains corresponds to a horizontal slice of a mountain range in which each mountain is a paraboloid. The circles of intersection of the plane and the paraboloids define the Laguerre froth [9]. In terms of bubbles, this means that the bubble walls move with a velocity that is proportional to $\sqrt{t-t_0}$ where $t-t_0$ is the time elapsed since nucleation. Such a model in two dimensions was studied by numerical methods in Ref. [10]. If the paraboloids are replaced by cones, the model comes closer to the present one.

A feature of our model of the first order phase transition is that bubbles cannot nucleate within pre-existing bubbles. This is appropriate to the case where the phases existing within bubbles are degenerate or nearly degenerate. However, in cases where a variety of non-degenerate bubbles can exist (for example, if the system has metastable vacuua), this assumption may have to be relaxed [11].

A two dimensional (dual) bubble lattice is constructed by connecting the centers of bubbles that have collided (Fig. 4). The three dimensional bubble lattice is similarly constructed and is shown in Fig. 5. The bubble lattice is almost fully triangulated though some violations of triangulation can occur. For example, if a tiny bubble gets surrounded by two large bubbles, the center of the tiny bubble will only be connected to the centers of the two surrounding bubbles and this can lead to plaquettes on the lattice that are not triangular. The characteristics of this bubble lattice hold the key to the percolation of phases and the formation of topological defects. In particular, the average number of vertices to which any vertex is connected is expected to play a crucial role. This number is called the

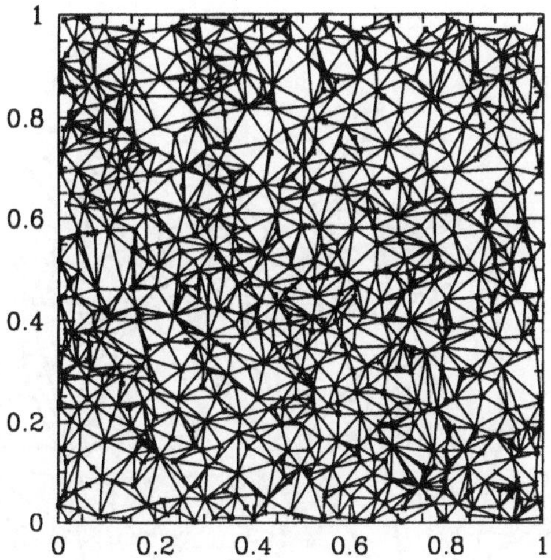


*Figure 4.* The crosses denote bubble centers that are in the + phase and the filled squares denote bubble centers that are in the - phase. If two bubbles collide, their centers are joined by straight lines. The figure then shows the "random bubble lattice" expected in a first order phase transition in two spatial dimensions.

"mean coordination number" of the lattice, and we now determine this quantity analytically.

First we consider the two dimensional case. We denote the number of points in the lattice by $P$, the number of edges by $E$ and the number of faces by $F$. Then the Euler-Poincaré formula [12] tells us

$$\chi = P - E + F \tag{5}$$

where, $\chi$ is the Euler character of the lattice and is related to the number of holes in the lattice (genus). In our case, the lattice covers a plane which we can compactify in some way, say by imposing periodic boundary conditions. Then $\chi$ is the genus of the compact two dimensional surface. For us it will only be important that $\chi = O(1)$. Next, if $\bar{z}$ is the (average) coordination number, we can see that

$$E = \frac{\bar{z}}{2} P \ ,$$

since, a given point is connected to $\bar{z}$ other points but each edge is bounded by two points. Also,

$$F = \frac{2}{3} E \ ,$$

since each line separates 2 faces but then each face is bounded by 3 lines. Now, using (5) gives

$$1 - \frac{\bar{z}}{2} + \frac{\bar{z}}{3} = \frac{\chi}{P} \simeq 0 \ ,$$

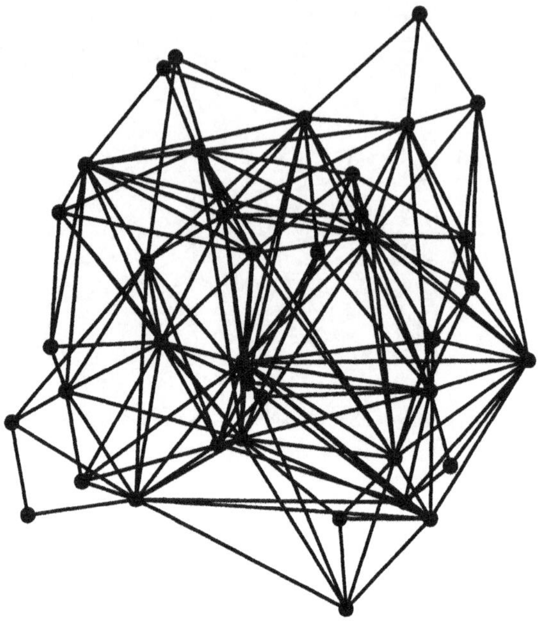

*Figure 5.* A portion of the three dimensional dual bubble lattice.

since $P$ is assumed to be very large. Therefore, in two dimensions, $\bar{z} = 6$, a result that first appeared in the botanical literature [13, 9].

In three dimensions the analysis to evaluate $\bar{z}$ is somewhat more complicated. The Euler-Poincaré formula now says

$$\chi = P - E + F - V \ , \tag{6}$$

where $V$ is the number of volumes in the lattice. Now, in addition to the usual coordination number $\bar{z}$, we also need to define a "mean face coordination number" $\bar{y}$ which counts the average number of faces sharing a common edge. In terms of $\bar{y}$ and $\bar{z}$, the relations between the various quantities for a triangulated three dimensional lattice are:

$$E = \frac{\bar{z}}{2}P \ , \quad F = \frac{\bar{y}}{3}E \ , \quad V = \frac{2}{4}F \ , \tag{7}$$

where the first equation is as in two dimensions, the second equation follows from the definition of $\bar{y}$ and the fact that the lattice is triangulated, and the last relation follows because a face separates two volumes and a volume is bounded by four faces that form a tetrahedron. Inserting these relations in (6) leads to:

$$\bar{z} = \frac{12}{6 - \bar{y}} \ , \tag{8}$$

where, as before, we assume that $P$ is very large and ignore the $\chi/P$ term. Note that the relation between $\bar{y}$ and $\bar{z}$ is purely topological and will hold for any triangulated lattice.

We now want to estimate $\bar{y}$. For this we work in a "mean field" approximation where we assume that the edge lengths are fixed. We consider two vertices $A$ and $B$ separated by a unit distance. We wish to find the number of points that can be connected to both $A$ and $B$, subject to the constraint that the connected points are at unit distance from each other. This will give the (average) number of faces that share the edge from $A$ to $B$ and hence will be the face coordination number $\bar{y}$. Let us choose $A$ to be at the center of a sphere of unit radius and $B$ to be at the North pole. Then the additional points $P_1,...,P_y$, have to lie on the circle at latitude 60 degrees to satisfy the distance constraint. Then one finds that the azimuthal angular separation of two sequential points $P_i$ and $P_{i+1}$ is 70.5 degrees. Therefore

$$\bar{y} = \frac{360}{70.5} = 5.1, \tag{9}$$

which then leads to [14]

$$\bar{z} = 13.4\ . \tag{10}$$

It is worth noting the ingredients that have entered into the analytic estimate of $\bar{z}$. The relation (8) is a topological statement about the lattice, but the estimate for $\bar{y}$ is geometric, depending on the assumption that the edges have fixed length. In principle, the edge lengths can fluctuate but our estimate for $\bar{y}$ will still be valid if the fluctuations average out.

In Fig. 6 we show the distribution of coordination number in our three dimensional simulations. The average coordination number is found to be $\bar{z} = 13.34 \pm 0.05$ and agrees quite closely with the mean field result. For comparison, Voronoi foam has $\bar{z} = 15.54$ and the Johnson-Mehl model has $\bar{z} > 13.28$ [15]. The reason why $\bar{z}$ is larger in the Voronoi model is that, in this model, the cells stop growing on collision in the direction of the collision, thus leading to anisotropic growth. It can be shown that anisotropy of the cells leads to a higher value of $\bar{z}$ [9].

The mean value of $z$ is not a good characteristic of the distribution of $z$ since the distribution is skewed (the modal value of $z$ is 7) and it is of interest to characterize the entire distribution of $z$. In the literature on domain physics, attempts to derive the distribution of coordination number are often based on maximizing the "entropy" of the lattice. The expression used for the entropy is the one proposed by Shannon [16, 17].

$$S = -\sum_n p_n \ln p_n \tag{11}$$

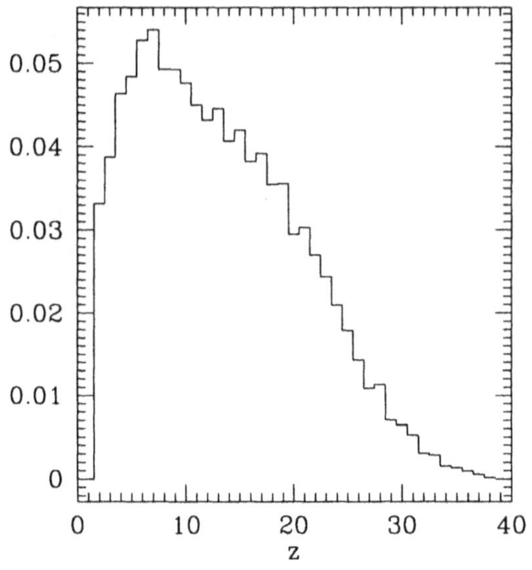


*Figure 6.* The coordination number relative frequency distribution for the three dimensional dual bubble lattice.

where, $p_n$ is the probability that a vertex will be connected to $n$ other vertices. In addition, one needs to insert the Euler-Poincaré constraint in the Shannon entropy via Lagrange multipliers. (There is also the issue of assigning *a priori* probabilities [17].) In two dimensions the constrained extremization is relatively straightforward, since the Euler-Poincaré constraint fixes the average coordination to be 6, *i.e.*

$$\sum_n np_n = 6 \quad \text{(in 2D)} \ .$$

In three dimensions, we may once again introduce the constraint that the average coordination number of the lattice is fixed (even though there is some freedom in choosing its value). Then, on extremizing $S$ in eq. (11) with respect to $p_n$, we find an exponential fall-off of the distribution. Indeed, the distribution shown in Fig. 6 has an exponential fall-off:

$$f(z) \sim \exp[-0.25z] \ , \quad z > 20 \ . \tag{12}$$

1.2.2. *Percolation on a random bubble lattice*

We now turn to the formation of defects on the bubble lattice. We put a $+$ phase on a bubble with probability $p$ and a $-$ phase with probability $1-p$ (as shown in Fig. 4 in the two dimensional case). We then find the size distribution of $+$ clusters and calculate the moments of the cluster distribution function after removing the largest cluster from the distribution [18]. That is, we calculate:

$$S_l(p) = \sum_{s \neq s_{max}} s^l n_s(p) \tag{13}$$

for $l = 0, 1, 2, ...$, where the sum is over cluster sizes ($s$) but does not include the largest cluster size, and $n_s(p)$ is the number of clusters of size $s$ divided by the total number of bubbles. In Fig. 7 we show the first three moments as a function of $p$, where the turning point in $S_2$ marks the onset of percolation. To understand this, first consider the behavior of the second moment for small $p$. As we increase $p$, there are fewer + clusters (as seen from the $S_0$ graph) probably due to mergers, but the merged cluster sizes are bigger (as seen from the $S_1$ graph). Since the second moment places greater weight on the size of the cluster than on the number density as compared to the lower moments, it grows for small $p$. For large $p$, however, as we increase $p$ further, the additional + clusters join the largest cluster of +'s and are not counted in the second moment. In fact, some of the smaller clusters also merge with the largest cluster and get removed from the sum in (13). This causes the second moment to decrease at large $p$. Hence, the second moment has a turning point and the location of this turning point at $p_c$ marks the onset of percolation. In three dimensions we find $p_c = 0.17 \pm 0.01$ (from Fig. 7), which is well under 0.5, while in two dimensions we find $p_c = 0.50 \pm 0.01$ which is consistent with 0.50. (The two dimensional version of Fig. 7 may be found in [6].)

It is interesting to compare the critical probabilities we have found with lattice based results for site percolation where the regular lattice has a coordination number close to that of the random bubble lattice. In two dimensions a triangular lattice has $\bar{z} = 6$ and $p_c = 0.5$. In three dimensions, a face centered cubic lattice has $\bar{z} = 12$ and $p_c = 0.198$ [18]. These values of the critical probabilities are fairly close to our numerical results.

1.2.3. *Open Problems*

*Exponents:* The value of the critical percolation probability is not universal. However, the percolation exponents are expected to be universal. These have not been evaluated for the random lattice and would be worth determining.

*Bias:* The rather low value of $p_c$ in three dimensions means that domain walls formed between degenerate vacua ($p = 0.5$) will percolate and almost all of the wall energy will be in one infinite wall. Furthermore, even if the vacua are not degenerate, *i.e.* there is bias in the system, infinite domain walls can still be produced. If the properties of the bubble lattice are insensitive to small biases, infinite domain walls would be produced for

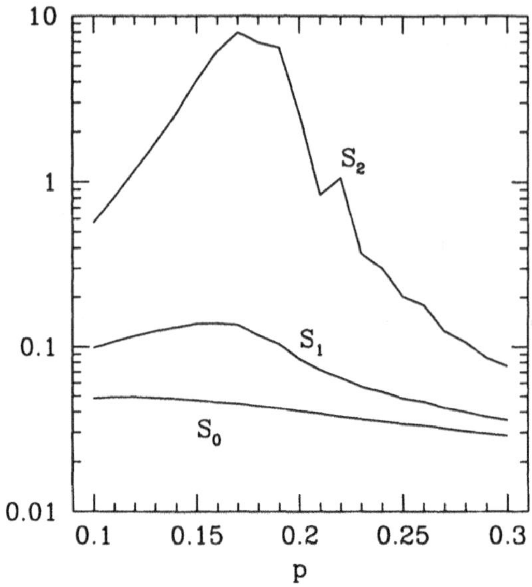


*Figure 7.* The zeroth, first and second moments of the cluster distribution function versus the probability $p$ in three dimensions.

$p > 0.17$. However, it is likely that the bubble lattice will depend on the bias in three ways. First, the nucleation rate of bubbles of the metastable vacuum will be suppressed compared to that of the true vacuum. Secondly, the velocity with which bubbles of the two phases grow can be different. Thirdly, bubbles of the true vacuum may nucleate within the metastable vacuum. In addition to these factors, bubbles may not retain their spherical shape while expanding due to instabilities in their growth. The effect of these factors on the percolation probability will be model dependent. For example, the bubble velocities will depend on the ambient plasma, and the nucleation rates on the action of the instantons between the different vacua. The effect of including these factors on the random bubble lattice and percolation has not yet been studied.

*Other defects:* The formation of topological strings on the random bubble lattice follows the algorithm described in Ref. [1]. It is found that about 85% of the strings in the simulation are infinite. This number should be compared with earlier static simulations of string formation which yield a slightly lower fraction (about 80%). The distribution of other types of defects should be determined.

*Phase equilibration:* The analysis described here neglects phase equilibration processes when domains of different phases collide. This may be justified if the time scale $\tau$ in eq. (4) is short compared to the typical time required for phase equilibration. In the case of domain walls, phase equilibration in two colliding bubbles can only occur by the motion of the phase separating wall across the volume of one of the bubbles. In this case, the neglect of phase equilibration is justified if the domain wall velocity is much smaller than the bubble wall velocity. It would be interesting to see how the results change if phase equilibration is important.

## 2. Interactions of Defects

Once a string network forms in any system, the strings start moving under their tension. Inevitably string collisions occur. What happens then?

It is a classic result that two strings intercommute (reconnect) upon intersection (Fig. 8). This conclusion seems to hold regardless of the details of the collision (angles and velocities), as well as the physical model (global or local strings)[3]. Intercommuting has been observed in computer simulations and experimentally. There are important supporting arguments that provide insight into intercommuting but there is no analytical proof that intercommuting of Abelian strings must necessarily occur[4].

The phenomenon of intercommuting is vital to the relaxation (coarsening) of the system. In cosmology, intercommuting provides a means for the string network to dissipate and prevent it from dominating the universe.

A question that has recently attracted some attention is the interaction of *different types* of defects arising in the same physical model [20]. Consider, for example, a phase transition in which both domain walls and point defects (global or local monopoles) can be formed. On formation, the domain walls will start moving under their own tension. Inevitably, they will collide with the point defects. What happens then? Do the point defects pass through the walls to the other side? Or do they undergo some microphysical transformation? The answer to these questions are very important for understanding the coarsening of the system.

Based on several different arguments, it was suggested in [20] that the monopoles do not pass through the domain walls. Instead they undergo a microphysical transformation and unwind on the walls. The arguments in support of this conjecture are:

(i) There is an attractive force between the monopoles and the walls since monopoles can save the expense of having to go off the vacuum in their

3If there are strong attractive forces between the strings and they are nearly parallel upon collision, they may form a bound state which smears out the collision region and the strings may then separate again, thus passing through each other without intercommuting [19].

4Certain non-Abelian strings do not intercommute but this is for topological reasons. Instead of intercommuting, such strings can get connected by another type of string upon collision.

(a) (b)

*Figure 8.* Two incoming strings in (a) collide, reconnect and emerge as in (b).

core by moving on to the wall. So the monopoles can form bound states with the walls. Then, as there is no topological obstruction to the unwinding of monopoles on the wall, the monopoles on the wall can continuously relax into the vacuum state.

(ii) The investigation of a similar system - Skyrmions and walls - has been dealt with in full detail in Ref. [21]. These authors find that the Skyrmion hits the wall, sets up traveling waves on the wall and dissipates. They also find that, even though it is topologically possible for the Skyrmion to penetrate and pass through the domain wall, this never happens. They attribute their finding to the coherence required for producing a Skyrmion. That is, the penetration of a Skyrmion may be viewed as the annihilation of the incoming Skyrmion on the wall and the subsequent creation of a Skyrmion on the other side. However, the annihilation results in traveling waves along the wall that carry off a bit of the coherence required to produce a Skyrmion on the other side. Hence, even though there is enough energy in the vicinity of the collision, a Skyrmion is unable to be created on the other side of the wall. I think that these considerations apply equally well to monopole-wall interactions and that monopoles will never pass through the wall - just as strings never pass through each other.

(iii) The interactions of vortices and domain walls separating the A and B phases of $^3$He have been studied and also observed experimentally. It is found that singular vortices do not penetrate from the B phase into the A phase [22].

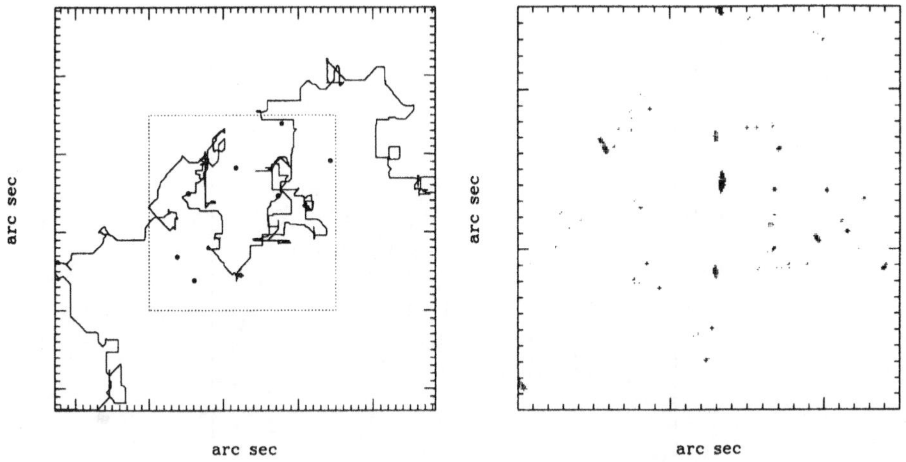


*Figure 9.* The figure on the left shows the projected string configuration along with the randomly located unlensed sources inside the dashed box. The figure on the right shows the resulting images of the sources The full string shown in the figure on the left was used to determine the lensing effects.

What is sorely needed is a direct check of this conjecture. (A very recent study [23] has confirmed monopole dissolution on walls in a particular model under some restricted conditions.) If confirmed, it would be good to be able to understand the result at a deeper level.

## 3. Observation of Cosmic Strings

If cosmic strings lie between us and some distant sources of light, they will cause distortions in the image of the source or cause multiple images to be produced. Two of the most distant cosmological sources are the cosmic microwave background radiation and quasars. The cosmic microwave background radiation comes to us from when the universe was only about a million years old, while the light from quasars started out about 1 billion years ago. (The present age of the universe is about 10 billion years.) A major difference between these two sources is that the CMBR is an almost uniform source, while the quasars are point-like. This leads to different kinds of observable signatures of strings in the two cases. Cosmic strings can imprint a pattern of anisotropy of the CMBR and they can gravitationally lens quasars and galaxies with a frequency given by the probability that a quasar will lie behind a string. Both effects are quite small - the anisotropy is at the level of about 1 part in $10^5$ and the probability of lensing is also of this order of magnitude. In either case, however, the signature of strings appears to be quite distinct and can lead to confirmation of their

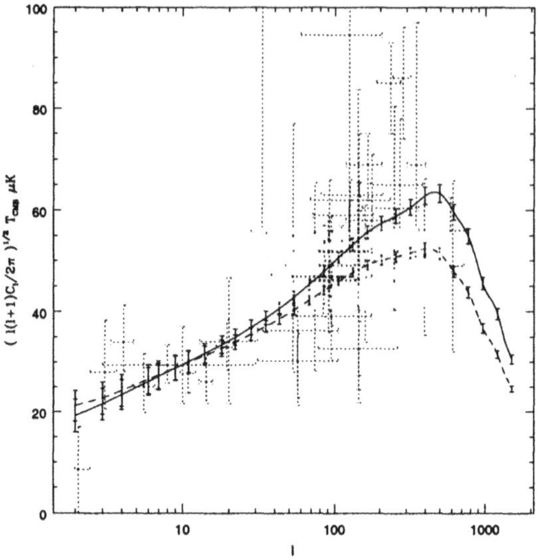


*Figure 10.* The graph shows the total angular power spectrum in CMBR anisotropy for wiggly (solid line) and smooth (dashed line) strings when $\Omega_{baryons} = .05$, $\Omega_{CDM} = 0.25$ and $\Omega_\Lambda = 0.7$ and using small values for the string velocities.

presence or absence at a certain energy scale.

### 3.1. GRAVITATIONAL LENSING

The most tricky part in studying the gravitational lensing of distant sources by strings is the actual construction of the string itself. In Ref. [24], the string was constructed by using flat spacetime simulations as proposed by Smith and Vilenkin [25]. Light from distant sources is then propagated in the gravitational field of this string. The results are shown in Fig. 9.

What seems most striking about the lensed sources is that they seem to trace out the string. This is due to the wiggly nature of the string since small sections of the projected string can act like very massive objects.

### 3.2. CMBR DISTORTIONS

Once again, the most difficult aspect of calculating the CMBR distortions is to find a reliable model of the network of strings. Direct simulations of the string network have provided invaluable information about certain properties of the network. The game is to convert this information into a model that can then be fed into the machinery to calculate the CMBR anisotropy. Recently I have been working [26] with Levon Pogosian to compute the anisotropy using the model developed in [27, 28]. (Some other

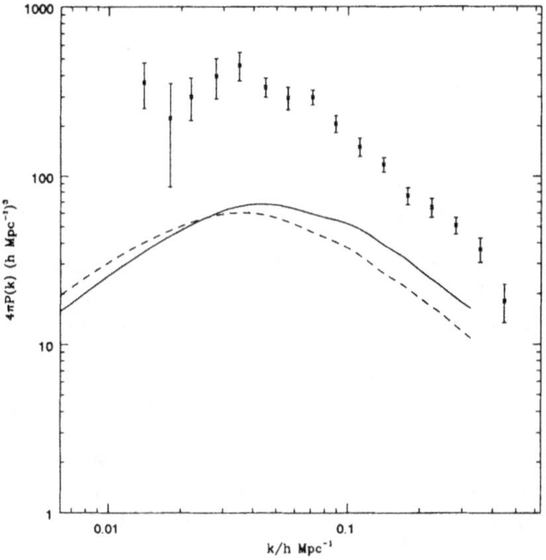


*Figure 11.* The graph shows the power spectrum of the density inhomogeneities for the same two models and the observed data points.

references to related literature may be found in [29].) The results depend on details of the string model as well as the cosmological model. In Fig. 10 I show our preliminary results for the power in the $l^{th}$ spherical harmonic of the CMBR anisotropy as a function of $l$ together with the observed data points. (Space does not permit a lengthier explanation of the graphs in this section but background details can be found in several textbooks, for example Ref. [30].) At the moment, the observations do not confirm or reject the hypothesis that cosmic strings may be responsible for the CMBR anisotropy.

The density perturbations produced by cosmic strings would lead to large-scale structure formation. Then one might compare the power spectrum of the density inhomogeneities produced by strings to those observed in the galaxy distribution. Here too, the details of the string and the cosmological model are all crucial. Within the limitations of the models used, the string predictions do not agree with observations at the level of a factor of about 2 in the amplitude of the density fluctuations (Fig. 11). However, further modeling of the string network and analysis is necessary before we can be sure of this result.

## Acknowledgments

This work was supported by the Department of Energy (USA).

## References


1. Vachaspati, T. and Vilenkin, A. (1984) Formation and evolution of cosmic strings, *Phys. Rev.* **D30**, 2036.
2. Scherrer, R.J. and Frieman J. (1986) Cosmic strings as random walks, *Phys. Rev.* **D33**, 3556.
3. Copeland E.J., Kibble T.W.B. and Steer D.A. (1998) The evolution of a network of cosmic string loops, *Phys. Rev.* **D58**, 043508.
4. Pogosian L. and Vachaspati T. (1997) Relaxing the geodesic rule in defect formation algorithms, *Phys. Lett.* **B423**, 45.
5. Scherrer R. J. and Vilenkin A. (1997) Cosmic string formation from correlated fields, *Phys. Rev.* **D56**, 647; Scherrer R. J. and Vilenkin A. (1998) Lattice-Free simulations of topological defect formation, *Phys. Rev.* **D58**, 103501.
6. Vachaspati T. (1997) Formation of topological defects, *ICTP 1997 Summer School Lectures on Cosmology*, hep-ph/9710292.
7. de Laix A.A. and Vachaspati T. (1999) On random bubble lattices, *Phys. Rev.* **D59**, 045017.
8. Borrill J., Kibble T.W.B., Vachaspati T. and Vilenkin A. (1995) Defect production in slow first order phase transitions, *Phys. Rev.* **D52**, 1934.
9. Rivier N. (1993) Order and disorder in packings and froths, *Disorder and Granular Media*, eds. D. Bideau and A. Hansen, North-Holland.
10. Telley H. (1989), Modelisation et simulation bidimensionnelle de la croissance des polycristaux, Ph. D. Thesis, EPFL, Lausanne, (unpublished).
11. Gleiser M., Heckeler A.F. and Kolb E.W. (1997) Modeling thermal fluctuations: phase mixing and percolation, *Phys. Lett.* **B405**, 121.
12. Nash C. and Sen S. (1983) Topology and Geometry for Physicists, Academic Press, London.
13. Lewis F.T. (1928), The correlation between cell division and the shapes and sizes of prismatic cells in the epidermis of Cucumis, *Anat. Record* **38**, 341.
14. Rivier N. (1982) Recent results on the ideal structure of glasses, *J. Physique Coll.* **43**, C9-91.
15. Meijring J.L. (1953) Interface area, edge length, and number of vertices in crystal aggregates with random nucleation, *Philips Res. Rep.* **8**, 270.
16. Shannon C.E. (1948) The mathematical theory of communication, *Bell Systems Technical Journal* **27**, 379; Shannon C.E. and Weaver W. (1949) *The Mathematical Theory of Communication*, University of Illinois Press.
17. Rivier N. (1985) Statistical crystallography: structure of random cellular networks, *Phil. Mag.* **B52**, 795.
18. Stauffer D. (1979) Scaling theory of percolation clusters, *Phys. Rep.* **54**, 1.
19. Bettencourt L.M.A. and Kibble T.W.B. (1994) Non-intercommuting configurations in the collisions of type-I U(1) cosmic strings, *Phys. Lett.* **B332**, 297.
20. Dvali G., Liu H. and Vachaspati T. (1998) Sweeping away the monopole problem, *Phys. Rev. Lett.* **80**, 2281.
21. Kudryavtsev A., Piette B., Zakrzewski W.J. (1997) Skyrmions and domain walls in (2+1) dimensions, hep-th/9709187, DTP-97/25.
22. Trebin H.-R. and Kutka R. (1995) Relations between defects in the bulk and on the surface of an ordered medium - a topological investigation, *J. Phys.* **A28**, 2005; Misirpashaev T. Sh. (1991) Topological Classification of defects at phase interface, *Sov. Phys. JETP* **72**, 973; Krusius M., Thuneberg E.V. and Parts U. (1994) A-B phase-transition in rotating superfluid He-3, *Physica* **B197**, 376.
23. Alexander S., Brandenberger R., Easther R. and Sornborger A. (1999) On the interaction of monopoles and domain walls, hep-ph/9903254.
24. de Laix A.A., Krauss L.M. and Vachaspati T. (1997) Gravitational lensing signature of long cosmic strings, *Phys. Rev. Lett.* **79**, 1968.
25. Smith A.G. and Vilenkin A. (1987) Numerical simulation of cosmic string evolution

in flat space-time, *Phys. Rev.* **D36**, 990.

26. Pogosian L. and Vachaspati T. (1999) Cosmic microwave background anisotropy from wiggly strings, astro-ph/9903361, *Phys. Rev.* **D**, in press.
27. Vincent G., Hindmarsh M. and Sakellariadou M. (1997) Correlations in cosmic string networks, *Phys. Rev.* **D55**, 573.
28. Albrecht A., Battye R. and Robinson J. (1997) The case against scaling defect models of cosmic structure formation, *Phys. Rev. Lett.*, **79**, 4736; Albrecht A., Battye R. and Robinson J. (1998) Detailed study of defect models for cosmic structure formation, *Phys. Rev.* **D59**, 023508.
29. Bouchet F.R., Bennet D.P. and Stebbins A. (1988) Microwave anisotropy patterns from evolving string networks, *Nature* **335** 410; Perivolaropoulos L. (1993) COBE vs cosmic strings: an analytical model, *Phys. Lett.* **B298**, 305; Perivolaropoulos L. (1995) Spectral analysis of microwave background perturbations induced by cosmic strings, *Ap. J.* **451**, 429; Allen B., Caldwell R.R., Dodelson S., Knox L., Shellard E.P.S. and Stebbins A. (1997) CMB anisotropy induced by cosmic strings on angular scales $> 15'$, *Phys. Rev. Lett.* **79**, 2624; Pen U.-L., Seljak U. and Turok N. (1997) Power spectra in global defect theories of cosmic structure formation, *Phys. Rev. Lett.* **79**, 1611; Avelino P.P., Shellard E.P.S., Wu J.H.P. and Allen B. (1998) Cosmic-string-seeded structure formation, *Phys. Rev. Lett.* **81**, 2008; Battye R., Robinson J. and Albrecht A. (1998) Structure formation by cosmic strings with a cosmological constant, *Phys. Rev. Lett.* **80**, 4847; Contaldi C., Hindmarsh M. and Magueijo J. (1999) The power spectra of CMB and density fluctuations seeded by local cosmic strings, *Phys. Rev. Lett.* **82**, 679; Copeland E.J., Magueijo J. and Steer D.A. (1999) Cosmological parameter dependence in local string theories of structure formation, astro-ph/9903174.
30. Peacock J.A. (1999) Cosmological Physics, Cambridge University Press.

# SHARDS OF BROKEN SYMMETRY

*Topological Defects as Traces of the Phase Transition Dynamics*

W.H. ZUREK, L.M.A. BETTENCOURT, J. DZIARMAGA
*Theoretical Division, Los Alamos National Laboratory*
*Los Alamos NM 87545, USA*

AND

N.D. ANTUNES
*Dépt. de Physique Théorique, Université de Genève*
*24 quai E. Ansermet, CH 1211, Genève 4, Switzerland*

We discuss the origin of topological defects in phase transitions and analyze their role as a "diagnostic tool" in the study of the non-equilibrium dynamics of symmetry breaking. Homogeneous second order phase transitions are the focus of our attention, but the same paradigm is applied to the cross-over and inhomogeneous transitions. The discrepancy between the results in $^3$He and $^4$He is discussed in the light of recent numerical studies. The possible role of the Ginzburg regime in determining the vortex line density for the case of a quench in $^4$He is raised and tentatively dismissed. The difference in the anticipated origin of the dominant signal in the two ($^3$He and $^4He$) cases is pointed out and the resulting consequences for the subsequent decay of vorticity are noted. The possibility of a significant discrepancy between the effective field theory and (quantum) kinetic theory descriptions of the order parameter is briefly touched upon, using atomic Bose-Einstein condensates as an example.

## 1. Introduction

The theory of the creation of topological defects appeals to models of critical dynamics and to our understanding of the processes which occur when phase transitions take place. Consequently, topological defects can be used as "symptoms", macroscopic manifestations of underlying physical processes, which in turn can help diagnose the nature of critical dynamics.

For first order phase transitions there is little doubt that nucleation - a process understood for over half a century - is an essentially accurate, universal yet simple model . A similarly simple model of the dynamics of second order phase transitions was proposed much more recently [Zurek 1984 [1]; 1985 [2]; 1993 [3]]. One of its implications is the ability to predict the size of the ordered patches of the new lower symmetry phase, right



*Y.M. Bunkov and H. Godfrin (eds.),*
*Topological Defects and the Non-Equilibrium Dynamics of Symmetry Breaking Phase Transitions, 77–103.*

from its inception. This allows one to calculate the initial density of topological defects through the estimate put forward in the seminal paper by Kibble [4, 5]. It may also lead to a revision of the scenarios for baryogenesis and chiral symmetry restoration [Gill 1997 [6]; Dziarmaga and Sadzikowski 1998 [7]], as well as other related phenomena (such as the A-B $^3$He transition, see Volovik 1996 [8], Bunkov and Timofeeskaya 1997 [9]). Some of the predictions based on the new paradigm have been successfully tested and refined in numerical experiments [Laguna and Zurek [10, 11]; Yates and Zurek [12]; Antunes, Bettencourt and Zurek [13], Stephens *et al.* [14]]. More importantly the prediction of copious vortex production in superfluid phase transitions has been experimentally verified in $^3$He by two very different strategies in two distinct parameter regimes [15, 16], (see also papers by Krusius and Godfrin in this volume). The situation in $^4He$ [17] and the initial indications from high temperature superconductors [18] are, however, at best inconclusive. Indeed there are still differences concerning analytic estimates of the initial density of defects in the underdamped case [19, 20], even in 1D. The aim of this paper is two-fold: We shall start with a brief summary of the paradigm on which the emerging understanding of second order phase transitions is based. We shall then explore its extrapolations and investigate the experimental, numerical and analytic evidence for and against this mechanism, in various settings. This paper is not really an introductory survey to the extent to which our lectures were. We have decided that the existing literature (including the reviews of Zurek [21] and Eltsov, Krusius and Volovik [22] as well as the other papers mentioned above and below) already serves this purpose. Rather we aim to perform a "reconnaissance by force" of what is likely to be the most interesting "proving grounds" for the ideas summarized briefly in the following section.

## 2. Critical dynamics and defect formation

Second order transitions fall into universality classes which are characterized by the behavior of the healing length $\xi$ and the relaxation time $\tau$ (among other quantities) as a function of the relative temperature

$$\epsilon(T) = \frac{T - T_c}{T_c}. \tag{1}$$

Thus, $\tau \sim |\epsilon|^{-\nu z}$ and $\xi \sim |\epsilon|^{-\nu}$ diverge in the vicinity of $\epsilon = 0$, where $\nu$ and $z$ are universal critical exponents. A very specific model which represents a large class of second order phase transitions is the so-called Landau-Ginzburg theory. There, the dynamics of the order parameter is thought to effectively obey a Langevin equation of the form:

$$\ddot{\varphi} + \eta\dot{\varphi} - c^2\nabla^2\varphi + \frac{1}{2}\left[\lambda\varphi^3 + m^2\phi\right] = \mathcal{O}(t, x). \tag{2}$$

Above, $\eta$ characterizes the viscosity in the system, while $c$, $\lambda$ are constant coefficients. The mass term $m^2$ can depend explicitly on time. Moreover, the correlation function for the noise

$$\langle \mathcal{O}(t,x)\mathcal{O}(t',x')\rangle = 2\eta\Theta\delta(x-x')\delta(t-t'), \tag{3}$$

includes the temperature parameter $\Theta$, which can vary. The change of $m^2$, eg. $m^2 = m_0^2\epsilon(t)$ or of $\Theta$ or both, may precipitate the phase transition.

We shall assume that in the vicinity of the critical temperature $\epsilon(t)$ obeys a simple relation

$$\epsilon(t) = t/\tau_Q. \tag{4}$$

In that case the dynamics of the order parameter can be approximately divided into the adiabatic and impulse regimes (Zurek [2, 3]), with the boundary which occurs at time $\hat{t}$ when the relaxation time of the order parameter equals the characteristic time on which $\epsilon(t)$ changes:

$$\tau(\epsilon(\hat{t})) = \frac{\epsilon(\hat{t})}{\dot{\epsilon}(\hat{t})} = \hat{t}. \tag{5}$$

The timescale on which the order parameter will react to changes of $\epsilon(t)$ depends on whether $\dot{\varphi}$ or $\ddot{\varphi}$ dominates. For the Landau-Ginzburg theory, in the two cases

$$\tau_{\dot{\varphi}} = \frac{\eta\tau_0^2}{|\epsilon|}, \qquad \tau_{\ddot{\varphi}} = \frac{\eta\tau_0}{|\epsilon|^2}, \tag{6}$$

respectively, where $\tau_0 = 1/m_0$, and $m_0$ is the mass term in Eq. (2), evaluated for $T = 0$ (i.e. for $\epsilon = -1$).

Using $\epsilon(t) = t/\tau_Q$, we can now solve for $\hat{t}$, to obtain:

$$\hat{t}_{\dot{\varphi}} = \pm\tau_0\sqrt{\eta\tau_Q}, \qquad \hat{t}_{\ddot{\varphi}} = \pm\tau_0^{2/3}\tau_Q^{1/3}. \tag{7}$$

To estimate the scale of the domains which could have become uniform through dynamics in the adiabatic regime, we should need:

$$\hat{\epsilon}_{\dot{\varphi}} = \pm\sqrt{\frac{\eta\tau_o^2}{\tau_Q}}; \qquad \hat{\epsilon}_{\ddot{\varphi}} = \pm\left(\frac{\eta\tau_o^2}{\tau_Q}\right)^{2/3}. \tag{8}$$

The characteristic scale is then given by $\hat{\xi} = \xi_0/\sqrt{|\hat{\epsilon}|}$, which yields:

$$\hat{\xi}_{\dot{\varphi}} = \xi_0\left(\frac{\tau_Q^2}{\eta\tau_0^2}\right)^{1/4}; \qquad \hat{\xi}_{\ddot{\varphi}} = \xi_0\left(\frac{\tau_Q}{\tau_0}\right)^{1/3}. \tag{9}$$

The initial density of defects can now be estimated using an argument due to Kibble [4], which will imply a $\hat{\xi}$-sized unit of defect per $\hat{\xi}$ sized volume, i.e. $n \sim 1/\hat{\xi}^2$ for vortex strings in two spatial dimensions.

The above calculation is based on the assumption that the order parameter approximates its equilibrium configuration until $-\hat{t}$, at which point it ceases to evolve dynamically (although noise and damping continue unabated). The dynamical evolution restarts at $+\hat{t}$, below the critical point, but by then it may be too late to undo non-trivial topological arrangements of $\varphi$ inherited from above $T_c$.

This same paradigm decides when the overdamped or underdamped estimates are relevant. For, in view of the above argument, it is essential to decide whether the dynamics of the order parameter is overdamped at $\hat{t}$, i.e. whether;

$$\eta \dot{\varphi}_{|\hat{t}} > \ddot{\varphi}_{|\hat{t}}. \tag{10}$$

This can be evaluated directly from Eq. (2) with the help of the above estimates for $\hat{t}$ and $\hat{\epsilon}$, and leads to the inequality:

$$(\eta \tau_0)^3 > \tau_0/\tau_Q. \tag{11}$$

Numerical studies have by now confirmed this paradigm. The scalings which we obtained follow theoretical predications both when the quench is induced by an explicit change of the mass term in Eq. (2) (Laguna and Zurek [10, 11], Yates and Zurek [12]) and when the temperature of the noise $\Theta$ is adjusted, but $m$ set to a constant (Antunes, Bettencourt and Zurek [13]). Moreover, the switch from the overdamped to the underdamped behavior occurs where expected, and with the consequences consistent with the scaling implied by the paradigm Eqs. (8)-(11) (Laguna and Zurek [10, 11], Yates and Zurek [12]). The same reasoning can be of course repeated using other values of critical exponents relevant for other cases (Halperin and Hohenberg [23]), which has been already done in some cases (Zurek [21]; Antunes, Bettencourt and Zurek [13]).

While the scalings accord well with the theoretical predictions, the specific density of defects $n$ is lower than the appropriate inverse power of $\hat{\xi}$;

$$n = \frac{1}{\left(f^2 \hat{\xi}\right)^2}, \tag{12}$$

where $f$ is always more than unity, and usually in the range $8 - 15$ [10, 11, 12, 13]. We shall return to its estimates later in this paper.

## 3. Crossover transitions

An interesting case of transitions which does not conform to the second-order universality class occurs when the critical scaling behavior of the healing length and of the relaxation time "tapers off" (i.e. is fully analytic) very near to $\epsilon = 0$. For instance a hypothetical relaxation time and healing length dependences;

$$\tau = \frac{\tau_0}{(|\epsilon| + \Delta)}; \qquad \xi = \frac{\xi_0}{\sqrt{|\epsilon| + \delta}}, \tag{13}$$

illustrate such a crossover transition.

Examples of crossover phenomena are ubiquitous. One of the most interesting cases comes from the study of the electroweak standard model, where, for Higgs masses not yet excluded by experiment, the transition appears to be a crossover [24]. A crossover transition may also occur in the presence of impurities, anisotropies, weak external fields, or finite size scaling, instead of the expected critical behavior of the ideal model. It also substitutes critical behavior when non-perturbative fluctuations exist in the spectrum of the theory which are favored entropically and can destroy long range order. An example of the latter is a $\lambda\phi^4$ theory or a (short-range) Ising model in one spatial dimension.

Thus, $\tau$ and $\xi$ in the one-dimensional cases investigated numerically [10, 11] as examples of the second order phase transition are expected to taper off in the immediate vicinity of the critical temperature. Presumably this occurs for very small values of $\Delta$ and $\delta$, so that the scaling behavior encountered in the vicinity of $\hat{\epsilon}$ is not affected. Nevertheless, it is interesting to investigate what does the paradigm predict in the case of such crossover transformations.

We follow the footsteps of the argument outlined in the preceding section, and obtain $\hat{t}$ by solving :

$$\tau(\epsilon(\hat{t})) = \hat{t}, \tag{14}$$

which now leads to the quadratic equation

$$|\hat{t}|^2 + |\hat{t}|\Delta\tau_Q - \tau_0\tau_Q = 0. \tag{15}$$

Consequently

$$|\hat{t}| = \frac{-\Delta\tau_Q + \tau_Q\sqrt{\Delta^2 + 4\tau_0/\tau_Q}}{2}, \tag{16}$$

and

$$|\hat{\epsilon}| = \frac{-\Delta}{2} + \frac{1}{2}\sqrt{\Delta^2 + 4\tau_0/\tau_Q}, \tag{17}$$

where we have picked the physically relevant root of Eq. (15).

We note that in the limit of a 'real' second order phase transition ($\Delta \to 0$) we recover the old result, Eq. (8), providing that the change of notation ($\tau_0$ now used to be $\eta\tau_0^2$ in Eq. (8)) is acknowledged. On the other hand, when the quench is very slow and $\Delta^2 >> 4\tau_0/\tau_Q$, $\epsilon \to \Delta\tau_0/\tau_Q$, which itself is small compared with $\Delta$ (and, presumably also $\delta$ since $\Delta \sim \delta$ can be expected). Consequently for relatively rapid quenches

$$\hat{\xi} = \frac{\xi_0}{\sqrt{\left(\frac{\tau_0}{\tau_Q}\right)^{1/2} + \delta}}, \tag{18}$$

which approaches Eq. (9) for sufficiently small $\delta$. In the other limit;

$$\hat{\xi} = \xi_0/\sqrt{\delta}, \tag{19}$$

and the size of the coherent domains of the order parameter saturates.

We note that the above discussion should be regarded more as an exercise in extending the paradigm rather than as a generically valid theory, applicable to all crossover phase transitions. In particular, in some cases second order transitions may change into crossovers when an external bias which influences the choices of the broken symmetry vacuum is introduced. In such cases the externally imposed (rather than spontaneous) symmetry breaking will favor a particular vacuum and will lead to a suppression of topological defect production (Dziarmaga 1998 [20], Dziarmaga and Sadzikowski 1998 [7]).

Moreover, in case of the crossover transitions the influence of the Ginzburg regime may need to be carefully examined as its role in the generation and survival of topological defects is still a subject of dispute.

## 4. Inhomogeneous Transitions

Homogeneous quenches are a convenient idealization and may be a good approximation in some cases. However, in reality, the change of thermodynamic parameters is unlikely to be ideally uniform:

1) Experiments carried out in $^3He$ [15, 16], where a small volume of superfluid is re-heated to normal state, and subsequently rapidly cools to the temperature of the surrounding superfluid, are a good example of an inhomogeneous quench: The normal region shrinks from the outside. Yet, topological defects are created, thus suggesting that the phases of distinct domains within the re-heated region are selected independently.

2) Another example are relativistic heavy ion collisions where, according to Bjorken scenario (Bjorken 1983 [25]), a finite volume of quark-gluon plasma can be created. The plasma expands in the direction of collision and cools

from the outside in the perpendicular direction. The phase transition in this case can be first or second order (or a smooth crossover) depending on the parameters of the collision.

3) Any generic experiment based on pressure and/or temperature quench is to some degree inhomogeneous because of finite velocity of sound and finite heat conductance.

The mass parameter $\epsilon(t,\vec{r})$, varying in both time and space, must be considered in defect formation. As a consequence, locations entering the broken symmetry phase first could communicate their choice of the new vacuum as the phase ordered region spreads in the wake of the phase transition front. When this process dominates, symmetry breaking in various, even distant, locations is no longer causally independent. The domain where the phase transition occurred first may impose its choice on the rest of the volume, thus suppressing or even halting production of topological defects. This happens if velocity of the critical front is less than certain characteristic velocity.

## 4.1. SECOND ORDER TRANSITION

The characteristic velocity in an overdamped transition can be estimated as follows: The freeze-out healing length is set at $\hat{t}$ as $\hat{\xi} = \xi_0\ (\tau_Q/\tau_0)^{1/4}$. At the same instant the relaxation time is $\hat{\tau} = (\tau_Q\tau_0)^{1/2}$. These two scales can be combined [2] to give a velocity scale

$$\hat{v} = \hat{\xi}/\hat{\tau} = v_0\ (\tau_0/\tau_Q)^{1/4}\ , \tag{20}$$

where $v_0 = \xi_0/\tau_0$.

The density of defects $N$ as a function of critical front velocity is expected to change qualitatively at $\hat{v}$. Above $\hat{v}$ the homogeneous estimates should hold. Below $\hat{v}$ the density should be suppressed. Kibble and Volovik [26] suggested that $N \sim v/\hat{v}$ for small $v < \hat{v}$. Dziarmaga, Laguna and Zurek [27] argued that $N$ is exponentially suppressed below $\hat{v}$. There is a lot of qualitative difference between the two proposals. The former option suggests that however one makes a quench one will always get some defects, the latter implies that if one's inhomogeneous quench is sufficiently slow one will get no defects at all. In what follows we will quantify what "sufficiently slow" means.

*Decay of the False Vacuum* As a simple warm up exercise, let us consider decay of a false symmetric vacuum to a true symmetry broken ground state in a one-dimensional dissipative $\varphi^4$ model

$$\partial_t^2\varphi + \eta\,\partial_t\varphi \; - \; \partial_x^2\varphi \; + \frac{1}{2}\,(\varphi^3 - \epsilon\varphi) = 0\;, \tag{21}$$

where $\varphi(t,x)$ is a real order parameter and $\epsilon$ measures the degree of symmetry breaking i.e. $m^2 = -\epsilon$. Without loosing generality, we look for a solution $\varphi(t,x)$ which interpolates between $\varphi(t,-\infty) = -\sqrt{\epsilon}$ and $\varphi(t,+\infty) = 0$. Such a solution can not be static. It is a stationary half-kink

$$\varphi(t,x) = -\sqrt{\epsilon}\left(1+\exp\left[\frac{\sqrt{\epsilon}}{2}\frac{(x-v_t t)}{\sqrt{1-v_t^2}}\right]\right)^{-1} \tag{22}$$

moving with characteristic velocity

$$v_t = \left[1+\left(\frac{2\eta}{3\sqrt{\epsilon}}\right)^2\right]^{-1/2} \overset{\eta\to\infty}{\approx} \frac{3\sqrt{\epsilon}}{2\eta}\;. \tag{23}$$

It is worth noting that the decay velocity $v_t$ increases with $\epsilon$.

*Shock Wave* Our shock wave inhomogeneous quench model consists of a sharp "pressure front" propagating with velocity $v$; that is,

$$\partial_t^2\varphi + \eta\,\partial_t\varphi \; - \; \partial_x^2\varphi \; + \frac{1}{2}\,(\varphi^3 - \epsilon(t,x)\,\varphi) = \mathcal{O}(t,x)\;\;, \tag{24}$$

where

$$\epsilon(t,x) \; = \mathrm{Sign}\,(t - x/v) \tag{25}$$

is the relative temperature and $\mathcal{O}(t,x)$ is a Gaussian white noise of temperature $\Theta$ with correlations given by Eq. (3).

There are two qualitatively different regimes:

1) $v > v_t$, the phase front propagates faster than the false vacuum can decay. The half-kink (22) lags behind the front (25); a supercooled symmetric phase grows with velocity $v - v_t$. The supercooled phase cannot last for long; it is unstable, and the noise makes it decay into the true vacuum.

2) $v < v_t$, the phase front is slow enough for a half-kink to move in step with the front, $\varphi(t,x) = H_v(x - vt)$. The symmetric vacuum decays into one definite non-symmetric vacua. The choice is determined by the boundary condition at $x \to -\infty$. No topological defects are produced in this regime. The stationary solution $H_v(x - vt)$ is stable against small perturbations [27].

These expectations are borne out by the numerical study of kink formation in [27]. Numerical results are presented in Fig. 1.

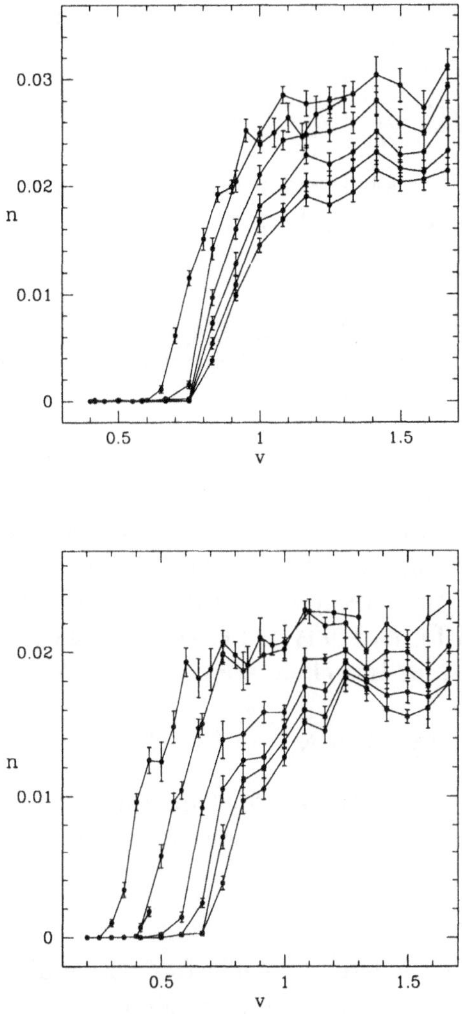


*Figure 1.* Up: Density of kinks $n$ as a function of velocity $v$ for the shock wave (25) with $\eta = 1$ (overdamped system). In this overdamped regime, the predicted threshold velocity is $v_t = 0.83$. The plots from top to bottom correspond to $\Theta = 10^{-1}, 10^{-2}, 10^{-4}, 10^{-6}, 10^{-8}, 10^{-10}$. At low $\Theta$, we get a clear cut-off velocity at $v \approx 0.8$, which is consistent with the prediction. Down: Density of kinks $n$ as a function of velocity $v$ for the linear inhomogeneous quench, Eq. (26), with $\tau_Q = 64$ and $\eta = 1$. The predicted threshold is $v_t = 0.77$. This cut-off is achieved for low $\Theta$. The plots from top to bottom correspond to $\Theta = 10^{-1}, 10^{-2}, 10^{-4}, 10^{-6}, 10^{-8}, 10^{-10}$.

*Linear Front* Let us consider now a system in which the inhomogeneous quench takes place via linear transition

$$\epsilon(t,x) = (t - x/v)/\tau_Q \tag{26}$$

In the absence of noise, the propagating front is followed by a stationary half-kink. This half-kink moves somewhat behind the front, its location is determined by the place where the threshold velocity (23) is equal to the front velocity, $v_t[\epsilon(t,x)] = v$. The distance between the front and the half-kink increases as $v^3$. This distance gives the size of the supercooled region. When the supercooled region is narrow then it is stable against small perturbations so that no defects are produced. If

$$v > v_t \equiv \left(1 + \frac{\eta^{3/2}\tau_Q^{1/2}}{11.7}\right)^{-1/2} \stackrel{\eta\to\infty}{\approx} \frac{3.42}{\eta}\left(\frac{\eta}{\tau_Q}\right)^{1/4} \equiv 4.07\,\hat{v}\ . \tag{27}$$

then the region is broad enough to be unstable [27] and the production of defects is no longer suppressed.

This prediction is confirmed by the numerical study of linear quenches in Ref. [27], compare Fig. 1. However, the threshold velocity apparently gradually decreases with increasing noise temperature $\Theta$. This decrease of the threshold for kink formation is due to the thermal nucleation of kinks. Quantitative estimates for this effect are given in [28].

### 4.2. FIRST ORDER TRANSITION

We assume the transition is strongly first order and that it goes by bubble nucleation. To be more specific we consider a toy model in 3 dimensions

$$\partial_t\varphi = \nabla^2\varphi - a\varphi + b\varphi^3 - c\varphi^5 + \mathcal{O}\ . \tag{28}$$

where $\varphi$ is real order parameter. The effective potential is of the $\varphi^6$ type. Provided that $b^2 > 4ac$, it has symmetric minimum at $\varphi = 0$ and two symmetry broken minima at $\varphi = \pm\varphi_m \equiv \pm\sqrt{(b+\sqrt{b^2-4ac})/2c}$. We assume that $b, c$ are constant and that symmetry breaking transition is driven by $a$ decreasing below its critical value $a_c = 3b^2/16c$. At $a = a_c$ all three minima are degenerate.

*Decay of the False Vacuum* Suppose that $a < a_c$. Let us consider decay of the false symmetric vacuum to the true symmetry broken phase in a one dimensional version of the model Eq. (28). We look for a solution which interpolates between $\varphi = \varphi_m$ for $x \to -\infty$ and $\varphi = 0$ for $x \to +\infty$. The solution is a stationary half-kink $H(x - v_t t)$ moving with velocity

$$v_t = \frac{-b + 2\sqrt{b^2 - 4ac}}{\sqrt{3c}} \tag{29}$$

which has an envelope function

$$H(x) = \frac{\varphi_m}{\sqrt{1 + \frac{\exp \alpha x}{2c}}} \ , \tag{30}$$

where $\alpha = \sqrt{4c/3}\varphi_m^2$. This way the false $\varphi = 0$ vacuum decays into the true $\varphi = \varphi_m$ vacuum in the absence of noise. The decay velocity $v_t$ is zero for $a = a_c$, it increases with increasing supercooling or with decreasing $a$.

*Shock Wave* In the shock wave model a sharp front propagates with velocity $v$

$$a = a_c - \Delta a \ \mathrm{Sign}(t - x/v) \ . \tag{31}$$

Similarly as for second order transitions there are two regimes:

1) $v > v_t$, the pressure front propagates faster than the false vacuum can decay. The half-kink lags behind the front. The supercooled phase in between them grows linearly with time. The phase is unstable, it decays by bubble nucleation just as for a homogeneous transition. Homogeneous estimates of defect density apply in this case.

2) $v < v_t$, the half-kink is faster. It moves in step with the front while its tail penetrating into the symmetric phase. There is no supercooled phase where bubbles could be nucleated. The symmetric phase goes smoothly into one of the symmetry broken phases.

*Linear Front* Let the inhomogeneous quench proceed by a linear front moving with velocity $v$

$$a = a_c - (t - x/v)/\tau_Q. \tag{32}$$

The half-kink follows the critical front staying at a certain distance behind it. The distance $D$ is such that the half-kink velocity $v_t$, which depends on the local value of $a$, is equal to the front velocity $v$, $v_t(a) = v$. With increasing $v$ the half-kink settles at increasing values of local $a$. Close to the critical front the radius of the critical bubble is infinite and at the same time the nucleation rate is infinitely small. As we go away from the front in the direction of the half-kink the critical radius shrinks. At a certain distance $L$ from the front the energy of the critical bubble becomes comparable to the temperature $\Theta$. At this point bubble nucleation becomes possible. If $L < D$ bubbles can be nucleated in the supercooled region between the front and the half-kink. If $L > D$ then there is no bubble nucleation and no defects can be born in the supercooled area.

Now we estimate the critical velocity such that $L = D$. The half-kink is located at such an $a$ that $v_t(a) = v$. $L = D$ providing that for this $a$ the energy of the critical bubble $E(a)$ is equal to temperature $\Theta$. The critical bubble is a metastable spherically symmetric static solution of Eq. (28) with, say, $\varphi_m$ vacuum inside and 0 vacuum outside its wall. Its energy can be easily estimated when the width of its wall is negligible as compared to its radius $R_c(a)$. An approximate solution is given by $H[r - R_c(a)]$, where the critical radius is

$$R_c(a) = \frac{\sqrt{12c}}{-b + 2\sqrt{b^2 - 4ac}} \ . \tag{33}$$

The energy of the critical bubble $E(a)$ has a negative volume contribution, $(4\pi R_c^3/3)V(\varphi_m)$, and a positive surface tension term, $(4\pi R_c^2) \int dx \ [H'(x)]^2$. When the solution of $v_t(a) = v$ is put into $E(a)$ and then the equation $E(a) = \Theta$ is solved, one obtains a critical velocity

$$v_{cr} = \left( \frac{\pi b(3b^2 - 6bc + 16c^2)}{4c^3\Theta} \right)^{1/3} \tag{34}$$

for $L = D$. For $v > v_{cr}$ bubbles can nucleate in between the half-kink and the front and thus the necessary condition for topological defects production is satisfied.

The formula for $v_{cr}$, Eq.(34), is still a crude lower estimate for the critical velocity. In fact it is not sufficient to nucleate some bubbles. Individual bubbles would coalesce with the half-kink without any chance to trap any nontrivial winding number. The bubbles should be nucleated in large numbers or have enough time to grow so that they can mutually coalesce before merging with the half-kink. Still, the argument which leads to $v_{cr}$ demonstrates that there is a threshold velocity for defect formation.

### 4.3. HIGHER DIMENSIONS

The theory can be generalized to higher dimensions and to a complex order parameter in a straightforward manner. Its major result is that a subthreshold inhomogeneous quench does not produce any variation of the order parameter in the direction normal to the front. This excludes any possibility of production of vortex loops or closed membranes entirely contained in the bulk, as well as of any pointlike defects. Some extended defects can grow into the bulk provided their seeds were created at this edge of the system where the symmetry was broken first. In first approximation such, say, vortices grow into the bulk, following the passing front, while keeping their direction normal to the front. In the end we do not get any chaotic tangle of strings and string loops but parallel "combed" vortices. There are two important perturbations to this "combed" picture:

1) Thermal fluctuations make the strings look more random but without backtracking and with string tension tending to smooth the small scale fluctuations. The ends of the strings and antistrings at the critical front are wandering around. Eventually an end of a string and of an antistring may meet so that the strings join into a half-loop with its both ends attached to the initial edge of the system. String tension shrinks the half-loop to the edge where it unwinds.

2) A much more efficient factor to remove vortices from the bulk are their mutual interactions. Global parallel string and antistring attract one another so that their ends at the critical front do not seek each other at random but tend to fuse in a deterministic way. This mechanism makes the number of strings in the bulk decay with increasing distance between the front and the initial edge.

The factors (1) and (2) lead to a picture in which the critical front initially draws some parallel strings and antistrings from the edge, then the strings recombine by joining ends and shrinking back to the edge. In the end only the net surplus of strings (or antistrings) is left in the bulk.

These ideas are supported by experiments:

1) Disclinations produced during a quench from disordered to nematic phase in liquid crystals [29]. This is a weakly first order transition. In early attempts to make cosmological experiments in liquid crystals the disclinations were observed to grow approximately combed, join ends and shrink to the initial edge. Later on it was realized that these quenches were not homogeneous enough [30].

2) Czochralski method of growing monocrystals, which is widely used to grow silicon monocrystals necessary for microchips. In this method, discovered in the thirties, a surface of liquid material is touched with a monocrystal template. As the template is slowly lifted up it drags a column of crystal out of the container. The top part of the column is cold while its bottom part is at the melting temperature - the transition is inhomogeneous. If the template is lifted slowly enough, then no defects of the crystal lattice are produced which might spoil the monocrystal.

To conclude this section: in an inhomogeneous quench there is a threshold velocity $v_t$ of the critical front. Above the threshold defects are produced like in a homogeneous quench. Below the threshold one gets no defects; instead a clean monocrystal or a "disoriented chiral condensate" is grown with a vacuum which may be uniform over significant distances, but which differs from the true vacuum.

## 5. Defect formation and the Ginzburg Regime

Recently a new $^4$He experiment [17] was devised, improving on the apparatus used earlier by McClintock *et al.* [31] to implement a superfluid transition in $^4$He through a sudden pressure quench. The corresponding results are rather surprising. They show no evidence for the formation of topological defects at the anticipated levels, contrary to expectations based both on the old experiment [31], the theory[1] and the $^3$He data [15, 16]. The discrepancy with the earlier $^4$He quench data is now seen as the evidence of mechanical stirring in the first version of the experiment. Nevertheless to address this discrepancy with $^3$He it was suggested [32] that because the Ginzburg regime in $^4$He extends over a broad range of temperatures around the $\lambda$-line, large scale fluctuations may be able to unwind and alter the configuration of the order parameter (in contrast to $^3He$) while the quench proceeds.

The Ginzburg temperature is defined through the loss of ability of the order parameter to hop, through thermal activation, over the potential barrier between broken symmetry vacua. Thus one might worry with Karra and Rivers [32] that when the defect densities are eventually measured, at a much later time, little or no string would have survived unwinding through thermal activation. In this section we investigate this possibility and more generally report a numerical study of the effect of thermal fluctuations on topological defect formation and evolution.

Originally the Ginzburg temperature $T_G$ was suggested to be the time of formation of topological defects [4], since, at lower temperatures, thermal fluctuations would be unable to overcome the potential energy barrier associated with the defect's topological stability.

In reality the situation is more complex. In equilibrium at any given temperature $T$ (including of course temperatures in the Ginzburg regime) a range of string configurations will exist. However, long strings can only exist in equilibrium strictly above $T_c$ [33, 34].

To freeze them out, i.e., to *form* them, energy (associated with the string tension) must be extracted from the system. This necessarily breaks time invariance and will lead to locally preferred nonequilibrium field configurations. Subsequently the system will order over larger and larger spatial scales, leading to mutual string annihilation.

The initial density of defects entering this stage of evolution is computed by the theory of section II. This density is set at an effective temperature $-\hat{\epsilon}$, which in $^4$He is well within the estimates for the width of the Ginzburg regime. By contrast, in $^3$He the Ginzburg temperature is small compared

1Although a factor $f \gtrsim 10$ in the formula for the string density $n \sim 1/(f\hat{\xi})^2$ could explain the new results and seems consistent with recent numerical studies [13].

to the typical $\hat{\epsilon}$. What happens to the initial densities of string when the system is exposed to temperatures in the Ginzburg regime for an extended amount of time?

In order to investigate this issue we need a quantitative definition of $T_G$. In tune with the arguments given above consider a volume of characteristic size $\xi(T)$, the correlation length, and a theory with two energetically degenerate minima of an effective potential $V(\phi)$, separated by a potential barrier $\Delta V$. The rate for the field to change coherently from one minimum to the other per unit volume due to thermal activation is $\exp[-\Delta V/k_B T]$. For an effective potential of the form (obtained, eg. perturbatively at 1-loop)

$$V(\phi) = -\frac{1}{2}m^2(T)\varphi^2 + \lambda\varphi^4 \tag{35}$$

$\Delta V = \frac{m(T)^4}{4\lambda}$. For a volume $\xi^3$, we define $T_G$ such that the probability of overcoming the potential barrier is of order unity:

$$T_G \,:\, \frac{\Delta V(T_G)}{T_G}.\xi^3(T_G) = 1 \quad \Leftrightarrow \quad \lambda T_G/m(T_G) = \frac{1}{4}. \tag{36}$$

This definition however has some caveats, for instance, an effective potential of the form Eq. (35) is only valid for the mean field and not on smaller scales. A more careful accounting of scales leads to different results [35], which show an enhancement of the hoping probability. Thus, the factor of 1/4 in Eq. (36) should not be taken at face value.

A more rigorous definition arises from the range of temperatures below $T_c$ for which fluctuations are large and consequently where perturbative finite temperature field theory fails to be useful. In order to set up a perturbative scheme at finite temperature from an initial 3+1 dimensional quantum field theory one implements dimensional reduction which is valid provided the temperature is high compared to all mass scales. As a consequence the coupling of the dimensionally reduced 3D field theory becomes dimensionful, i.e. $\lambda \to \lambda T = \lambda_3$. In order to proceed one has to identify an appropriate dimensionless coupling. This is done by taking $\lambda T/m(T)$. The Ginzburg regime is entered when this 3D effective coupling becomes strong, in the vicinity of the critical point, namely

$$T_G \,:\, \lambda T_G/m(T_G) = 1. \tag{37}$$

To compute $T_G$ one needs the scaling of $m(T)$ in the critical domain. We write $m^2(T) = m_0^2\epsilon^\nu$, with $\epsilon$ being the reduced temperature $\epsilon = |\frac{T-T_c}{T_c}|$.

Thus $\epsilon_G = -0.18$ for $\nu = 0.5$. This mean-field estimate produces an upper bound in $T$ for $T_G$ (and lower bound for $\beta = 1/T$). For realistic 3D exponents, $\nu = 0.67$, we obtain $\epsilon_G = -0.25$. The first criterion, based on

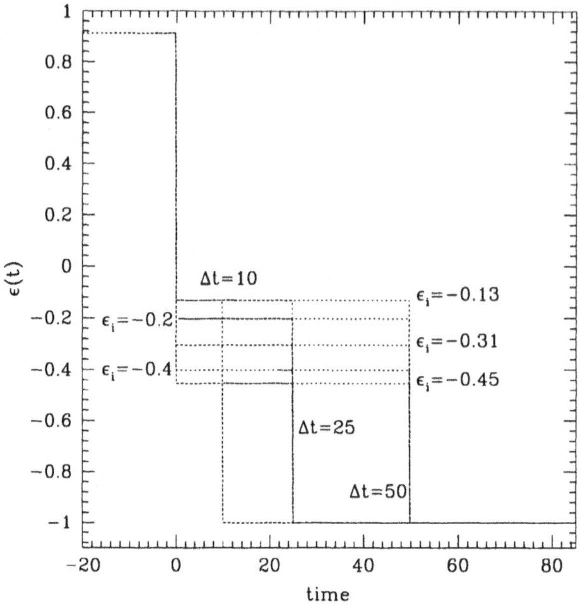


*Figure 2.* Temperature trajectories for testing the effect of exposure to the Ginzburg regime on string densities. The system is first thermalized at a high temperature and then placed in contact with a heat bath at an intermediate temperature $\epsilon_i$ below $T_c$, for a time interval $\Delta t$.

the hopping of a correlation sized volume, results in higher values of $T_G$. This brings about a relatively large uncertainty in the value of $T_G$, which is $18 - 25\%$ below $T_c$.

## 5.1. STRINGS SURVIVE THE GINZBURG REGIME

In order to investigate the role of the Ginzburg temperature in the *dynamics* of defect formation we deliberately expose the system to a heat bath at temperature $\epsilon_i$, within the Ginzburg regime and below. We repeat this procedure for a range of time intervals $\Delta t$, after which the bath temperature is taken to zero. This set of temperature trajectories is shown in Fig. 2. We are attempting to emulate the worst case scenario of an experimental quench where the temperature or pressure are dropped monotonically but where the system makes a long stopover within the Ginzburg regime.

We would expect that, if the Ginzburg regime indeed produced enhanced decay of strings, then the string densities measured at later times should be smaller the longer the time the system spent within the range $T_c \geq T \geq T_G$.

We have measured the final string densities at a time $t >> \Delta t$.

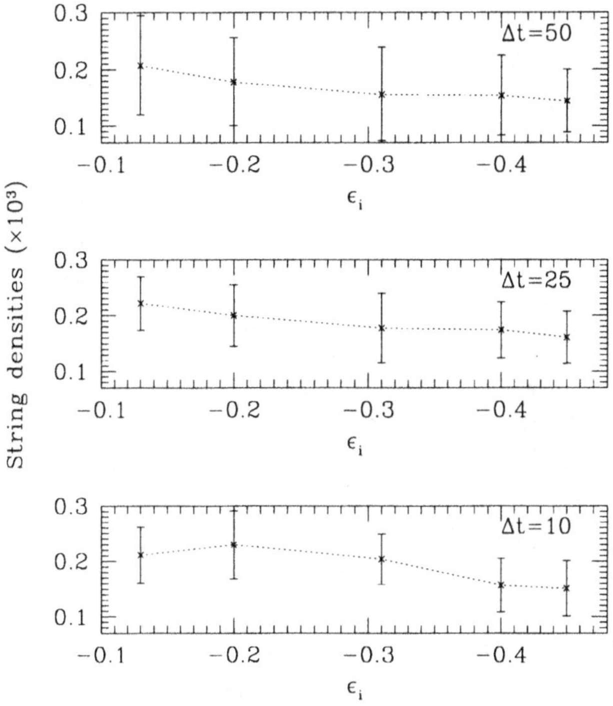


*Figure 3.* The string density measured at a later time $t >> \Delta t$ vs. intermediate temperature $\epsilon_i$. From top to bottom the three plots correspond to $\Delta t = 10, 20, 50$, during which the system remained in contact with a heat bath at $T_i$. There is no visible role played by intermediate temperatures within the Ginzburg regime.

Our results for the final string densities as a function of intermediate temperature $\epsilon_i$ and $\Delta t$ are summarized in Figure 3. There is no apparent effect of the Ginzburg regime in reducing string densities at formation.

If any trend is visible from Figure 3 it is the opposite, namely that the lower $\epsilon_i$, the less string is measured at later times. This is consistent with the relaxation of the string network, resulting in vortex annihilation controlled by the string tension (which is smaller near $T_c$) and with results the thermodynamics of vortex strings [33, 34]

## 5.2. MEMORY OF THE ORDER PARAMETER CONFIGURATION NEAR $T_C$

An independent test on the possible role of thermal fluctuations in affecting string densities consists in reheating a quenched system to a temperature around $T_c$ (both below and above it) and cooling it again. This process tests the memory of the order parameter as well as that of other related quantities

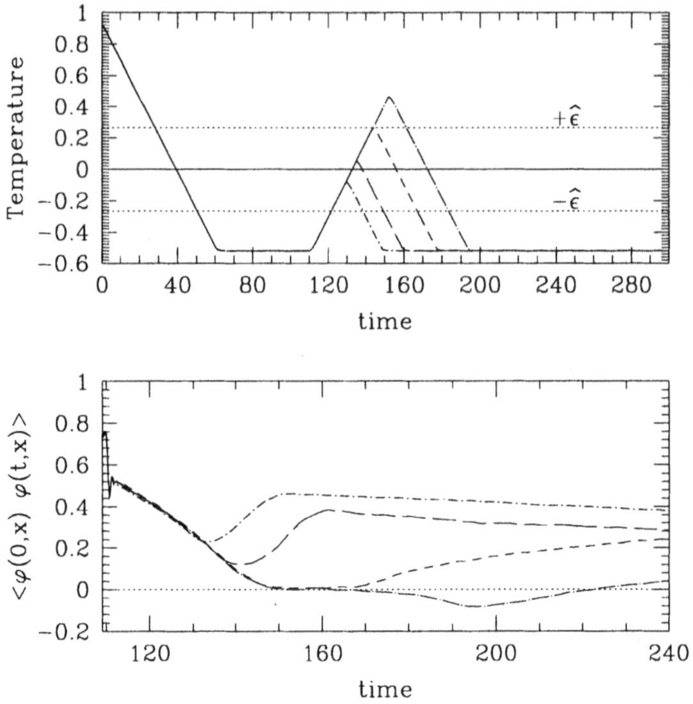


*Figure 4.* FIG. 4. a. Dependence of the bath temperature $\epsilon$ in time. After being quenched in temperature ($\tau_Q = 80$) the system is reheated at the same rate to a temperature $\epsilon_f = 0.469, 0.256, 0.061, -0.068$ (top to bottom) 'and cooled again. b. The correlation function between the field at the time just before reheating and at later times, $\langle \varphi_i(t_{\rm rh}, x)\varphi_i(t + t_{\rm rh}, x)\rangle$ is plotted. There is a universal short time transient for the decorrelation of the field over small scales while the long time tails of the correlation function describe change over the mean fields. All four trajectories cross the Ginzburg regime, but only those reaching or crossing $+\hat{\epsilon}$ display a significant memory loss.

[see also Yates and Zurek 98 [12]], such as defects. These temperature ($\epsilon(t)$) trajectories are illustrated in Fig. 4a.

We are particularly interested in investigating under what circumstances thermal fluctuations can affect the large scale configuration of the order parameter.

We define the unequal-time correlation function

$$\langle \varphi(x, t_{\rm rh})\varphi(x, t + t_{\rm rh})\rangle = N \sum_{j=1}^{2} \sum_{i} \varphi_j(x_i, t_{\rm rh})\varphi_j(x_i, t + t_{\rm rh}), \tag{38}$$

where $N$ is an irrelevant normalization factor. This correlator has several interesting properties. For short times it displays a characteristic time, which describes the decay of correlations over very small spatial scales. This is the initial transient in Fig. 4b. For later times the residual correlation comes from the motion of the order parameter (the field volume average). This average can be either positive or negative but, if thermal, will converge to zero at and above $T_c$.

Now, we are interested in determining whether the final field configuration over large spatial scales is correlated to the configuration prior to reheating. Fig. 4 shows that only if one crosses $T_c$, by more than $+\hat{\epsilon}$, is the memory of the initial quenched configuration erased (see in particular the two trajectories reaching higher temperatures in comparison to the others). For these trajectories the field correlations reach zero and after reheating evolve to a value manifestly different from that prior to reheating.

For trajectories within the Ginzburg regime, that do not cross $T_c$, the change in the configuration of the order parameter as measured by Eq. (38) is small. In particular the field configuration existing before reheating is approximately recovered as the fields are cooled. The same is true for the string densities, including those of long strings.

Thus we are led to conclude that even prolonged exposure of a quenched field configuration to the Ginzburg regime has little consequences in changing the order parameter configurations emerging at $-\hat{\epsilon}$, and associated string densities. In addition we have shown that to truly destroy a quenched field configuration existing below $-\hat{\epsilon}$, one has to expose the system to temperatures well above $T_c$. In particular for any particular quench trajectory, a temperature of $T \sim T_c + \hat{\epsilon}$, must be reached and maintained for a time $\sim \hat{t}$ in order to erase memory of the initial configuration.

These results fully support the theory of section II for the critical dynamics of second order transitions and all known thermodynamic results for vortex strings in $O(N)$ theories. Thus we expect the results of this section to carry over from our models to the Lancaster $^4$He experiments. The results of Ref. [17] in these experiments cannot therefore be attributed to the effects of Ginzburg regime in $^4He$.

In the next section we offer an alternative explanation.

## 6. What is being observed in the $^3$He and $^4$He experiments

The several experiments in Helium, and more recently in superconductors, testing the theory of defect formation rely on substantially different processes to induce the phase transition and measure defects.

In this section we analyze, in the light of our own theoretical results, how experimental procedures can lead to the detection of substantially different defect densities.

Two particular factors play a decisive role in the value of the topological defect density measured - the time and procedure of measurement after the quench and the initial/final state of the system.

### 6.1. THE LANCASTER EXPERIMENTS IN $^4HE$

In the Lancaster experiments in $^4$He the defect density is measured through the attenuation of a second sound signal (a heat pulse). This probe can only detect densities above a certain threshold (if the theory of section II is used $f \leq 10$ would be required, which is at odds with the results of the numerical studies [10, 12] and especially [13]). Moreover, the density at formation has to be extrapolated from the data obtained at relatively late times - the signal is noisy shortly after the quench [17].

After being formed by the critical dynamics of the phase transition vortex strings decay away, as the system orders and cools. This decay has been modeled by Vinen's Equation

$$\dot{n} = -\gamma n^2, \tag{39}$$

where $n$ is the length of string per unit volume, i.e. the string length density, $\gamma > 0$. This model has been observed in the same experiment to describe very well the decay of vorticity induced initially through a fluid flow.

Vorticity created thermally is potentially different from that formed under an external flow. We know from several theoretical and numerical indications that a thermal distribution of vortices close to the transition is comprised of both long strings and small loops, see Fig. 5.

These two populations decay very differently in the wake of the quench. Without any mechanism for stabilization the loops tend to disappear in a fast transient. In contrast the long strings loose some of their small scale structure but survive, and will ultimately set the decay pattern described by the Vinen equation.

It is the surviving long strings - eventually measured - that will provide the experimental signal in the $^4$He experiments. This is shown in Figs. 5 and 6. As the system is quenched from higher temperatures or pressures, an initial string network comprised of strings of all lengths looses its loops and settles to a much slower decay trend dominated by long strings.

The crucial question then is whether enough long string would persist at the time of measurement to yield a positive signal.

We have performed a very similar procedure in a numerical "experiment" [13]. We observed that at long times string densities could be measured that agreed extremely well with all features of the theory. Our definition of *long* times was intimately connected to the completion of the phase transition expressed in the expectation values of the order parameter $\langle|\phi|\rangle \simeq 0.9 - 0.975$. The effective $f$ measured then was in the range $f = 11 - 16$. All our indications are that the Lancaster $^4$He experiment

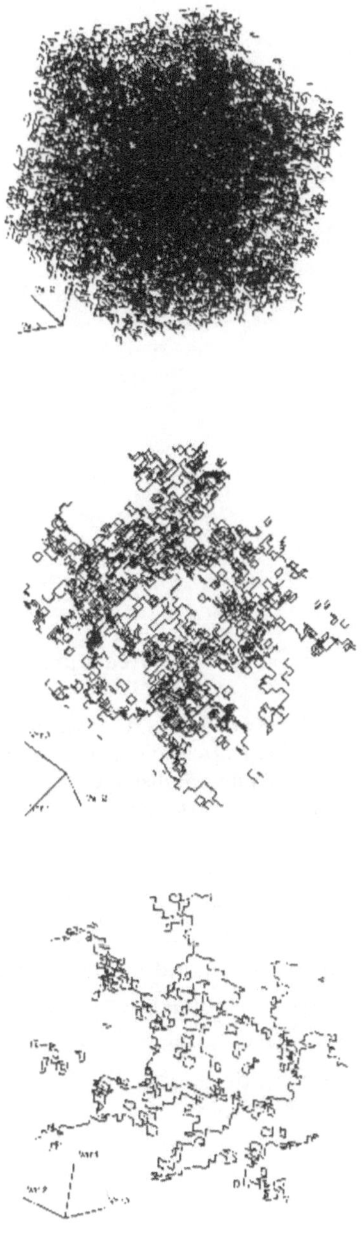

*Figure 5.* The decay of vorticity under a quench. Initially the defect network includes both long strings and small loops. At late times the network coarsens and only long strings remain, see also Fig. 6.

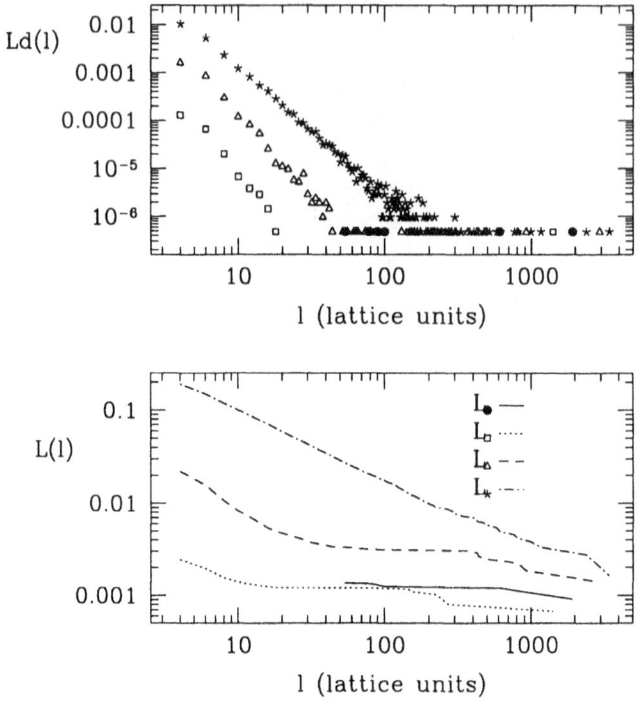


*Figure 6.* FIG. 6. String length $l$ distributions Ld(l), taken between $+\hat{t}$ and the 'time of formation' ($\langle|\phi|\rangle = 0.95$), for $\tau_Q = 64$. Data sets denoted by ($\star$,$\triangle$,$\square$, $\bullet$) correspond to increasingly later times. Lines show the integral distributions, eg. $L_\bullet(l) = \sum_l^\infty l' Ld_\bullet(l')dl'$. It is clear that at late times only long strings survive.

performs its measurements much later (up to 2 orders of magnitude) than we do, thus leading to even smaller string densities. Such string densities would evade detection under the second sound experimental probe, which may lead to the negative result reported recently [17].

In order to measure a positive signal in these circumstances the measurements would have to be made sooner, after a much faster quench, or with a higher sensitivity.

## 6.2. EXPERIMENTS IN $^3$HE

In contrast to the experiments in $^4$He described above which appear sensitive to the "infinite" string, experiments in $^3$He have pursued two independent strategies both of which allow one to stabilize and detect defect loops of various sizes: either vorticity is stabilized and amplified by the flow and then measured directly using nuclear magnetic resonance (the Helsinki experiment) or it is inferred from a certain amount of missing energy (the

Grenoble and Lancaster $^3$He experiments). Of these two procedures the first is more direct - vorticity formed during the quench is forced to migrate to the center of the container, through the existence of a subcritical rotation velocity, where it is detected. This permits loops of string of length larger than a known threshold to survive decay and results naturally in a higher density, i.e. the defect density is measured effectively very shortly after the transition takes place and need not be limited the the "infinitely long string". As a consequence, much smaller values of $f$ and larger string densities are measured than in the $^4$He experimental setting.

Both remaining $^3$He experiments end at a region of the phase diagram far from the transition - $^3$He being very much colder than in the Helsinki experiment. A lower effective temperature results in the effective absence of damping mechanisms which in turn leads to the preservation of even small vortex loops. Dissipation mechanisms rely on the presence of quasiparticles. Thus, when the medium is very cold, energy dissipation will slow exponentially and vortices can be stabilized by a coherent flow resulting from their motion through the superfluid. We expect therefore the string population in all $^3$He experiments to be mostly in the form of relatively small loops. In Helsinki, the largest loops are stabilized by the slow rotation of the whole system, and their density can be extrapolated to the smaller loops, leading to the total consistent with the Grenoble and Lancaster experiments, where - one may guess - all of the loops survive for a long time in the absence of dissipation.

The lifetime of these loops is thus expected to be much longer than that of thermal loops formed at a quench through the $\lambda$-line in $^4He$. As a result the long time decay of vorticity may also be very different in these two cases as the former corresponds to an ensemble of moving loops, at relative distances much larger than their typical radius, but the latter contains strings of all sizes, where the mean distance between strings is comparable to their length.

This conjectured picture, supported in part by numerical studies, leads to the conclusion that both experimental settings in $^3$He should lead to a positive result, compatible with a relatively small value of $f$ relevant for all loops (we get $f \simeq 4$, when $n \sim 1/(\hat{\xi} f)^2$ is used to fit early data in Fig. 6), whereas in $^4$He the smallness of the signal at the time of first measurement makes the detection more difficult and at present below the sensitivity threshold.

## 7. Discussion

The mechanism we described early on in this paper is based on the analysis of the behavior of the order parameter $\varphi$. The order parameter is clearly a phenomenological entity and the equations that govern its evolution are approximate and in many cases postulated rather than derived. On the other hand the underlying physics is usually very specific. It may, for instance, involve atoms of some particular isotope such as $^4He$. Thus, in principle, one could formulate an exact microscopic theory of particular second order phase transformations. However, in all of the experimentally accessible cases discussed above such a fundamental theory is simply too complicated to lead to useful conclusions. The superfluid transition in $^4$He is a good example: Strong interactions in $^4$He make it impossible to proceed rigorously all the way starting at the microscopic level. Analysis of related issues in the field theoretic context is also difficult (D. Boyanovski, H. J. de Vega, R. Holman, these proceedings [36]). Recently however a new system has become experimentally accessible: Atomic Bose-Einstein condensates (BEC's) undergo the second-order phase transition at much lower densities. Natural approximation schemes can be therefore suggested, and the exact microscopic theory can be studied in greater detail than for the "old" superfluids.

We shall not attempt to review the theoretical or experimental situation in BEC's. Good reviews already exist (see eg. Ketterle et al. [37]). Our aim is simply to point out that questions concerning the formation of topological defects can be posed and analyzed within a much more fundamental formalism, which is explicitly quantum. The approximations start from the Schrödinger, equation and lead in a controlled manner to master equations for the density operator of the condensing system. Further approximations result in a quantum kinetic theory. Preliminary analysis of these issues [38] allows one to recover the key scaling relations and the key predictions we have described in section II. Indeed, time dependent Landau-Ginzburg theory follows as an approximation to some of the terms which one obtains from the microscopic treatments. On the other hand, the microscopic theory contains additional terms, which alter predictions concerning the formation of topological defects. Limited studies [38] indicate that the predicted densities of the vortex lines or of the winding numbers would be smaller than those based on the scalings of the order parameter (see section II of this paper). Moreover, corrections seem to be more significant as the ratio $\tau_0/\tau_Q$ decreases.

The possibility of experimental studies of defect formation in BEC quenches nevertheless exists and may lead to exciting insights into the problem.

Superconductors may be the other useful testing ground. Indeed, one experiment has been reported at this meeting [18], with the claim of conflicting results, which seemed to depend on the geometry. Rapid cooling produced no detectable signal in a high-temperature film, although it is far from clear whether any was expected. The original claim that the effect was ruled out at the "$\sim 10^3$ level" was based on an overly optimistic prediction, which did not recognize that the total flux expected to arise in the experimentally studied geometry scales as $n^{1/4}$, i.e. only with the *fourth root* of the total number of defects (rather than with the square root).

The revised prediction is close to the claimed sensitivity of the experiment, and given the uncertainties in the critical exponents of the high-temperature superconductor, as well as the possibility of imperfect trapping of the defects, etc. it is unfortunately impossible to extract useful constraints from the existing negative experiment.

One of the experiments carried out by the same group, in the loop geometry has, on the other hand, yielded positive results. This experiment also operates near the edge of detectability. It detects the flux induced by a loop which is artificially broken into a large number N of superconducting sections, which are then rapidly reconnected. The predicted flux should have a Gaussian distribution with a random direction and intensity corresponding to $\sigma \sim \sqrt{N}$ flux quanta (Zurek 1996 [21]). SQUIDs used in the experiment are not sensitive to the flux at 1 $\sigma$ level, but large signals appear to have been indeed found (Raz Carmi, talk at Les Houches, Feb. 1999 [18]).

The available experimental results can be therefore described as confusing. In liquid crystals the results seem to be the perhaps least ambiguous, but they concern a (weakly) first order transition. In superfluids, $^3$He is still the strongest case for the mechanism, especially since the results between all the experiments (carried out in quite different parameter regimes, and using very different techniques) are consistent. Further experimental and numerical studies to investigate the role of rotation in stabilizing the vortex loops and to explore the implications for the A-B phase transition, etc., are nevertheless essential.

The existing $^4$He data are clearly disappointing, but not at odds with a more conservative theoretical estimate. Moreover as we have argued above there may be a way to reconcile estimates of vortex line density obtained from $^3$He and $^4$He experiments, even without any special appeal to the Ginzburg regime [32] or to quantum kinetic theory.

Finally the first experimental reconnaissance into quenches in superconductors is preliminary in its nature and ambiguous in its results.

In the meantime, numerical studies have confirmed and refined the basic indications of the theory of order parameter dynamics.

One may be therefore justified in the expectation of an exciting but uncertain future. A lot is at stake, including the understanding of the phase transition dynamics, nature of the order parameter and other collective observables of quantum many body systems and perhaps even the relation between the quantum and the classical.

## References


1. Zurek, W. H. (1984), 'Experimental Cosmology: strings in superfluid Helium', Los Alamos preprint LA-UR-84-3818.
2. Zurek, W.H.(1985) Cosmological experiments in superfluid Helium, Nature **317** 505.
3. Zurek, W.H. (1993) Cosmic strings in laboratory superfluids and topological remnants of other phase transitions, Acta Physica Polonica B **24** 1301.
4. Kibble, T.W.B. (1976) Topology of cosmic domains and strings, J. Phys. A **9**, 1387.
5. Kibble, T.W.B. (1980) Some implications of a cosmological phase-transition, Phys. Rep. **67**, 183.
6. Gill, A.J. (1997), Baryon Density in the Central Region of a Heavy-Ion Collision, hep-ph/9706327.
7. Dziarmaga, J. and Sadzikowski, M. (1999) Antibaryon density in the central rapidity region of a heavy ion collision, Phys.Rev.Lett.**82**, 4192.
8. Volovik, G. (1996) Cosmology, Particle Physics and Superfluid 3He, Czech J. Phys. **46**, 3048.
9. Bunkov, Y.M., and Timofeevskaya, O.D. (1998) "Cosmological" scenario for A-B phase transition in superfluid 3He Phys. Rev. Lett. **80** 1308.
10. Laguna, P. and Zurek, W.H. (1997) Density of kinks after a quench: When symmetry breaks, how big are the pieces?, Phys. Rev. Lett. **78**, 2519.
11. Laguna, P. and Zurek, W.H. (1998) Critical Dynamics of Symmetry Breaking: Quenches, Dissipation and Cosmology, Phys. Rev. **D58**, 5021.
12. Yates A. and Zurek, W. H. (1998) Vortex formation in two dimensions: When symmetry breaks, how big are the pieces?, Phys. Rev. Lett. **80**, 5477.
13. Antunes, N.D., Bettencourt, L.M.A. and Zurek, W.H. (1999) Vortex String Formation in a 3D U(1) Temperature Quench, Phys. Rev. Lett. **82** 2824.
14. Stephens, G.J., Calzetta, E.A., Hu, B.L. and Ramsey, S.A. (1999) Defect Formation and Critical Dynamics in the Early Universe, Phys.Rev. D **59** 045009.
15. Baüerle, C., *et al.* (1996) Laboratory simulation of cosmic-string formation in the early universe using superfluid He-3, Nature **382**, 332.
16. Ruutu, V.M.H. *et al.* (1996) Vortex formation in neutron-irradiated superfluid He-3 as an analog of cosmological defect formation, Nature **382**, 334; Ruutu, V.M.H.*et al.* (1998) Defect Formation in Quench-Cooled Superfluid Phase Transition, Phys. Rev. Lett. **80**, 1465.
17. Dodd, M. E. (1998), *et al.*, Non-Appearance of Vortices in Fast Mechanical Expansions of Liquid 4He Through the Lambda Transition, Phys. Rev. Lett., **81**, 3703.
18. Carmi, R., Polturak, E. (1999) Search for spontaneous nucleation of magnetic flux during rapid cooling of $YBa_2Cu_3O_{7-\delta}$ films through $T_c$, To appear in Phys. Rev. B
19. Lythe G.D. (1998) Stochastic PDEs: domain formation in dynamic transitions, Anales de Fisica Vol. 4 55.
20. Dziarmaga, J. (1998) Density of kinks just after a quench in an underdamped system, Phys. Rev. Lett. **81**, 1551.
21. Zurek, W.H. (1996) Cosmological experiments in condensed matter, Phys. Rep. **276** 177.
22. Eltsov, V.B., Krusius, M., and Volovik, G.E. (1998) Superfluid 3He: a Laboratory Model System of quantum field theory, cond-mat/9809125.
23. Hohenberg, P.C. and Halperin, B.L. (1977) Theory of dynamic critical phenomena, Rev. Mod. Phys. **43**, 435.
24. Kajantie, K., Laine, M., Rummukainen, K. and Shaposhnikov, M. (1996) Is there a hot electroweak phase transition at $m$(H) greater than or similar to $m$(W)?, Phys. Rev. Lett. **77**, 2887–2890.
25. Bjorken, J.D. (1983) Highly relativistic nucleus-nucleus collisions : the central rapidity region, Phys. Rev.**D 27**, 140.
26. Kibble, T.W.B. and G.E. Volovik (1997) On Phase Ordering Behind the Propagating Front of a Second-Order Transition, Pis'ma v ZhETF **65**, 96.

27. Dziarmaga, J., Laguna, P. and Zurek, W.H. (1998) Symmetry Breaking with a Slant: Topological Defects after an Inhomogeneous Quench, cond-mat/9810396, to appear in Phys.Rev.Lett.
28. Kopnin, N. B. and Thuneberg, E.V., in preparation.
29. Chuang, I., Dürrer, R., Turok, N. and Yurke, B. (1991) Cosmology in the laboratory : defect dynamics in liquid-crystals, Science **251**, 1336; Bowick, M.J., Chander, L., Schiff, E.A., and Srivastava, A.M. (1994) The cosmological Kibble mechanism in the laboratory : string formation in liquid-crystals, *ibid.* **263**, 943.
30. Yurke, B. private communication.
31. Hendry, P.C. *et al.* (1994) Generation of defects in superfluid He-4 as an analog of the formation of cosmic strings, Nature **368**, 315.
32. Karra G. and Rivers R. J. (1998), A reexamination of quenches in Helium 4 and Helium 3, Phys. Rev. Lett. **81**, 3707.
33. Antunes, N.D., Bettencourt, L.M.A. and Hindmarsh, M. (1998) The Thermodynamics of Cosmic String densities in U(1) Scalar Field Theory, Phys. Rev. Lett. **80** 908.
34. Antunes, N.D. and Bettencourt, L.M.A. (1998), The length distribution of vortex strings in U(1) equilibrium scalar field theory, Phys. Rev. Lett. **81** 3083.
35. Bettencourt, L.M.A. (1995) Coarse-Grained Fluctuation Probabilities in the Standard Model and Subcritical Bubbles, Phys. Lett. B **356** 297 (1995).
36. Boyanovsky, D., de Vega, H. and Holman, R. (1999) Non-equilibrium phase transitions in condensed matter and cosmology: spinodal decomposition, condensates and defects, these proceedings.
37. Ketterle, W., Durfee, D.S., Stamper-Kurn, D.M. (1999) Making, probing and understanding Bose-Einstein condensates, cond-mat/9904034.
38. Anglin, J.R. and Zurek, W.H. (1998) Winding up by a quench: vortices in the wake of rapid Bose-Einstein condensation, quant-ph/9804035.

# THE 'GRENOBLE' COSMOLOGICAL EXPERIMENT

*The Kibble-Zurek Scenario in Superfluid $^3He$*

C. BÄUERLE, YU.M. BUNKOV, S.N. FISHER[1] AND H. GODFRIN
*Centre de Recherches sur les Très Basses Températures*
*Centre National de la Recherche Scientifique*
*BP166 38042 Grenoble Cedex 09, France*

We describe in this lecture neutron irradiation experiments performed on superluid $^3$He-B at ultralow temperatures. The measurement of the thermal energy released by the neutron-$^3$He nuclear reaction reveals the existence of an energy deficit. We propose that this deficit can be ascribed to the formation energy of vortices, these topological defects being created by the Kibble-Zurek mechanism as the region heated to the normal state by the neutron irradiation recrosses the superfluid transition. A quantitative agreement is found with Zurek's predictions for the density of topological defects created during a rapid quench through a symmetry-breaking phase transition.

## 1. Introduction

The lectures presented by Tony Leggett illustrated that $^3$He is one of the most sophisticated systems that condensed matter physicists understand well. It displays a normal-superfluid phase transition characterised by a simultaneous breaking of orbital, spin and gauge symmetries. The superfluid phases are described by a complex matrix order parameter field which allows the existence of a large variety of collective modes and topological defects.

Moreover, the symmetries broken in $^3$He are thought to be very similar to those broken in the cosmological phase transitions of the early Universe, as described in Tom Kibble's lectures. It is therefore appealing to use this wonderful model system to test Zurek's scenario of the creation of topological defects during a rapid phase transition.

Indeed, these ideas motivated the realisation of two experiments, in Helsinki and Grenoble[1, 2], which provide complementary information on the dynamics of the second order phase transition and the formation of

1 Present address: Department of Physics, Lancaster University, Lancaster LA1-4YB, United Kingdom.



*Y.M. Bunkov and H. Godfrin (eds.),*
*Topological Defects and the Non-Equilibrium Dynamics of Symmetry Breaking Phase Transitions*, 105–120.

topological defects. This lecture will concentrate on the experimental investigation of Zurek's scenario in superfluid $^3$He performed in Grenoble.

## 2. Superfluid $^3$He

The phase diagram of superfluid $^3$He is shown in Figure 1. The external parameters which define the thermodynamic state of the system are the pressure P, the temperature T and the magnetic field B. Owing to its large zero point energy and weak Van der Waals interactions, $^3$He remains in the liquid state down to the absolute zero of temperature for pressures below about 3.4 Mpa. At higher pressures there is a first order transition to the solid state. The normal liquid to superfluid phase transition is of second order and occurs at a transition temperature $T_c$ which is close to 1 mK at zero pressure and increases to around 2.5 mK at the solidification pressure.

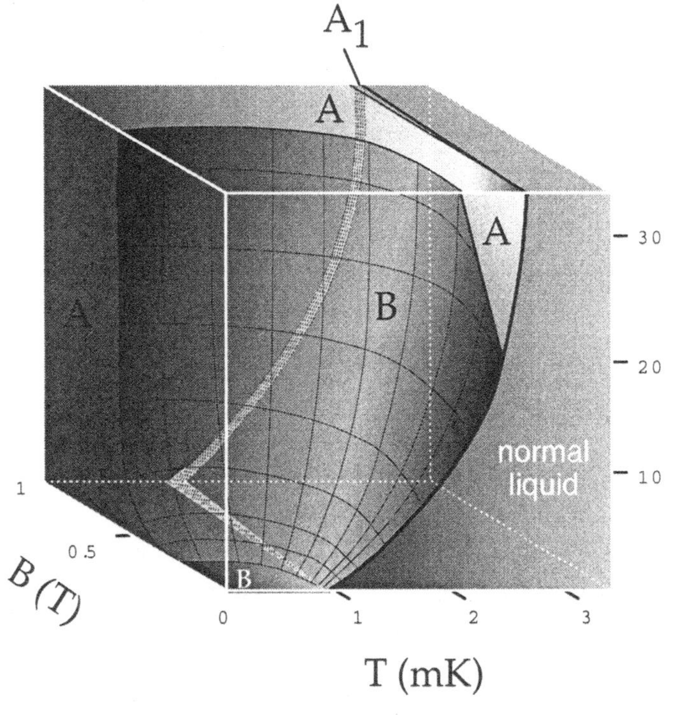


*Figure 1.* The phase diagram of $^3$He in the milliKelvin temperature range.

The superfluid transition takes place, in simple terms, due to the pairing $^3$He atoms (to form Cooper pairs) which can then Bose-condense forming the superfluid condensate. The complexity of the $^3$He condensate (in contrast to that of a conventional superconductor) arises from the fact that $^3$He

Cooper pairs have a spin angular momentum $S=1$ and an orbital angular momentum $L=1$. The extra degrees of freedom associated with the directions of these angular momenta lead to the existence of different superfluid phases with rather exotic properties.

In low magnetic fields, the B-phase superfluid occupies the majority of the phase diagram and the A-phase is stable only at relatively high temperatures and pressures. The effect of the magnetic field is to split the A transition, leaving a narrow range of temperature just below $T_c$ for the $A_1$-phase, but also favouring the A-phase at the expense of the B-phase over a greater part of the phase diagram.

The experiments described below have been performed in low magnetic fields (small on the scale of Figure 1) and at temperatures as low as $0.1\,T_c$, i.e. deep in the B-phase region of the phase diagram. The measurements were also made at different transition temperatures by changing the pressure. Results for three pressures (0, 0.6 and 1.94 MPa) were obtained before the cell finally exploded while pressurising to 2.2 MPa (the plastic experimental cell was not designed to work at such high pressures).

The choice of the B-phase for the measurements reported here was motivated by both theoretical and technical considerations. In many ways, the B-phase is much simpler than the A-phase. In low magnetic fields, the B-phase consists of an equal mixture of Cooper pairs with the three allowed values (+1, 0 and -1) for the projections of the spin (**S**) and orbital (**L**) angular momenta, giving zero total angular momentum $\mathbf{J} = \mathbf{L}+\mathbf{S}$. The order parameter of the superfluid B-phase has the form $A_{\mu\nu} = 1/\sqrt{3}\, e^{i\phi}\, R_{\mu\nu}(\mathbf{n}, \theta)$. The matrix $R_{\mu\nu}$ (which describes long range order in the *relative* orientation of **L** and **S**) corresponds to a rotation around a vector **n** by an angle $\theta = 104$ degrees. In the language of group theory, the transition from the normal state to the B-phase is described as the symmetry breaking transition: $SO(3)_{\mathbf{L}} \times SO(3)_{\mathbf{S}} \times U(1)_{\phi} \rightarrow SO(3)_{\mathbf{L+S}}$.

The primary excitations of the superfluid are quasiparticles (note that we use this denomination, loosely speaking, to describe quasiparticles and quasi-holes). Quasiparticles can be very loosely considered as unbound $^3$He atoms. The quasiparticle excitation spectrum has an energy gap $\Delta$ equal to one half of the pairing energy. For the B-phase in low magnetic fields, the energy gap is isotropic with a magnitude close to that predicted by the weak coupling (BCS) expression $\Delta = 1.76\, k_B T_c$ (although there is a definite deviation from this expression at higher pressures due to strong coupling corrections). Consequently, at low temperatures, the density of thermally excited quasiparticles is proportional to the Boltzmann factor, $e^{-\Delta/k_B T}$. Needless to say, at temperatures of the order of 0.1 $T_c$ the quasiparticle density is extremely small (Fig. 2).

The B-phase superfluid at temperatures of around 0.1 $T_c$ provides us

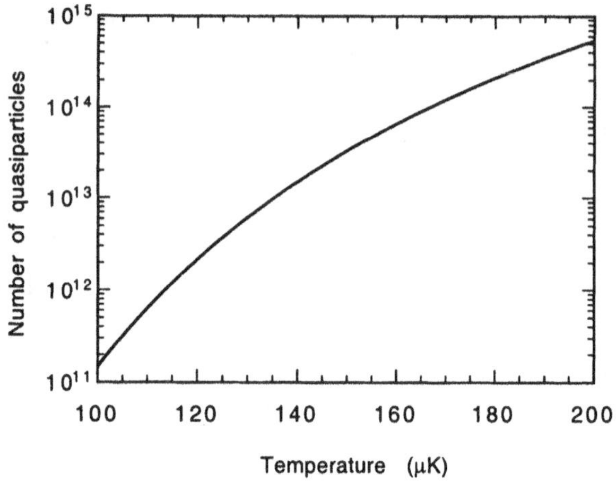


*Figure 2.* The density of thermally excited quasiparticles in $0.1\,\mathrm{cm}^3$ of superfluid $^3$He-B as a function of temperature.

with a vacuum which can not easily be matched either in Nature or in other laboratories. At such ultra-low temperatures the few surviving thermal quasiparticles are in the ballistic regime: their mean free path exceeds the dimensions of a typical experimental cell by several orders of magnitude. Furthermore, the thermal energy content of superfluid $^3$He-B under these conditions is only on the order of $1\,\mathrm{keV}/cm^3$. The system is therefore extremely sensitive to depositions of tiny amounts of energy. The resulting increase of the quasiparticle density can be measured by using a vibrating wire resonator immersed in the fluid. This simple technique offers a highly sensitive and versatile probe of the excitations of the superfluid.

## 3. The vibrating wire technique

The vibrating wire resonator consists of a simple superconducting wire loop, which acts as a high quality mechanical resonator. The wire motion is excited by placing it in a magnetic field and passing an alternating current through it. The resulting Lorentz force drives the wire into oscillation. The motion of the wire through the magnetic field induces an a.c. voltage across the wire proportional to the amplitude of its motion. This voltage is amplified and measured using a lock-in amplifier. Owing to the natural restoring force of the wire loop, the amplitude of oscillation reaches a maximum when the driving frequency coincides with the mechanical resonance frequency.

The frequency width of the wire resonance is a measure of the damping forces opposing the wire motion. In vacuum the resonance is extremely

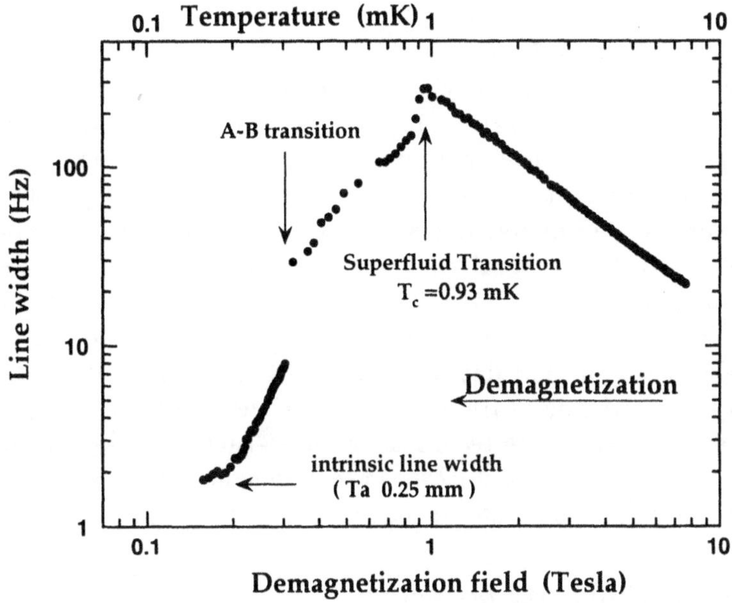


*Figure 3.* The vibrating wire resonance width as a function of temperature in liquid $^3$He observed during cooling by adiabatic demagnetization.

sharp and its quality factor is only limited by frictional forces within the wire itself. The width of the resonance increases when the wire is placed in a viscous medium due to the additional damping. When placed in the B-phase, one observes a dramatic increase of the width as the temperature is increased from $0.1\,T_c$ to $T_c$ (Fig. 3). At higher temperatures the width of the wire resonance simply measures the viscosity of the fluid (the wires are often referred to as viscometers). At lower temperatures, viscosity has no meaning since the quasiparticle excitations become ballistic. The viscosity of liquid $^3$He at the transition temperature is similar to that of oil, resulting in very strong damping at temperatures close to $T_c$. The response of the vibrating wire in the superfluid at low temperatures is not trivial. As might be expected, the scattering of the quasiparticles by the wire is responsible for damping. Consequently, the resonance width has temperature dependence which is dominated by the Boltzmann factor $e^{-\Delta/k_B T}$ at low temperatures. The magnitude of the damping at low temperatures is, however, larger by three orders of magnitude compared to the predictions of a naive ballistic gas model. In fact, due to the quantum nature of the superfluid, the moving wire is predominantly pushed by particles incident from the front and pulled by holes incident from the rear. This results in a strong amplification of the damping force exerted on the wire motion. A quantitative theory of the damping can be obtained by considering the effect of superflow on the dispersion relation of the quasiparticles in the vicinity of the moving wire[3].

Since the frequency width of a vibrating wire resonator is a rapidly varying function of the temperature, it provides a very convenient thermometer for studies performed in superfluid $^3$He [3, 4, 5]. Furthermore, since the wire couples directly to the quasiparticle excitations, the problems faced by conventional thermometers resulting from poor thermal coupling is not an issue. In practice, several vibrating wires of different diameters are incorporated into the experimental cell. Small diameter (4.5 $\mu$m) wires are very sensitive at low temperatures, but their damping is too large measure accurately in the normal liquid, so much larger (125 $\mu$m diameter) wires are used for this temperature regime. One of the most spectacular applications of the vibrating wire technique is the so-called 'black-body radiator', originally used for quasiparticle beam experiments, but may also be used as a superfluid $^3$He bolometer or particles detector[6].

## 4. The superfluid $^3$He bolometer

A schematic of the superfluid $^3$He bolometer used for these measurements is shown in Figure 4.

The bolometer consists of a copper box with a volume of $\sim 0.1cm^3$ containing two vibrating wire resonators. A small, 60 $\mu$m diameter, hole in the wall of the box connects the superfluid $^3$He inside the bolometer with the surrounding bath of superfluid $^3$He. The heat leaks into the superfluid inside the bolometer must be kept extremely small (in the femtoWatt range) to enable the system to cool to the lowest temperatures (approaching 100 $\mu$K) in spite of the very weak thermal connection.

The principle of the bolometer is simple. Energy deposited inside the bolometer from whatever source (for instance by an incident particle) will generate an excess of quasiparticles. The increase of the quasiparticle density (determined by the volume of the bolometer) increases the frequency width of the vibrating wire resonance (which is monitored continuously by measuring the amplitude of the wire motion at its resonant frequency). The vibrating wire response is relatively fast, and the signal rises within a few seconds. This sharp rise is followed by a slow recovery as the excess quasiparticles 'leak out' through the bolometer hole. The time constant for the recovery is around one minute and is determined simply by the size of the bolometer hole. The bolometer is therefore particularly adapted to the detection of rare events of small energy. Data obtained in our typical laboratory conditions is shown in Figure 5. These 'background' particle events are due to cosmic rays and particles originating from the natural radioactivity of the surroundings.

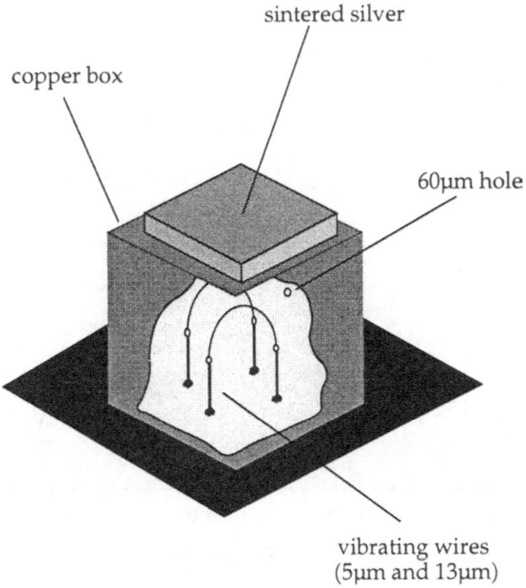


*Figure 4.* The superfluid $^3$He bolometer.

The amplitude of the signal is proportional to the energy deposited within the box. In order to establish an accurate relation between the increase of the vibrating wire resonance width and the deposited energy it is essential to calibrate the bolometer. Due to the large thermal boundary resistance between a solid and superfluid $^3$He it is difficult to transfer heat to the $^3$He (i.e. to create quasiparticles) using a conventional electrical heater. However, it is possible to use a vibrating wire for this purpose by driving it with a large amplitude. Measuring the voltage component in phase with the current yields directly the dissipation, thus providing an accurate way to apply known amounts of energy into the superfluid. We have used continuous and pulsed heating techniques in order to achieve a calibration of the bolometer[5]. These measurements, which also constitute the basis for the development of a sub-milliKelvin temperature scale, provide us with an accurate energy calibration over the range 100 keV to 10 MeV. We shall see later that this calibration plays a crucial role in the experiment described here.

## 5. The 100 $\mu$K refrigerator

As can be inferred from Fig. 2, the bolometer sensitivity increases exponentially at low temperatures. Present cryogenic techniques allow us to cool liquid $^3$He down to 100 $\mu$K, a record temperature presently achieved in Lan-

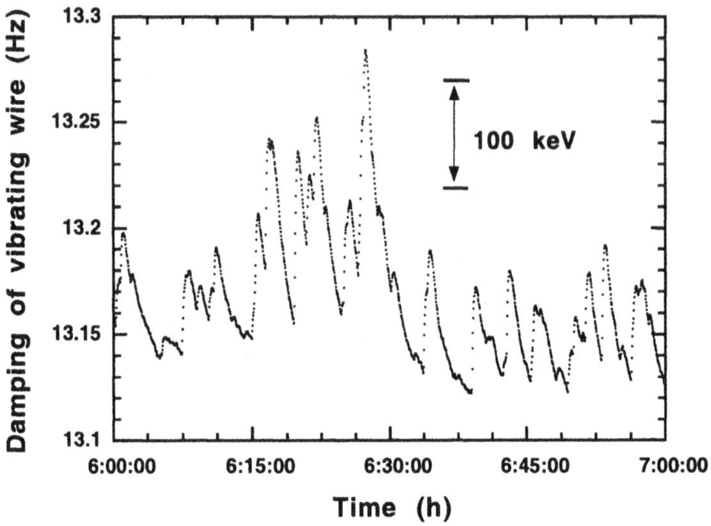


*Figure 5.* Background radiation events detected with a superfluid $^3$He bolometer.

caster and Grenoble. The cryogenic set-up consists of a Dewar filled with liquid $^4$He at 4.2 K, used as the first cryogenic stage. A vacuum chamber immersed in the liquid helium contains the coldest parts of the system: the dilution refrigerator unit and the nuclear demagnetisation stage. The cooling process used in dilution refrigerators consists of mixing the two helium isotopes ($^3$He and $^4$He) in the fluid state. The 'Grenoble type' dilution refrigerator used in this experiment allows us to pre-cool the nuclear stage down to temperatures in the milliKelvin range. Copper nuclear stages, which consist of copper rods, wires or plates specially annealed to achieve extremely high thermal conductivities, are widely used now for refrigeration in the sub-milliKelvin range. Their cooling principle is based on the adiabatic demagnetisation of the copper nuclear spins. In contrast to dilution refrigeration, which provides continuous cooling, nuclear demagnetisation is a one-shot process. Initially, the nuclear stage is thermally connected to the dilution refrigerator to absorb the heat generated as the magnetic field, provided by a superconducting magnet, is increased to about 8 Teslas. The combination of a large magnetic field and low temperatures polarises the nuclear spins thus reducing the spin entropy.

The stage is then thermally disconnected from the dilution refrigerator by means of a superconducting heat switch. The magnetic field B is then slowly reduced. Since the spin entropy is a function of B/T and the nuclear stage is now thermally isolated, the temperature T of the copper nuclear spins decreases proportionately with the magnetic field during the demagnetisation. While extremely low temperatures can be achieved in the nuclear spin system, it is not so easy to use the technique as a refrigerator to cool other systems. In particular, the coupling between the cold nuclear spins and liquid $^3$He is very poor.

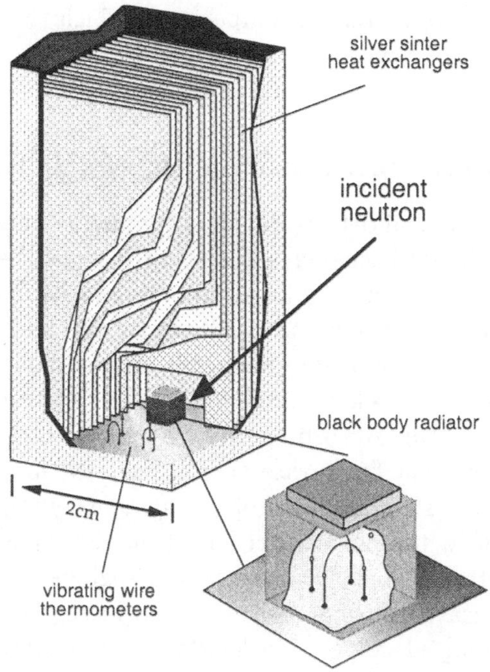


*Figure 6.* The 'Lancaster type' nuclear demagnetisation stage and the superfluid $^3$He bolometer.

The thermal path includes the coupling between the spins and the conduction electrons in the copper, the coupling between the copper electrons and those in the sintered silver powder (used as a large area heat exchanger to improve thermal coupling), and finally the coupling between the silver sinter and liquid $^3$He. The latter coupling is exceptionally poor for superfluid $^3$He at low temperatures due to the small quasiparticle density, so rather specialised cryogenic techniques are required to cool $^3$He to such low temperatures. The experiment described here incorporated a 'Lancaster type' nuclear stage which is specifically designed for cooling superfluid $^3$He to the lowest temperatures (see Figure 6).

The nuclear stage consists of two nested cells called the outer and inner cells. Both cells contain copper refrigerant immersed in liquid $^3$He. The outer cell contains copper powder which cools to a temperature of around 1mK with a relatively large heat capacity. This acts as a thermal shield to reduce heat leaks into the inner cell. The inner cell refrigerant consists of around 70 0.1mm thick copper plates covered with a thin layer of sintered silver powder. This arrangement, with the fluid in close contact with the nuclear refrigerant, maximises the thermal coupling. As shown in Figure 6, the bolometer is located inside the inner cell, in a small volume cut out of the copper plates. These techniques allow us to cool liquid $^3$He somewhat below 100 $\mu$K (the minimum temperature which the current vibrating wire thermometers can measure is limited to 100 $\mu$K due to their intrinsic damping).

## 6. The Kibble-Zurek mechanism in superfluid $^3$He

The creation of topological defects during a symmetry breaking phase transition has already been discussed extensively from a theoretical point of view during this School. We have heard about Tom Kibble's mechanism of cosmic string formation during the GUT transition, and also Zurek's scenario introducing the concept of dynamically driven defect creation. These very general ideas, as Zurek pointed out, could be tested in condensed matter systems like $^4$He. According to this model, crossing the normal-superfluid transition at a finite rate should lead to the formation of vortex lines. The critical slowing down of the system dynamics determines a rate limited cut-off in the correlation length in the vicinity of the second order phase transition. Causally disconnected regions of the low temperature phase are generated during the transition, having a typical size of the order of the rate limited correlation length. As the system evolves, inhomogeneities in the order parameter result in a tangle of topological defects (line defects are commonly known as vortices in superfluid He or cosmic strings in the Universe). The predictions of this model[7] are summarised in Figure 7.

The density of topological defects resulting from the transition depends on several intrinsic parameters. The time $\tau_0$ is the response time of the ordered phase at low temperatures and $\xi_0$ is the zero temperature coherence length (giving the minimum length scale for variations in the order parameter). The defect density is also controlled by the external parameter $\tau_Q$ which is a measure of the time scale for cooling through the phase transition (the quench time). A simple formula, derived by Woijzek Zurek, provides an estimate of the separation $\beta$ between vortices expected from this scenario: $\beta = \xi_0 \, (\tau_Q/\tau_0)^{1/4}$.

## 7. The 'Grenoble' cosmological experiment

Zurek's formula suggests that a substantial number of vortices should be created in $^3$He during a sufficiently rapid temperature quench through the

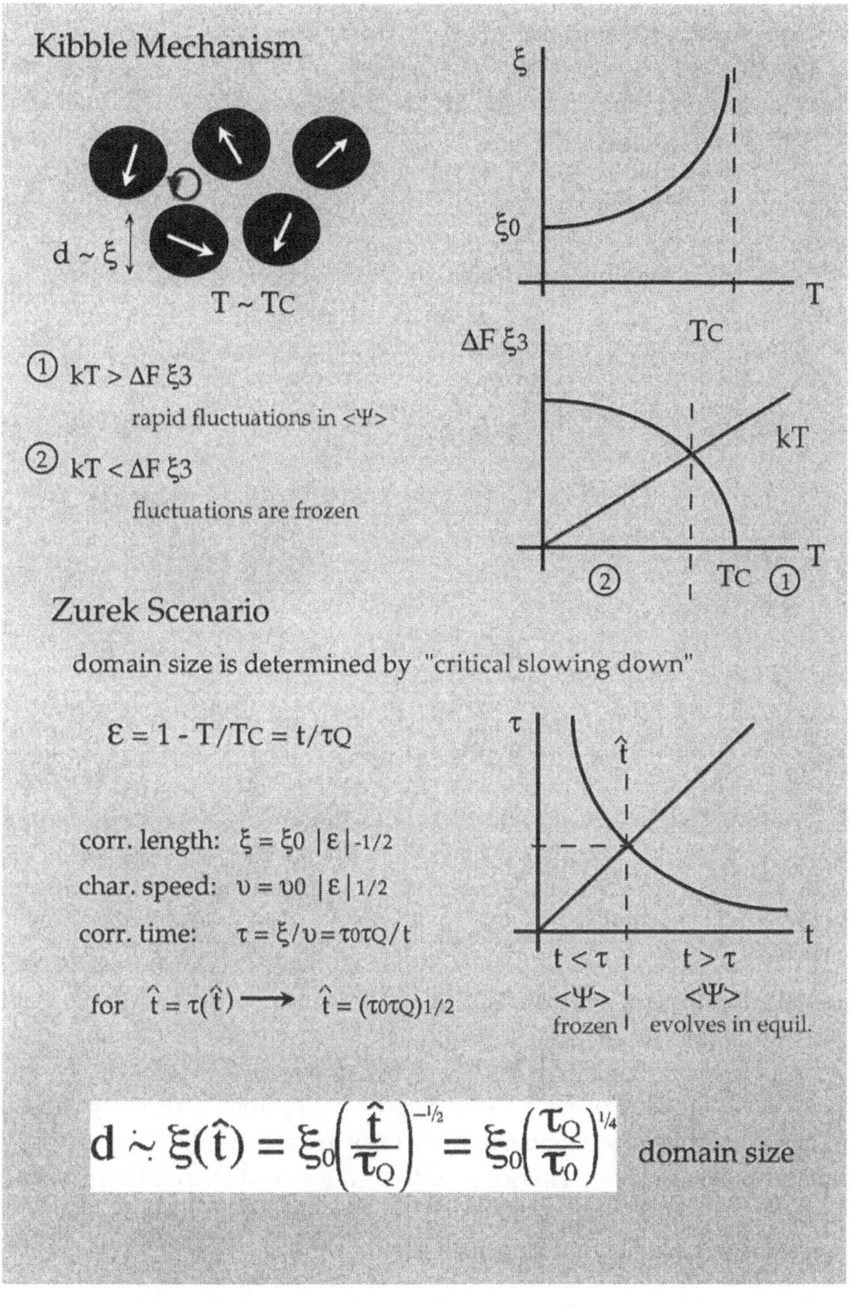


*Figure 7.* Kibble-Zurek mechanism of topological defects creation during a symmetry-breaking phase transition.

normal-superfluid phase transition. However, there are two major problems faced by the experimentalist. First, it is impossible to achieve sufficiently high cooling rates for a macroscopic sample of superfluid $^3$He. The second difficulty is that vortex tangles are very difficult to detect. Fortunately, there exists a particularly convenient method to generate a very localised hot spot in superfluid $^3$He. The $^3$He nucleus has a very high cross section for low energy neutrons via the capture process: n + $^3$He $\rightarrow$ $^3$H + p. Each capture process deposits an energy of 764 keV shared between the proton and triton, which are thermalised within a distance on the order of 30$\mu$m. Since this 'hot spot' of normal liquid is rather small and is surrounded by very cold superfluid, the cooling is very rapid and is completed after about 1$\mu$s. This is sufficiently rapid to generate a significant density of vortices as predicted by Zurek's formula. For the measurements described here, the superfluid inside the bolometer reached a base temperature of about 160 $\mu$K, limited by the femtowatt heat leak from the bolometer walls and the small size of the bolometer hole. A small flux of low energy neutrons was applied to the bolometer using a standard AmBe neutron source placed a couple of meters away from the refrigerator.

Both the Helsinki and Grenoble groups used neutrons to generate the phase transition, but very different experimental techniques were used for observing the resulting vortex tangle. The Helsinki[2] group worked at relatively high temperatures (not very far below $T_c$) and used a rotating cryostat. The rotation expands some of the vortex loops generated by the neutron events, which could then be detected using nuclear magnetic resonance techniques. The observations verified that the statistical distribution of the sizes of the large vortex loops generated by the neutron irradiation followed the theoretical prediction.

In the present experiment, performed at much lower temperatures, the extremely high energy sensitivity of the vibrating wire technique enables us to establish a detailed energy balance for the neutron capture process. The total energy deposited by the neutron capture is accurately known and the total amount of heat deposited in the system is directly measured from the vibrating wire response. If vortices are created in the process, they should store some of this energy resulting in a deficit in the thermal energy deposited. This supposes, of course, that the vortex lifetime is longer than the measuring time, a condition which is fulfilled at ultra-low temperatures.

We have analysed a large number of neutron capture events detected by the calibrated vibrating wire bolometer. The results are summarised in figure 8. Here we show histograms of the number of neutron events versus the measured thermal energy deposited for three different pressures. Clearly, a substantial shift is observed in the peak position with respect to the total 764 keV energy deposited by the capture process.

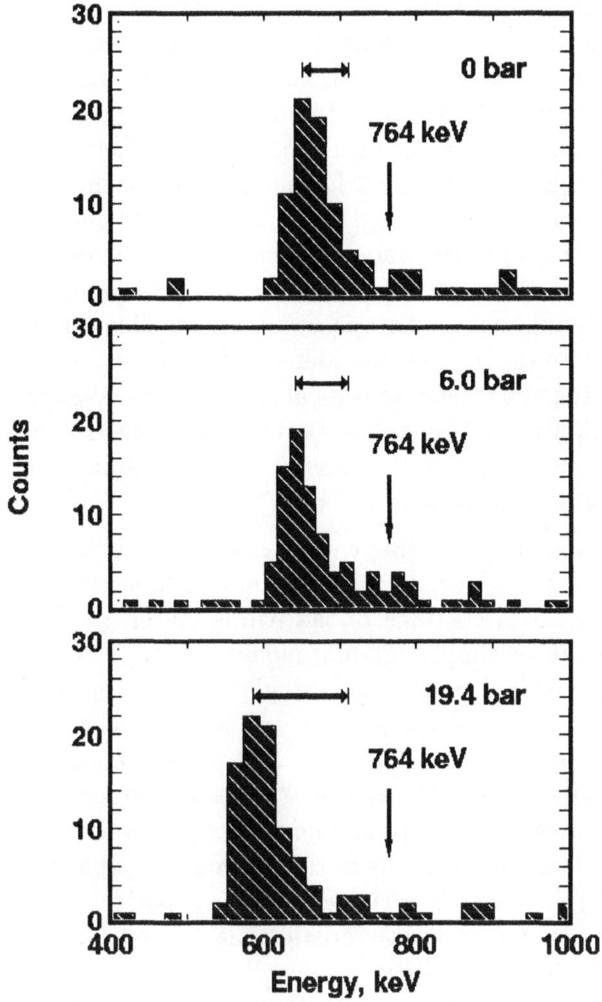


*Figure 8.* The energy spectra recorded in superfluid $^3$He irradiated by neutrons at ultralow temperatures. The energy deficit with respect to the 764 keV deposited by the neutrons can be attributed in part to the formation of a vortex tangle by the Kibble-Zurek mechanism (see text).

Before we can attribute the measured energy deficit to the creation of vortices, we must first consider what other processes could reduce the energy available for thermalisation. There are other defects that could be generated in the superfluid, however these have a negligible energy compared to vortex lines. So any other source of energy deficit must involve the loss of energy from the interior of the bolometer (energy transfer to walls of the bolometer for instance). The thermal boundary resistance between the bolometer wall and the superfluid is quite enormous at these temperatures, so any loss via thermal excitations is out of the question. This leaves scintillation. Experiments performed in liquid $^4$He have shown that helium is quite a good scintillator, losing energy from particle collisions via the emission of UV photons to which helium is transparent. The energy loss due to UV photons accounts for 6 to 8% of the total energy deposited by high energy electrons in $^4$He, and we estimate that the effect should be on the order of 3% for neutron irradiation. It has been pointed out by Tony Leggett that the dimer formation processes responsible for the scintillation may be sensitive to the isotope considered, and that a pressure dependence might also be expected. Since, at present, there exists no experimental data on the scintillation from $^3$He we use a estimated value of 25 keV for the energy lost as ultraviolet radiation. This is a relatively small correction to the energy deficit, as can be seen from Table 1.

A large fraction of the missing energy can therefore be attributed to the creation of vortices in the superfluid. We now compare our measurements with the theoretical predictions of the Kibble-Zurek mechanism. We can do this using a rather simple thermal model. We suppose that the energy $E = (764 - 25)$ keV is deposited at a single point in the fluid. The energy quickly thermalises generating a sphere of hot normal fluid which subsequently cools via quasiparticle diffusion with a diffusion constant $D$ (taken to be that of bulk liquid $^3$He at the transition temperature $T_c$). As the sphere cools, the volume of normal fluid first expands before finally shrinking to zero. The maximum radius of the normal fluid sphere is found to be $R \simeq 0.4(E/CT_c)^{1/3}$ where $C$ is the liquid heat capacity just above $T_c$. The cooling takes place with a characteristic time given by $\tau_Q \simeq R^2/4D$. The values obtained for the maximum radius $R$ and the cooling time $\tau_Q$ using this model are shown in Table 1. The strong pressure dependence of these variables is largely attributed to the pressure dependence of $T_c$.

The energy per unit length of a vortex is mainly due to the kinetic energy of the circulating superflow and is given by $E_L \simeq (\rho/4\pi)(\hbar/2m_3)^2 \ln(\beta/\xi_0)$ (about 0.5keV/mm at zero pressure). This enables us to convert total vortex energy into total vortex length as shown in table 1. In order to obtain the average vortex spacing we simply assume that this vortex length is evenly distributed over a volume equal to the maximum volume of the normal fluid sphere. The result of such a simple analysis, shown in Table 1, is that the average vortex spacing is found to be around 8 coherence lengths and rather insensitive to pressure. Such a result seems rather intriguing considering that all the variables which enter the calculation are rather strongly pressure dependent. The estimate of the theoretical prediction of

TABLE 1. Experimental data and results

| $^3$He Pressure | 0 | 0.6 | 1.94 | MPa |
|---|---|---|---|---|
| Energy deficit | 110 | 120 | 175 | keV |
| Vortex Energy E | 85 | 95 | 150 | keV |
| Radius of hot spot | 28 | 18 | 12 | $\mu$m |
| Quench time $\tau_q$ | 0.29 | 0.56 | 1.25 | $\mu$sec |
| Coherence length $\xi_0$ | 77 | 39 | 21 | nm |
| $\tau_0 = \xi_0 / v_F$ | 1.3 | 0.78 | 0.54 | nsec |
| Vortex length L | 22 | 23 | 32 | cm |
| $\beta / \xi_0$ experiment | 8.3 | 8.3 | 7.1 | - |
| $\beta / \xi_0$ theory | 3.8 | 5.2 | 6.9 | - |

the Kibble-Zurek mechanism in table 1 is obtained directly from $\tau_Q$ and the known properties of the bulk superfluid.

The theoretical prediction of the vortex spacing in coherence length units is seen to be pressure dependent and about a factor of two smaller than the values inferred from our measurements. However, this is a surprisingly good agreement considering the large simplifications made in the analysis of the experiment. The major difficulty lies in the lack of knowledge concerning the processes by which the initial reaction energy is converted into thermal energy and ultraviolet radiation. Assuming a spherical hot spot which cools via quasiparticle diffusion is certainly a major simplification. Alternative cooling models have been discussed in the literature and during this School[8]. The diffusive cooling scenario, on the other hand, seems to have some support from very recent theoretical calculations[9]. Furthermore, the Kibble-Zurek theory is itself a rather simplified model and should only be expected to give the correct scale for the vortex spacing. Clearly, more theoretical and experimental work is required to improve our quantitative understanding of these process.

Leaving the finer details aside, the good semi-quantitative agreement obtained so far in these pioneering experiments provides strong support for the Kibble-Zurek scenario. As predicted, the rapid superfluid phase transition generated by a neutron capture creates a very significant density of topological defects.

## References

1. Bäuerle, C., Bunkov, Yu.M., Fisher, S.N., Godfrin, H. and Pickett, G.R. (1996) Laboratory simulation of cosmic string formation in the early Universe using superfluid $^3$He, *Nature*, **Vol. 382**, pp. 332-334.

2. Ruutu, V.M.H., Eltsov, V.B., Gill, A.J., Kibble, T.W.B., Krusius, M., Makhlin, Yu.G., Plaçais, B., Volovik, G.E. and Wen, Xu (1996) Vortex formation in neutron-irradiated superfluid $^3$He as an analogue of cosmological defect formation, *Nature*, **Vol. 382**, pp. 334-336.
3. Fisher, S.N., (1991) The mechanical properties of superfluid $^3$He-A and $^3$He-B in the ballistic regime, *Ph.D. Thesis, Lancaster University.*
4. Fisher, S.N., Guénault, A.M, Kennedy, C.J. and Pickett, G.R. (1989) Beyond the two-fluid model: Transition from linear behavior to a velocity-independent force on a moving object in $^3$He-B, *Phys. Rev. Lett.*, **Vol. 63**, pp. 2566–2569 and references therein.
5. Bäuerle, C., Bunkov, Yu.M., Fisher, S.N. and Godfrin, H. (1998) Temperature scale and heat capacity of Superfluid $^3$He-B in the 100 $\mu$K range, *Phys. Rev. B*, **Vol. 57**, pp. 14381-14386.
6. Bradley, I.D., Bunkov, Yu.M., Cousins, D.J., Enrico, M.P., Fisher, S.N., Follows, M.R., Guénault, A.M., Hayes, W.M., Pickett, G.R. and Sloan, T. (1995) Potential dark matter detector? The detection of low energy neutrons by superfluid $^3$He, *Phys. Rev. Lett.*, **Vol. 75**, pp. 1887-1891.
7. Zurek, W.H. (1996) Cosmological experiments in Condensed Matter systems, *Physics Reports*, **Vol. 276**, pp. 178–221 and references therein.
8. Yip, S.K. and Leggett, A.J. (1990) Nucleation and growth of $^3$He-B in the super-cooled A-phase, in *Helium Three, Modern Problems in Condensed Matter Sciences*, **Vol. 26** edited by W.P. Halperin and L.P. Pitaevskii (North Holland, Elsevier Science Publishers B.V., New York, pp. 523.
9. Kopnin, N.B. and Thuneberg, E.V. (1999) Time-dependent Ginzburg-Landau analysis of inhomogeneous normal-superfluid transitions, *Phys. Rev. Lett.*, **Vol. 83**, pp. 116–119 and references therein.

# "AURORE DE VENISE" - COSMOLOGICAL SCENARIO OF THE A-B PHASE TRANSITION IN SUPERFLUID $^3$HE

YU. M. BUNKOV
*Centre de Recherches sur les Très Basses Températures CNRS, BP166, 38042, Grenoble, France.*

These lectures will review symmetry-breaking phase transitions in the special case of the breaking of combined symmetries and the subsequent formation of different states with different ordering. The two most interesting systems with these futures are the Universe and superfluid $^3$He. In both cases the energy potential below the transition temperature has a complicated multidimensional profile with many local minima and saddle points, which correspond to different states. Consequently, not only topological defects, but also islands of different metastable states can be created. Using $^3$He we can analyse the properties and experimental consequences of such transitions and, in particular, the first order phase transition between the two low symmetry states.

## 1. Introduction

Since in the main part of this article we will be considering superfluid $^3$He, we will start by describing the Universe. In this Advanced Studies Institute we have learnt, that the Universe is just a model (not yet well understood) for superfluid $^3$He. In our current work we have studied the probability of creation of two different states (A and B states) after a fast superfluid transition in $^3$He. Later, we were very surprised when we found that a similar problem may be worth considering for the Universe. The early Universe cools through a cascade of symmetry breaking phase transitions. These transitions may have produced many types of topological defects: domain walls, cosmic strings (vortices) and monopoles [1, 2]. In this article we raise the question that not only topological defects but also islands of different metastable states can appear and survive for some time after a fast transition. For example, the standard cosmological model proposes a transition from the state with symmetry $SU(5)$ to a lower symmetry state $SU(3) \times SU(2) \times U(1)$. But, in addition, a number of other low symmetry states can be produced at this transition. It can be shown [3]



*Y.M. Bunkov and H. Godfrin (eds.),*
*Topological Defects and the Non-Equilibrium Dynamics of Symmetry Breaking Phase Transitions*, 121–137.

that the ground state energy of these states lies between the energy of the $SU(3) \times SU(2) \times U(1)$ state and the $SU(4) \times U(1)$ state. However, we learn from the physics of the superfluid transition in $^3$He, that it is not only the ground state energy that determines the final state of the system. There is also the question of the creation probability of different states, which is related to the potential energy profile between the high-energy state with unbroken symmetry (the false vacuum in cosmological language) and the low energy states with different broken symmetries (true vacuums). We note that there be many different low energy states.

Usually, discussions about the cosmological transitions in the Universe do not go beyond the simple model of a gauge transition with the energy profile in the shape of a "Mexican hat". In this case the order parameter is described by one complex number or two Higgs fields with a single minima and a degenerate phase. In condensed matter this transition corresponds to the $^4$He superfluid transition or to the superconducting transition in conventional superconductors. The order parameter in $^3$He is described by 9 complex numbers, giving a manifold of 18 dimensions. The $SU(5)$ transition can be represented by a 25 dimensional manifold. It is quite near to $^3$He, but not to $^4$He. The difference being that multidimensional manifolds have an energy profile with a very complicated form and not the simple "Mexican hat" profile. There can be many local minima and saddle points in a multidimensional energy profile each corresponding to different states. We know practically nothing about the energy profile for the cosmological transitions. However, for the superfluid transition in $^3$He we know the energy profile very well, at least for the Ginzburg-Landau approximation near $T_C$, thus we can calculate the creation probability of the different states [4]. The energy profile of the order parameter in $^3$He has two minima, corresponding to the A and B states, as well as a few saddle points. In addition, in a magnetic field a third minima appears, corresponding to the $A_1$ state.

It is impossible to show graphically the 18-dimensional energy profile. However, to illustrate the main idea, let us deform the two dimensional "Mexican hat" potential picture to obtain two different states with different ground state energies, as shown in Fig. 1. Now one can see the A and B states. On this figure the B state has lower energy. At the transition the system moves downhill from the central part (false vacuum). The direction of this movement is determined by some small perturbation, which has, possibly, an equal probability for any direction. However, the profile of the energy potential is not symmetric. As shown in Fig. 1 there is a very clear asymmetry to this profile. Consequently, the probability of moving down to state A and to state B are different and is not determined by the ground state energy but by the asymmetry of the profile near the high symmetry

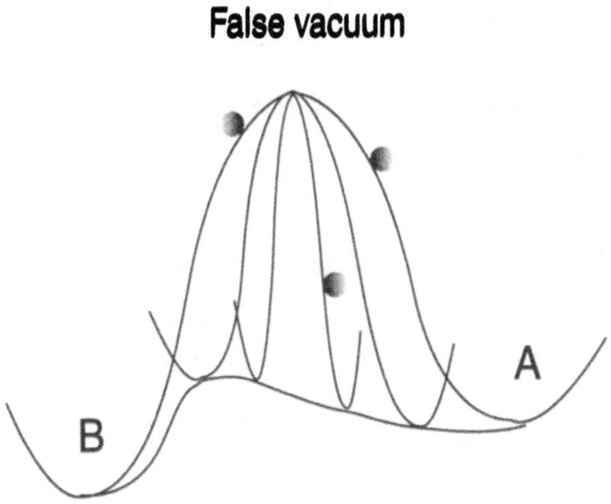


*Figure 1.* The schematic 2D presentation of energy asymmetry for the 18D manifold of the superfluid $^3$He order parameter.

state (false vacuum).

At a fast symmetry breaking phase transition the causally independent parts of the system can go to the different states. The consequences of these processes for cosmology have not yet been discussed. For example we do not know the creation probability of the $SU(3) \times SU(2) \times U(1)$ state in relation to the creation probability of the alternative states. We are not even certain that our state is not a metastable state! And if it is, how long the other states survived after the transition. Can the enlargement of $SU(3) \times SU(2) \times U(1)$ state to the regions, initially occupied by other states, be considered as an alternative to the inflatory model? We can not give an answer to all of these questions because we do not know the energy profile for a $SU(5)$ multidimensional manifold below the transition. However, we can answer these questions for superfluid $^3$He.

Experimentally the conditions for a fast superfluid transition in $^3$He can be achieved by a nuclear reaction of a neutron with a nucleus of a $^3$He atom in the superfluid state at ultra low temperature. This reaction locally deposits energy, creating a fireball of normal $^3$He, which then expands and cools down extremely rapidly. At the following very rapid superfluid transition the creation of topological defects (vortices) has been demonstrated recently to be in accordance with the Kibble-Zurek scenario for the cosmological analogue (see [7] in this book). We discuss here the extension of this scenario to the case when alternative symmetries may be broken and different states nucleated independently. We use the results of our calculations of the nucleation probability of the various states of superfluid $^3$He during a superfluid transition to explain the first order transition from supercooled A phase to the B phase, triggered by a nuclear reaction. We have named this new scenario "Aurore de Venise", owing to the islands of the city Venice. It is an alternative to the well-known "Baked Alaska" scenario of the A-B transition.

## 2. The $^3$He Ginzburg-Landau energy potential

Superfluid $^3$He has an order parameter which describes the simultaneous spin, orbital and gauge symmetries which are broken at the superfluid transition. The rotational and gauge symmetries of $^3$He are usually represented by a 3×3 matrix of complex numbers $A_{ai}$ which is known as the order parameter. Above the transition all the elements of the matrix are zero, representing full symmetry. Below the transition, some of these quantities become non-zero. The symmetry of the order parameter after the transition corresponds to the manifold of symmetries which remain unbroken. In the case of superfluid $^3$He there are 13 possibilities (13 states) corresponding to the various symmetries of the order parameter [8]. The free energy of these states can be expressed in the framework of the phenomenological theory of Ginzburg and Landau by:

$$\begin{aligned} F = & -\alpha A^*_{ai}A_{ai} + \beta_1 A^*_{ai}A^*_{ai}A_{bj}A_{bj} + \beta_2 A^*_{ai}A_{ai}A^*_{bj}A_{bj} \\ & +\beta_3 A^*_{ai}A^*_{bi}A_{aj}A_{bj} + \beta_4 A^*_{ai}A_{bi}A^*_{bj}A_{aj} + \beta_5 A^*_{ai}A_{bi}A_{bj}A^*_{aj} \end{aligned} \tag{1}$$

where $\alpha = \alpha_0(1 - T/T_c)$, which changes sign at the transition temperature $T_c$ and the quantities $\beta_i$ are functions of pressure (and also of temperature through the so-called "strong correction") and depend on the details of the microscopic interaction.

The different possible symmetries of the order parameter correspond to local minima and saddle points in this 18-dimensional energy surface. In superfluid $^3$He we know there are two stable states, the A and B phases. The energy balance between the A and B phases is determined by the relationship between the parameters $\beta_i$. At zero pressure, the B phase corresponds to the absolute minimum, while at pressures above 20 bar there is a region of temperature where the A phase becomes the preferred state.

These two states have different order parameter symmetries. In the B phase, relative spin ($_S$) - orbit ($_L$) symmetry $SO(3)_{S+L}$ remains unbroken (such that $A_{ai}$ resembles a rotation matrix). In the A phase (the "axial" state) the symmetry of the spin system is reduced to a gauge symmetry ($U_S$), which couples to the orbital motion to yield a combined symmetry of the orbital rotation and gauge ($_G$) fields $U_S \times U_{L+G}$.

## 3. The fast superfluid transition in $^3$He

The superfluid transition in $^3$He can be considered as the closest condensed matter analogy to the cosmological transitions. This analogy has been utilised in the experimental test of the Kibble cosmological mechanism of cosmic string creation. According to the Kibble mechanism [2], at the transition separate regions of the Universe are independently nucleated with a random orientation of the gauge field in each region. The size of these initial regions (domains) depends strongly on the rapidity with which the transition is traversed. According to Zurek [9] the fundamental distance between the independently-created coherent domains (in the language of [9] the distance $Z$ between the ensuing vortices ) is of the order of $Z = \xi_0(\tau_Q/\tau_0)^{1/4}$, where $\xi_0$ is the zero temperature coherence length, $\tau_0 = (\xi_0/v_F)$ is the characteristic time constant of the superfluid and $\tau_Q$ is the characteristic time for cooling through the phase transition. As the domains grow and make contact with their neighbours, the resulting gauge field cannot be uniform. The subsequent order-parameter "glass" forces a distribution of topological defects leading to a tangle of quantized vortex lines. The first quantitative tests of defect creation during a gauge symmetry transformation have been recently performed in superfluid $^3$He by the Helsinki [5] and Grenoble [6] groups as described in paper [7] of this book. In the Grenoble experiment the $^3$He was cooled down to an extremely low temperature, about 1/20 of $T_C$. Then the superfluid $^3$He was heated locally by neutron irradiation via the nuclear reaction:

$$^3\text{He} + \text{n} \Rightarrow {}^3\text{H}^- + \text{p}^+ + 764 \text{ keV}$$

The energy released by the neutron reaction heats a small region of the liquid $^3$He (about 30$\mu$m in diameter) into the normal state. This region cools rapidly through the superfluid transition owing to the rapid outflow of quasiparticles into the surrounding superfluid. For the experimental conditions it has been proposed that quasiparticles from the heated region disperse outwards, meaning that the hot bubble is cooled rapidly from its sides and that the cooling rate is so fast that the order parameter of the surrounding superfluid $^3$He cannot follow the changing temperature front fast enough (see [10] for theoretical details). Consequently, internal regions of the hot volume transit into the superfluid phase independently in accordance with the Zurek cosmological scenario. In the Grenoble experiment the excess number of quasiparticles created by the reaction has been counted and it was found that a significant fraction of the energy released by the reaction does not appear in the thermal reservoir of quasiparticles. This energy deficit agrees well in magnitude with the energy expected to be trapped as topological defects (in this case vortices) as calculated from Zurek's scenario for the Kibble mechanism.

Now let us take into account the possibility of nucleation of different states. The difference of the ground state energy between these states is relatively small. Let us say that it is negligible on the time scale of the transition! This means that regions, which independently enter the superfluid state, should not only have different orientation of the order parameter but may also correspond to states with different symmetries [11]. According to the Zurek scenario, regions on a distance scale of $Z$ undergo the superfluid transition separately. We can consider these regions as independent elementary samples of $^3$He. (Later we shall analyse the influence of the gradient energy between the different regions.) In [4] the creation of the superfluid phases in a single region during a rapid cooling were calculated numerically. For this the time dependent Ginzburg-Landau equation was applied in the form:

$$-\tau\frac{\partial}{\partial t}A_{ai}+\alpha_0\frac{T_c-T(t)}{T_c}A_{ai}-(\beta_1A^*_{ai}A_{bj}A_{bj}+\beta_2A_{ai}A^*_{bj}A_{bj}$$
$$+\beta_3A^*_{bi}A_{aj}A_{bj}+\beta_4A_{bi}A^*_{bj}A_{aj}+\beta_5A_{bi}A_{bj}A^*_{aj})=0 \qquad (2)$$

The initial conditions were the temperature $T = T_c$ and a small independent random perturbation of all 18 numbers of the $A_{ai}$ matrix. The temperature was then reduced over some time ( $10^{-9}$ – $10^{-7}$ s ) and the time dependence of the order parameter during this "downhill" process was calculated.

Both the symmetry of the order parameter $A_{ai}$ and the energy during this time-evolution was monitored. It was found that both the A and the B phases can develop. The final state depends on the starting $^3$He-A perturbation of the order parameter and the profile of the 18-dimensional potential surface. It does not depend on velocity of cooling or the final temperature. That is because in the formalism of the Ginzburg-Landau theory, the $\beta$'s do not depend on temperature. This expression corresponds to real $^3$He only near $T_c$. In other words, our results can only be applied to real $^3$He for relatively high temperatures. For low temperatures more complicated theories should be considered. Nevertheless, our results demonstrate a new explanation of the A-B phase transition which we will discuss later.

Other metastable states may develop transiently after the application of an initial perturbation with the exact symmetry of one of these states. However, the trajectory of $A_{ai}$ in these cases is unstable and any small

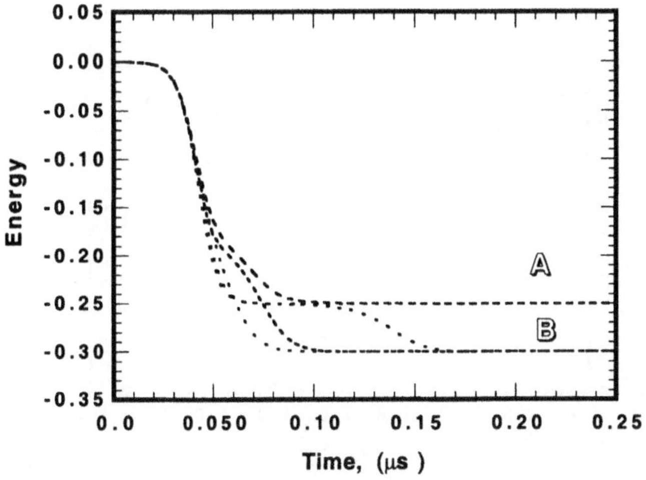


*Figure 2.* The time evolution of the free energy density during several different superfluid phase transitions after a small random perturbation. The temperature was reduced from $T = T_c$ to $T = 0$ in a time of $10^{-8}$s (after [4]).

perturbation away from the exact symmetry leads to the more stable A or B states.

It is important to note that, although according to Zurek the cooling rate determines the dimensions of the independent regions, the trajectory of the order parameter for a single coherent region is rate independent and is only determined by the profile of the Ginzburg-Landau potential. At zero pressure, when we have weak coupling with $\beta_i = (-1, 2, 2, 2, -2)$, the B phase corresponds to the absolute energy minimum. Our computer simulation shows that, even under these conditions, nucleation of the A phase has a high probability. It was found that the probability of B phase creation to be $54\% \pm 1\%$, while that of the A phase creation is 46%. It is difficult to visualise the trajectory of the order parameter in 18 dimensional space, but we can monitor the energy during the transition. On Fig. 2 we show typical trajectories of the superfluid $^3$He free energy after rapid cooling. In some cases the trajectory approaches a saddle point on the energy surface. The behaviour here is clarified by reducing the rate of energy change.

In order to study the influence of gradient energies on the development of the order parameter we consider a one-dimensional spatial sample of Zurek length $Z$ divided into 100 points. We chose $Z$ to agree with the Grenoble experiment at zero bar (about $8\xi_0$). Two different perturbations were applied, one for the first 50 points and the second for the remaining 50 points. The development of the $A_{ai}$ matrix during the "downhill" process

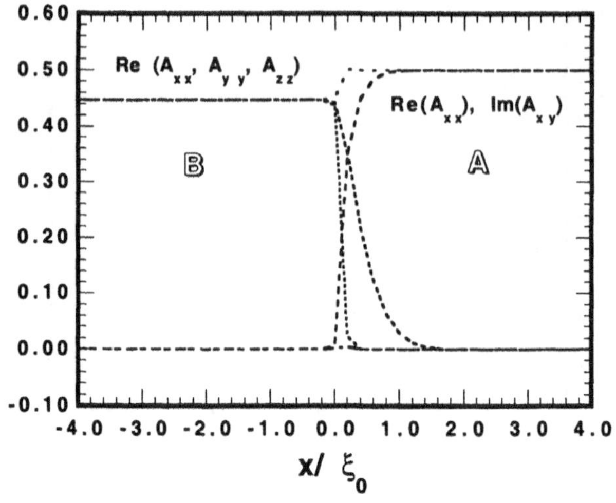


*Figure 3.* Computer simulation of the spatial distribution of non-zero terms of the order-parameter 0.25 $\mu$s after nucleation of the B phase at the left hand side and of the A phase at right hand side of the sample (after [4]).

is calculated at each point, taking into account the gradient of energy. The results of these calculations are shown in Fig. 3. We should mention here that this technique was first discussed in ref. [12] for calculations of the profile of the order parameter on the boundary.

In [4] it was found that the boundary between the two different states remains almost stationary during the main part of the "downhill" process. However, towards the end of this process the boundary begins to move in the energetically favourable direction. This result looks very natural, since the boundary displacement is determined by the energy difference and the time dependence of the energy is very similar for the two different symmetries at the beginning of the "downhill" process, as seen in Fig. 2.

Even for the case, when there is no energetic barrier between the two phases, like between the B-phase and so named planar phase at zero pressure, the states are fully developed first, and then the less energetically favorable state begins to shrink.

Our calculations for this case are shown in Fig. 4.

In the frame of the Ginzburg-Landau approximation the $\beta_i$ parameters depend only on the pressure. There are a number of theories which suggest somewhat different values for these parameters. We have used the parameters, calculated by Sauls and Serene [13] for the strong coupling model and parameters calculated for the spin fluctuation model [14]. In Fig. 5 the probability of A phase nucleation as a function of pressure is shown along with the energy balance between the A and B phases. It is important to

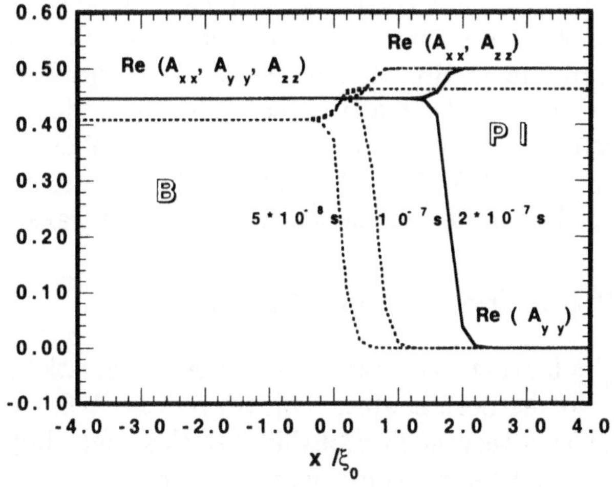


*Figure 4.* Computer simulation of the spatial distribution of non-zero terms of the order-parameter 0.005 $\mu$s, 0.01 $\mu$s and 0.02 $\mu$s after nucleation of the B phase at the left hand side and of the planar phase at right hand side of the sample.

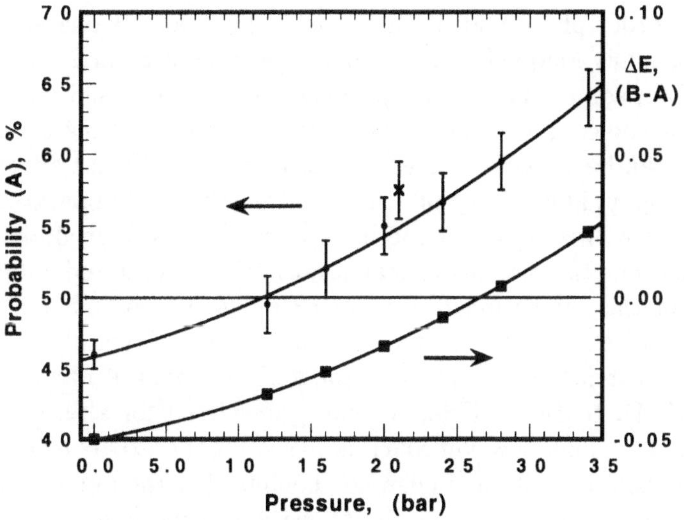


*Figure 5.* The probability of A state nucleation as a function of pressure for temperatures near $T_c$ and the difference of energy (F) between the A and B states. The probability calculated for the spin fluctuation model is shown by (×) (after [15]).

notice that the probability of A state nucleation may become greater than 50% even in the region where the B phase is stable.

All experimentalists who work with superfluid $^3$He have noticed the crucial asymmetry of the A-B transition. If one is cooling $^3$He at a pressure above 20 bar, the A phase may survive as a supercooled metastable state far below the equilibrium A-B transition line. On the other hand, on warming it is difficult to get superheated B phase. This observation may supply the critical jigsaw piece of information for the long-running puzzle of the A-B transition in superfluid $^3$He. We will use the results, discussed in this section, to explain our view of the nature of the A-B transition.

## 4. The A-B transition in superfluid $^3$He

There is no doubt that neutrons, or some other radiation, which locally heats $^3$He up to the normal state, can trigger the A-B transition. Many experiments show a very small probability of this triggering. In a new scenario, we have called "Aurore de Venise", we consider the fast local cooling of $^3$He through $T_c$ with the simultaneous and independent creation of many seeds of superfluid states with different order parameters. The triggering to the B phase can be explained as a formation of big enough cluster of B phase seeds. We consider the statistical conditions required for the A phase to be triggered to the B phase.

The first order phase transition between the A and B phases of superfluid $^3$He has many unique features which makes this system a very interesting test bed for some fundamental properties of transitions. The superfluid $^3$He-A can be cooled down to the "hypercooled" state and the transition to the B-phase can be triggered experimentally at very different experimental conditions. The unique purity of $^3$He at extremely low temperatures gives a hope to have well-controlled experimental conditions. Indeed, after many years of experimental and theoretical studies, the general understanding of the nature of this transition is far away from being complete. This may be particularly related to the fact that all the experimental results have been treated in relation to only one theoretical scenario, namely "Baked Alaska" (BAM). In this article, we have considered an alternative theoretical approach. This new approach is based on the existence of causally independent local transitions under fast cooling. For the name of scenario, we have taken into account that the transitions appear at the number of small islands, looking like a map of Venice.

The nucleation of B-phase in supercooled A-phase does not follow the usual scenario of classical nucleation by thermal fluctuations. That is a result of the extremely low temperatures of the transition and the relatively high surface tension, which gives the critical dimensions of a bubble of B-phase of order 0.5 $\mu$m. The estimated time scale of transitions due to thermal fluctuation is much longer than the lifetime of our Universe. (See the comprehensive discussion in [16])

Another mechanism of B-phase nucleation can be by quantum tunnelling through energetically forbidden regions of configuration space. This type of scenario has been proposed for first order phase transitions by Langer [17] and discussed in detail in [18, 19]. For cosmology and particle physics this process was named the "instanton" mechanism (See review [20]). For the A-B transition this mechanism also gives extremely small probability of nucleation [16].

A more interesting scenario is related to non-topological solitons, proposed for particle physics by Coleman [21] and named Q-balls. The order parameter of superfluid $^3$He is described by a matrix of 9 complex numbers, with different symmetries for the A and B phases. The transition can be considered as a Q-ball type excitation of the A-phase order parameter, which can transfer to the B-phase. The energy barrier in this case can be much lower than the case for the classical solution. This mechanism of the A-B transition has been discussed by Hong [22]. Later Leggett and Yip [16] criticised this scenario, mainly on a basis of the spontaneous transition. From our point of view, this mechanism requires more detailed consideration, particularly as a scenario for the transition in non-stationary conditions, such as found after local heating by radiation. These circumstances have not yet been carefully considered.

Finally, it is widely accepted now, that the A-B transition follows a neutron, or some other radiation impact, which locally heats the $^3$He into the normal state. The subsequent fireball of normal $^3$He cools by the expansion of excitations. The early work can be summarized by the following scheme [16]. If the fireball is cooling by diffusion, the locus of $T_C$ first expands and then contracts. The superfluid $^3$He-A just follows the locus, so that there is no possibility of the creation of a new phase. A solution was proposed by Leggett, where an inverse temperature distribution, on the front of fireball, sheltered the internal space. In this shelter, a new phase, the B-phase, could be created in the surrounding normal phase which then expands to above the critical dimension required for the B-phase to be stable relative to the A-phase. If this size is not reached the B-phase would be 'swallowed' by the developing A-phase. This is the "Baked Alaska" scenario. For a decade this scenario was considered to be the only valid explanation of the transition.

However, there are some strong contradictions in this scenario, first considered in [23]. According to our estimations, the formation of an inverse temperature front is very unlikely in superfluid $^3$He. A cosmic radiation event creates energetic quasiparticles. These non-equilibrium quasiparticles travel out from the site of the event and thermalise, creating many new quasiparticles. In the "Baked Alaska" scenario the non-equilibrium quasiparticles should not thermalise locally but should first fly on some distance from the event site, keeping the internal part relatively cold. Let us consider this conditions in details. We will even make a few assumptions in favour of the"Baked Alaska" scenario.

The mean free path of non-equilibrium quasiparticles increases quadratically with decreasing quasiparticle energy. The energetic quasiparticles excites new quasiparticle - quasihole pairs. Owing to the hypothetical shell type distribution of quasiparticles, we should assume that at time $t$ after the event all non-equilibrium quasiparticles have near the same energy $E(t)$ and are distributed somewhere on the distance of a mean free path $l_{(E)}$ from the primary event :

$$l_{(E)} \sim (1/p_F)(E_F^2/E^2(t)), \tag{3}$$

where $E_F$ is Fermi energy. (we put $\quad = 1$). The clouds of these non-thermalised quasiparticles should form the shell of The "Baked Alaska" scenario with a thickness significantly less then $l_{(E)}$. Let us analyse this possibility. The number of non-equilibrium quasiparticles is $N = E_0/E(t)$, where $E_0$ is the energy of event. These quasiparticles are distributed on the Fermi sphere. The density of states near the Fermi sphere is :

$$\Delta n = (V/\pi^2)k_F^2\Delta k, \tag{4}$$

All these $N$ quasiparticles should be distributed in momentum space

$$\Delta k/k = 1/2\, E(t)/E_F. \tag{5}$$

From these conditions we can find the volume $V$ in space, which should be occupied by the $N$ of quasiparticles. This volume corresponds to a sphere of radius

$$R_{(E)} \sim (1/p_F)(E_0E_F/E^2(t))^{1/3}, \tag{6}$$

A hypothetical sharp energy distribution of quasiparticles leads to the requirement of an even bigger spatial volume for the quasiparticles!

For the "Baked Alaska" conditions the quasiparticles should be distributed only in the shell, that means that the condition $l_{(E)} >> R_{(E)}$ should be fulfilled. This condition can be rewritten as

$$(E_0/E_F)^{1/3}(E(t)/E_F)^{4/3} << 1. \tag{7}$$

If we take $E_0$ to be $10^9 E_F$, as is the case for neutrons, then this condition is not fulfilled until the energy of the quasiparticles corresponds to $T_c$. One can possibly see the "Baked Alaska" quasiparticles distribution after small events with energy of order 10 keV. But these events cannot explain the A-B phase transition as we will discuss later.

In consequence, one cannot consider the fate of one isolated quasiparticle in an energetic event and the dominant process of quasiparticles thermalisation is mutual scattering among the excited quasiparticles within the hot spot leading to fast local thermalisation. This does not lead to the outward-moving wall of ballistic quasiparticles as conjectured in the "Baked Alaska" process. Thus, it is likely that the quasiparticles remain inside the hot bubble and expand by the usual diffusion process.

The inverse temperature front may be formed after a small energy events about 10 keV. But superfluid $^3$He is very sensitive to cosmic rays, as was first shown experimentally by the Lancaster group [24]. Even without a neutron radiation source there is a large fraction of high energy events. It is unlikely that the transition is triggered by so small events and not triggered by large events. We should exclude also the events with non-localised energy deposition, since in this case it is even more difficult to create a sharp front of quasiparticles.

The next condition, that was considered in favour of the "Baked Alaska" scenario was enlargement of the hot spot above the critical dimensions of B phase, so that it became stable against the surrounding A-phase. But again, event with energy above a tenth keV create a hot spot bigger than the critical radius for B phase transition. That means that "Baked Alaska" enlargement of the hot spot is not necessary.

And finally, one should take in mind that non-equilibrium quasiparticles carry the Fermi momentum and a relatively small energy. Owing to momentum and energy conservation, the new pairs of quasiparticles cannot keep information about the momentum of the primary quasiparticle and can easily return to the centre of fireball.

All this shows that the "Baked Alaska" scenario is not very realistic and we need to consider another scenario to explain the A-B transition. But firstly let us revise the experimental results.

## 5. Experimental results

Experimental investigations of the A-B transition have been performed in Helsinki [25] and in Stanford University [26]. The program of Stanford experiments was well thought out and gives very clear results. They found that two different mechanisms are responsible for the transition. One is determined by the quality of the surface of the cell walls. The quality of the walls can not be independently controlled, but after special procedures, which makes the walls smooth and clean, the lifetime of overcooled A-phase becomes very long and is determines by radiation. In this case it is a bulk transition mechanism. The surface triggering remains a mystery. In this article we consider triggering the bulk transition by radiation.

The experimental results are well fitted by equation [26],

$$\tau = C_0 \exp a(R_c(T)/R_0)^n, \qquad (8)$$

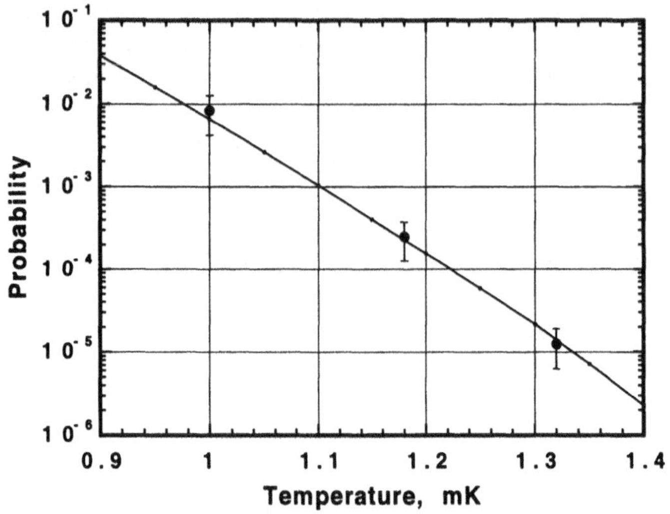


*Figure 6.* The experimental probability of the A-B transition after one neutron impact (data from [26]) and the calculated probability of the formation of the 50 correlated seeds of B phase.

where $\tau$ is the average lifetime of overcooled A-phase, $R_c(T)$ is the critical radius of the B-phase bubble as function of temperature, $R_0$ is the critical radius at zero temperature and $C_0$, $a$ and $n$ are fitting parameters.

This type of equation came originally from the theory of "Baked Alaska". The theory predicts $n$ to be about 3 - 5, while the best fit to the data gives 3/2. Other fitting parameters are not predicted by the theory. A magnetic field dependence was also measured, which scaled well with the changes of $R_c$ expected from the field dependence of the free energies. In other words the experiment proves the importance of the temperature and field dependence of the critical radius $R_c(T, H)$ for the B-phase transition process, but this is a property of superfluid $^3$He, not of the "Baked Alaska" mechanism itself.

A related problem is an extremely small probability of transition, as shown in Fig. 6. The Stanford experiment found that for some conditions thousands of neutrons were needed to trigger the transition! Neutrons always deposit an energy of about 700keV which creates a macroscopic hot spot of dimensions much bigger than the critical radius for B phase nucleation. How can we explain for this case the stochastic nature of the B phase transition?

Taking all this into account, we cannot conclude that the experiments make any solid confirmation for the "Baked Alaska" mechanism. For example, in [26] the authors argue about hot spots of order or smaller than

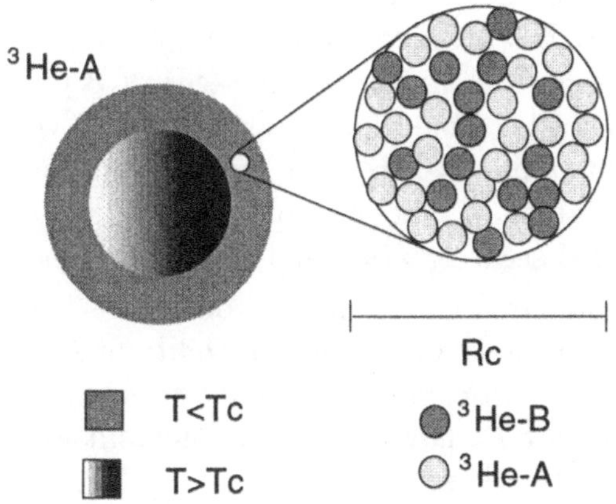


*Figure 7.* A schematic presentation of the A and B phase seeds which appears at diffusion cooling of the hot spot.

$R_c(T)$. Only in this case can the small probability of triggering be explained. However, there is a wide distribution of energy deposited in $^3$He after a cosmic event, with hot spots much bigger then $R_c(T)$. Why are they not also responsible for triggering a transition? Definitely, $R_c(T)$ plays a crucial role, but from our point of view this role is very different.

## 6. The "Aurore de Venise" scenario

In the conditions of fast cooling we do not need a normal shell to protect the interior of the hot bubble from the influence of the outside state. It was shown [10] that at cooling the diffusion temperature front moves so fast that A-phase lags far behind and seeds of new states appear independently of the surrounding $^3$He state. In the case of neutron heating, the dimensions of the hot spot ($R_h$) are considerably larger than the critical radius ($R_c$). There are a huge number of seeds of A and B phase nucleated inside the hot spot. This situation is shown in Fig. 7. By our estimations, at 29 bar, $R_h = 30\mu\text{m} \quad and R_0 = 0.5\mu\text{m}$, while the radius of independently created seeds $R_s = 0.07\mu\text{m}$.

We can suggest that the subsequent development of the very complicated mixuture of A and B phases is determind, firstly, by the relative densities of the seeds of these two phases, secondly by the seeds domain boundary surface energy and only in the third place by the energy balance between these phases. The conditions for equal probability of nucleation of the A and B phases are different from those corresponding to the equilibrium of their free energies. This is the reason for the asymmetry of the A-B transition. When the B state is energetically preferable, but the A state has higher probability of nucleation, the A state seeds percolate inside the hot spot. Consequently B state seeds shrink due to the A-B surface tension. In order to pass through the transition the seeds of B phase should percolate up to the critical cluster dimensions. This is possible when the number of B seeds at the region of dimensions $R_c$ is greater than the number of A seeds. There are two parameters that play an important role in these conditions; the probability of B seeds creation and the dimensions of $R_c$.

In order to compare the predictions of the new scenario with experimental results, we need to know the probability of A and B phase creation at low temperature. Unfortunately, we only know the probability of A and B seed formation near $T_c$, however, we can guess it for lower temperatures from the shape of $^3$He phase diagram.

Our current calculations show that the probability of A-B transition after neutron impact corresponds well to the probability of the correlated formation of about 50 seeds of B phase, as shown in Fig. 6. This means that the transition will trigger when about 50 seeds of B phase are created with a short distance between each other. It is difficult to go inside the dynamics of the interaction between the different seeds after transition. Indeed, the correlation of 50 seeds looks like a reasonable result, since it makes up a big fraction of B seeds inside a $R_c$ volume, which permits the percolation of the B state inside $R_c$ which triggers the transition.

Unfortunately, we can not yet produce quantitative relations for our theory. After consulting world specialists on percolation we have not found an analogous 3D problem which has been previously solved. Indeed, a very fast change of the probability of the B phase transition with cooling is a natural property of a 3D system near the point of percolation. Thus further investigations of the process of percolations are desirable.

In conclusion, we should say that our scenario explains in a very natural way, many features of A-B transition including the asymmetry of transition, its stochastic nature and its very small probability even after energetic radiation.

This work is a part of ESF Scientific Network "Topological defects" We are grateful to I. Bradley, D. J. Cousins, A. J. Gill, H. Godfrin, G. R. Pickett, O. D. Timofeevskaya and G. E. Volovik for many stimulating discussions.

## References

1. Zel'dovich, Ya. B., Kobzarev, I. Ya., Okun', L. B. (1975) Cosmological consequences of a spontaneous breakdown of a discrete symmetry, *Sov. Phys JETP* **40**, 1–5.
2. Kibble, T. W. B., (1976) Topology of cosmic domains and strings, *J. Phys.* **A9**, 1387–1398

3. Linde, A. (1990) *Particle physics and inflationary cosmology* Harwood acad., Switzerland.
4. Bunkov, Yu. M., Timofeevskaya, O. D., (1998) "Cosmological" scenario for A-B phase transition in superfluid $^3$He, *Phys. Rev. Lett.* **80**, 4927-4930.
5. Ruutu, V.M.H., Eltsov, V.B., Gill, A.J., Kibble, T.W.B., Krusius, M., Makhlin, Yu.G., Plaçais, B., Volovik, G.E. and Xu, W. (1996) Vortex formation in neutron-irradiated superfluid He–3 as an analogue of cosmological defect formation, *Nature* **382**, 334–336.
6. Bäuerle, C., Bunkov, Yu.M., Fisher, S.N., Godfrin, H. and Pickett, G.R. (1996) Laboratory simulation of cosmic string formation in the early Universe using superfluid He–3, *Nature* **382**, 332–334.
7. Bäuerle, C., Bunkov, Yu.M., Fisher, S.N., Godfrin, H. (1999) *The 'Grenoble' cosmological experiment. The Kibble-Zurek Scenario in Superfluid $^3$He.* This book.
8. Vollhardt, D., Wölfle, P. (1990) *The superfluid phases of helium 3*, Taylor and Francis, London.
9. Zurek, W.H. (1985) Cosmological experiments in superfluid helium, *Nature* **317**, 505–508.
10. Kibble, T.W.B., Volovik, G.E. (1997) On phase ordering behind the propagating front of a second order transition, *JETP letters* **65**, 102-107.
11. Volovik, G.E. was the first to point out these circumstances, in an unpublished version of article [5].
12. Schopohl, N. (1987) Spatial Dependence of the Order Parameter of Superfluid He at the A-B Phase Boundary *Phys. Rev. Lett.* **58**, 1664-1667.
13. Sauls, J.A., Serene, J.W. *Phys. Rev.* **B24**, 183 (1981).
14. Anderson, P.W., Brinkman, W.F. (1978) Theory of Anisotropic Superfluidity in $^3$He, *The physics of liquid and solid helium* , ed. by K. H. Bennemann, J. B. Ketterson, Wiley, New York, p 177-286.
15. Bunkov, Yu. M., Timofeevskaya, O. D., (1998) A and B phases nucleation at a fast superfluid transition in $^3$He. Can it be relevent for cosmology? *J. Low Temp. Phys.*, **110**, 45-50.
16. Leggett, A.J., Yip, S.K. (1990) Nucleation and Growth of $^3$He-B in the super-cooled A-phase, *Helium Three* , ed. by V. M. Agranovich, A. A. Maradudin, North-Holland, Amsterdam, p. 523-608.
17. Langer, J.S., (1969) *Ann. Phys.* **54**, 258.
18. Lifshitz, E.M., Kagan, Yu.M. (1972) Quantum kinetics of phase transition at temperatures close to absolute zero *Sov. Phys. JETP* **35**, 206-214.
19. Iordanskiy, S.V., Finkelshtain, A.M. (1972) Effect of quantum fluctuations on the lifetime of metastable states in solid *Sov. Phys. JETP.* **35**, 215-221.
20. Coleman, S., (1977) The uses of Instantons, *Int Summer School of Subnuclear Physics*, Ettore Majorana, Erice.
21. Coleman, S., (1985) *Nucl. Phys. B* **262** 263.
22. Hong, D.K., (1988) Q-balls in superfluid $^3$He *J. Low Temp. Phys.* **71** 483-494.
23. Bunkov, Yu. M., Timofeevskaya, O. D., (1999) Reply *Phys. Rev. Lett.* **82**, 3926-3927.
24. Bradley, D.I., Bunkov, Yu.M., Cousins, D.J., Enrico, M.P., Fisher, S.N., Follows, M.R., Guénault A.M., Hayes, W.M., Pickett, G.R., Sloan, T. (1995) Potential dark matter detector? The detector of low energy neutrons by superfluid $^3$He-B *Phys. Rev. Lett.* **75** 1887-1890.
25. Hakonen, P. J., Krusius, M., Salomaa, M. M., Simola, J. T. (1985) Comment on "Nucleation of $^3$He-B from the A phase: a cosmic ray effect?" *Phys. Rev. Lett.*, **54**, 245–248 .
26. Schiffer, P., Osheroff, D. D., Leggett, A. J. (1995). Nucleation of the AB transition in superfluid $^3$He: Experimental and theoretical consideration, *Progress in Low Temperature Physics* **14**, ed. by W. P. Halperin, Elsevier, Amsterdam.

# NON-EQUILIBRIUM PHASE TRANSITIONS IN CONDENSED MATTER AND COSMOLOGY: SPINODAL DECOMPOSITION, CONDENSATES AND DEFECTS

D. BOYANOVSKY$^{(A,B)}$, H.J. DE VEGA$^{(B,A)}$ AND R. HOLMAN$^{(C)}$
*(A) Department of Physics and Astronomy, University of Pittsburgh, Pittsburgh, PA 15260 USA*
*(B) LPTHE† Université Pierre et Marie Curie (Paris VI) et Denis Diderot (Paris VII), 4, Place Jussieu 75252 Paris, France*
*(C) Department of Physics, Carnegie Mellon University, Pittsburgh, PA 15213, USA*

These lectures address the dynamics of phase ordering out of equilibrium in condensed matter and in quantum field theory in cosmological settings, emphasizing their similarities and differences. In condensed matter we describe the phenomenological approach based on the Time Dependent Ginzburg-Landau (TDGL) description. We study the *exact* solution to the dynamics after a quench in this limit in Minkowski space time and in radiation dominated Friedman-Robertson-Walker Cosmology. There are some remarkable similarities between these very different settings such as the emergence of a scaling regime and of a dynamical correlation length at late times that describe the formation and growth of ordered regions. In quantum field theory and cosmology this length scale is constrained by causality and its growth in time is also associated with coarsening and the onset of a condensate. We provide a density matrix interpretation of the formation of defects and the classicalization of quantum fluctuations.

## 1. Phase Ordering Kinetics: an interdisciplinary fascinating problem

The dynamics of non-equilibrium phase transitions and the ordering process that occurs until the system reaches a broken symmetry equilibrium state play an important role in many different areas. Obviously in the physics of binary fluids and ferromagnets (domain walls) superfluids (vortex formation), and liquid crystals (many possible defects) to name but a few in

†Laboratoire Associé au CNRS UMR 7589.



*Y.M. Bunkov and H. Godfrin (eds.), Topological Defects and the Non-Equilibrium Dynamics of Symmetry Breaking Phase Transitions*, 139–169.

condensed matter, but also in cosmology and particle physics. In cosmology defects produced during Grand Unified Theory (GUT) or the Electro-weak (EW) phase transition can act as seeds for the formation of large scale structure and the dynamics of phase ordering and formation of ordered regions is at the heart of Kibble's mechanism of defect formation[1, 2, 3, 4]. Current and future measurements of Cosmic Microwave Background anisotropies will determine the nature of the cosmological phase transitions that influenced structure formation[5]. Also at even lower energies, available with current and forthcoming accelerators (RHIC and LHC) the phase transitions predicted by the theory of strong interactions, Quantum Chromodynamics (QCD) could occur out of equilibrium via the formation of coherent condensates of low energy Pions. These conjectured configurations known as 'Disoriented Chiral Condensates' are similar to the defects expected in liquid crystals or ferromagnets and their charge distribution could be an experimental telltale of the QCD phase transitions[6]. Whereas the GUT phase transition took place when the Universe was about $10^{-35}$ seconds old and the temperature about $10^{23}K$, and the EW phase transition occured when the Universe was $10^{-12}$ seconds old and with a temperature $10^{15}K$, the QCD phase transition took place at about $10^{-5}$ seconds after the Big Bang, when the temperature was a mere $10^{12}K$. This temperature range will be probed at RHIC and LHC within the next very few years. The basic problem of describing the process of phase ordering and the competition between different broken symmetry states on the way towards reaching equilibrium is common to all of these situations. The tools, however, are necessarily very different: whereas ferromagnets, binary fluids or alloys etc, can be described via a phenomenological (stochastic) description, certainly in quantum systems a microscopic formulation must be provided. In these lectures we describe a program to include ideas from condensed matter to the realm of quantum field theory, to describe phenomena on a range of time and spatial scales of unprecedented resolution (time scales $\leq 10^{-23}$ seconds, spatial scales $\leq 10^{-15}$ meters) that require a full quantum field theoretical description.

## 2. Main ideas from Condensed Matter

Before tackling the problem of describing phase ordering kinetics in quantum systems starting from a microscopic theory, it proves illuminating to understand a large body of theoretical and experimental work in condensed matter systems[7]-[10]. Although ultimately the tools to study the quantum problem will be different, the main physical features to describe are basically the same: the formation and evolution of correlated regions separated by 'walls' or other structures. Inside these regions an ordered phase exists which eventually grows to become macroscopic in size. Before attempting to describe the manner in which a given system orders after being cooled through a phase transition an understanding of the relevant time scales is required. Two important time scales determine if the transition occurs in or out of equilibrium: the relaxation time of long wavelength fluctuations (since these are the ones that order) $\tau_{rel}(k)$ and the inverse of the cooling rate $t_{cool} = T(t)/\dot{T}(t)$. If $\tau_{rel}(k) << t_{cool}$ then these wavelengths are in local thermodynamical equilibrium (LTE), but if $\tau_{rel}(k) >> t_{cool}$ these wavelengths fall out of LTE and freeze out, for these the phase transition occurs in a quenched manner. These modes do not have time to adjust locally to the temperature change and for them the transition from a high temperature phase to a low temperature one occur instantaneously. This description was presented by Zurek[11] analysing the emergence of defect networks after a quenched phase transition. Whereas the short wavelength modes are rapidly thermalized (typically by collisions) the long-wavelength modes with $k << 1/\xi(T)$ with $\xi(T)$ the correlation length (in the disordered phase) become *critically slowed down.* As $T \to T_c^+$ the long wavelength modes relax very slowly, they fall out of LTE and any finite cooling rate causes them to undergo a quenched non-equilibrium phase transition. As the system is quenched from $T > T_c$ (disordered phase) to $T << T_c$ (ordered phase) ordering *does not* occur instantaneously. The length scale of the ordered regions grows in time (after some initial transients) as the different broken symmetry phases compete to select the final equilibrium state. A *dynamical* length scale $\xi(t)$ typically emerges which is interpreted as the size of the correlated regions, this dynamical correlation length grows in time to become macroscopically large[7, 8, 9, 10].

The phenomenological description of phase ordering kinetics begins with a coarse grained local free energy functional of a (coarse grained) local order parameter $M(\vec{r})$[7, 8] which determines the *equilibrium* states. In Ising-like systems this $M(\vec{r})$ is the local magnetization (averaged over many lattice sites), in binary fluids or alloys it is the local concentration difference, in superconductors is the local gap, in superfluids is the condensate fraction etc. The typical free energy is (phenomenologically) of the Landau-Ginzburg form:

$$\begin{aligned} F[M] &= \int d^d\vec{x} \left\{ \frac{1}{2}[\nabla M(\vec{x})]^2 + V[M(\vec{x})] \right\} \\ V[M] &= \frac{1}{2}\, r(T)\, M^2 + \frac{\lambda}{4} M^4 \;\; ; \;\; r(T) = r_0(T - T_c) \end{aligned} \tag{1}$$

The equilibrium states for $T < T_c$ correspond to the broken symmetry states with $M = \pm M_0(T)$ with

$$M_0(T) = \begin{cases} 0 & \text{for } T > T_c \\ \sqrt{\frac{r_0}{\lambda}}(T_c - T)^{\frac{1}{2}} & \text{for } T < T_c \end{cases} \tag{2}$$

Below the critical temperature the potential $V[M]$ features a non-convex region with $\partial^2 V[M]/\partial M^2 < 0$ for

$$-M_s(T) < M < M_s(T) \;\; ; \;\; M_s(T) = \sqrt{\frac{r_0}{3\lambda}}(T - T_c)^{\frac{1}{2}} \quad (T < T_c) \tag{3}$$

this region is called the spinodal region and corresponds to thermodynamically unstable states. The lines $M_s(T)$ vs. $T$ and $M_0(T)$ vs. $T$ [see eq.(2)] are known as the classical spinodal and coexistence lines respectively.

The states between the spinodal and coexistence lines are metastable (in mean-field theory). As the system is cooled below $T_c$ into the unstable region inside the spinodal, the *equilibrium* state of the system is a coexistence of phases separated by domains and the concentration of phases is determined by the Maxwell construction and the lever rule.

**Question:** How to describe the *dynamics* of the phase transition and the process of phase separation?

**Answer:** A phenomenological but experimentally succesful description, Time Dependent Ginzburg-Landau theory (TDGL) where the basic ingredient is Langevin dynamics[7]-[10]

$$\frac{\partial M(\vec{r},t)}{\partial t} = -\Gamma[\vec{r}, M]\,\frac{\delta F[M]}{\delta M(\vec{r},t)} + \eta(\vec{r},t) \tag{4}$$

with $\eta(\vec{r},t)$ a stochastic noise term, which is typically assumed to be white (uncorrelated) and Gaussian and obeying the fluctuation-dissipation theorem:

$$\langle \eta(\vec{r},t)\eta(\vec{r}',t')\rangle = 2\,T\,\Gamma(\vec{r})\,\delta^3(\vec{r}-\vec{r}')\delta(t-t') \;\; ; \;\; \langle \eta(\vec{r},t)\rangle = 0 \tag{5}$$

the averages $\langle \cdots \rangle$ are over the Gaussian distribution function of the noise. There are two important cases to distinguish: **NCOP:** Non-conserved order parameter, with $\Gamma = \Gamma_0$ a constant independent of space, time and order parameter, and which can be absorbed in a rescaling of time. **COP:** Conserved order parameter with

$$\Gamma[\vec{r}] = -\Gamma_0\,\nabla^2_{\vec{r}}$$

where $\Gamma_0$ could depend on the order parameter, but here we will restrict the discussion to the case where it is a constant. In this latter case the average over the noise of the Langevin equation can be written as a conservation law

$$\begin{aligned} \frac{\partial M}{\partial t} &= -\nabla \cdot J + \eta \Rightarrow \frac{\partial}{\partial t}\langle \int d^3\, \vec{r} M(\vec{r},t)\rangle = 0 \\ \vec{J} &= \vec{\nabla}_{\vec{r}}\left[-\Gamma_0 \frac{\delta F[M]}{\delta M}\right] \equiv \vec{\nabla}_{\vec{r}}\mu \end{aligned} \tag{6}$$

where $\mu$ is recognized as the chemical potential. Examples of the NCOP are the magnetization in ferromagnets, the gap in superconductors and the condensate density in superfluids (the total particle number is conserved but not the condensate fraction), of the COP: the concentration difference in binary fluids or alloys. For a quench from $T > T_c$ deep into the low temperature phase $T \to 0$ the thermal fluctuations are suppressed after the quench and the noise term is irrelevant. In this situation of experimental relevance of a deep quench the dynamics is now described by a deterministic equation of motion,

for **NCOP**:

$$\frac{\partial M}{\partial t} = -\Gamma_0 \frac{\delta F[M]}{\delta M} \tag{7}$$

for **COP**:

$$\frac{\partial M}{\partial t} = \nabla^2 \left[\Gamma_0 \frac{\delta F[M]}{\delta M}\right] \tag{8}$$

which is known as the Cahn-Hilliard equation[7, 8]. In both cases the equations of motion are purely diffusive

$$\frac{dF}{dt} = \int d^3r \, \frac{\delta F[M]}{\delta M(\vec{r},t)} \frac{\partial M(\vec{r},t)}{\partial t} = -\Gamma_0 \begin{cases} \int d^3r \left(\frac{\delta F}{\delta M}\right)^2 & \text{NCOP} \\ \int d^3r \left(\vec{\nabla}\frac{\delta F}{\delta M}\right)^2 & \text{COP} \end{cases} \tag{9}$$

and in both cases $\frac{dF}{dt} < 0$. Thus, the energy is always diminishing and there is no possibility of increasing the free energy. Thus overbarrier thermal activation cannot be described in the absence of thermal noise, which is clear since thermal activation is mediated by large thermal fluctuations. The fact that this phenomenological description is purely dissipative with an ever diminishing free energy is one of the fundamental differences with the quantum field theory description studied in the next sections.

### 2.1. CRITICAL SLOWING DOWN IN NCOP:

Critical slowing down of long-wavelength fluctuations is built in the TDGL description. Consider the case of NCOP and linearize the TDGL equation above the critical temperature for small amplitude fluctuations near $M = 0$.

Neglecting the noise term for the moment and taking the Fourier transform of the small amplitude fluctuations we find

$$\frac{dm_k(t)}{dt} \approx -\Gamma_0 \left[k^2 + r_0(T - T_c)\right] m_k(t) \tag{10}$$

showing that long-wavelength small amplitud fluctuations relax to equilibrium $m_k = 0$ on a time scale given by

$$\tau_k \propto \left[k^2 + r_0(T - T_c)\right]^{-1} \tag{11}$$

As $T \to T_c^+$ the long-wavelength modes are critically slowed down and relax to equilibrium on very long time scales. Therefore a TDGL description leads to the conclusion that if the cooling rate is finite, the long-wavelength modes will fall out of LTE and become quenched. As the temperature falls below the critical, these modes will become unstable and will grow exponentially.

2.2. LINEAR INSTABILITY ANALYSIS:

Let us consider now the situation for $T << T_c$ and neglect the thermal noise. The early time evolution after the quench is obtained by linearizing the TDGL equation around a homogeneous mean field solution $M_o(t)$. Writing

$$M(\vec{r}, t) = M_o(t) + \frac{1}{\sqrt{\Omega}} \sum_{\vec{k} \neq 0} m_k(t)\, e^{i\vec{k}\cdot\vec{r}} \tag{12}$$

where $\Omega$ is the volume of the system, and considering only the linear term in the fluctuations $m_k(t)$ the linearized dynamics is the following: **COP:** for $M_o(t)$ the conservation gives

$$\frac{dM_o(t)}{dt} = 0$$

since $M_o$ is the volume integral of the order parameter [see eq.(6)] and for the fluctuations we obtain

$$\frac{dm_k(t)}{dt} = \omega(k)\, m_k(t) \quad ; \quad \omega(k) = -\Gamma_0\, k^2 \left[k^2 + \left.\frac{\partial^2 V[M]}{\partial M^2}\right|_{M_o}\right] \tag{13}$$

In the spinodal region $\left.\frac{\partial^2 V[M]}{\partial M^2}\right|_{M_o} < 0$ there is a band of unstable wave vectors $k^2 < \left|\left.\frac{\partial^2 V[M]}{\partial M^2}\right|_{M_o}\right|$ for which the frequencies are positive and the fluctuations away from the mean field grow exponentially.

**NCOP:** separate the $\vec{k} \neq 0$ from the $\vec{k} = 0$ in the linearized equation of motion:

$$\frac{dM_o(t)}{dt} = -\Gamma_0 \left.\frac{dV[M]}{dM}\right|_{M_o} \tag{14}$$

$$\frac{dm_k(t)}{dt} = -\Gamma_0 \left[\frac{\delta F[M]}{\delta M}\right]_{M_o(t)} m_k(t) = -\Gamma_0 \left[k^2 + \left.\frac{\partial^2 V[M]}{\partial M^2}\right|_{M_o}\right] \tag{15}$$

whereas the first equation (14) determines that $M_o(t)$ rolls down the potential hill towards the equilibrium solution, the second equation also displays the linear instabilities for the same band of wave vectors as in the COP in the spinodal region $|M_o(t)| \leq M_s(T)$ [see eq. (3)] for which the fluctuations grow exponentially in time. Thus in the linearized approximation for both NCOP and the COP the spinodal instabilities are manifest as exponentially growing fluctuations. These instabilities are the hallmark of the process of phase separation and are the early time indications of the formation and growth of correlated regions which will be understood in an exactly solvable example below.

### 2.3. THE SCALING HYPOTHESIS: DYNAMICAL LENGTH SCALES FOR ORDERING

The process of ordering is described by the system developing ordered regions or domains that are separated by walls or other type of defects. The experimental probe to study the domain structure and the emergence of long range correlations is the equal time pair correlation function

$$C(\vec{r}, t) = \langle M(\vec{r}, t) M(\vec{0}, t)\rangle \tag{16}$$

where $\langle \cdots \rangle$ stands for the statistical ensemble average in the initial state (or average over the noise in the initial state before the quench) and will become clear(er) below. It is convenient to expand the order parameter in Fourier components

$$M(\vec{r}, t) = \frac{1}{\sqrt{\Omega}} \sum_{\vec{k}} m_k(t)\, e^{i\vec{k}\cdot\vec{x}}$$

and to consider the spatial Fourier transform of the pair correlation function

$$S(\vec{k}, t) = \langle m_{\vec{k}}(t) m_{-\vec{k}}(t)\rangle \tag{17}$$

known as the **structure factor** or power spectrum which is experimentally measured by neutron (in ferromagnets) or light scattering (in binary fluids)[12]. The scaling hypothesis introduces a dynamical length scale $L(t)$ that describes the typical scale of a correlated region and proposes that

$$C(\vec{r},t) = f\left(\frac{|\vec{r}|}{L(t)}\right) \Rightarrow S(\vec{k},t) = L^d(t)\, g(kL(t)) \tag{18}$$

where $d$ is the spatial dimensionality and $f$ and $g$ are scaling functions. Ultimately scaling is confirmed by experiments and numerical simulations and theoretically it emerges from a renormalization group approach to dynamical critical phenomena which provides a calculational framework to extract the scaling functions and the deviations from scaling behavior[7]. This scaling hypothesis describes the process of phase ordering as the formation of ordered 'domains' or correlated regions of typical spatial size $L(t)$. For NCOP typical growth laws are $L(t) \approx t^{1/2}$ (with some systems showing weak logarithmic corrections) and $L(t) \approx t^{1/3}$ for scalar and $\approx t^{1/4}$ for vector order parameter in the COP case[7, 9, 10].

## 2.4. AN EXACTLY SOLVABLE (AND RELEVANT) EXAMPLE: THE LARGE $N$ LIMIT

We consider the case where the order parameter has $N$-components and transforms as a vector under rotations in an N-dimensional Euclidean space, i.e. $\vec{M}(\vec{r},t) = (M_1(\vec{r},t), M_2(\vec{r},t), \cdots, M_N(\vec{r},t))$. For $N = 1$ an example is the Ising model, for $N = 2$ superfluids or superconductors (where the components are the real and imaginary part of the condensate fraction or the complex gap respectively), $N = 3$ is the spin one Heisenberg antiferromagnet, etc. For $N = 1$ the topological defects are domain walls (topological in one spatial dimension), for $N = 2$ they are vortices in $d = 2$ and vortex lines in $d = 3$, for $N = d = 3$ the topological defects are monopoles or skyrmions which are possible excitations in Quantum Hall systems and also appear in nematic liquid crystals[2]. For $N \to \infty$ and fixed $d$ no topological defects exist. However the exact solution of the large $N$ model gives insight and is in fairly good agreement with growth laws for fixed $N$ systems which had been studied experimentally and numerically[7, 10]. In quantum field theory the non-equilibrium dynamics of phase transitions has been studied in Minkowsky and cosmological space-times[15, 16, 17, 18, 19]. In cosmological space-times it has been implemented to study the collapse of texture-like configurations[5, 13, 14] (see later). The large $N$ limit is an exactly solvable limit that serves as a testing ground for establishing the fundamental concepts and that can be systematically improved in a consistent $1/N$ expansion. It provides a consistent formulation which is *non-perturbative*, renormalizable and numerically implementable and has recently been invoked in novel studies of non-equilibrium dynamics in quantum spin glasses and disordered systems[20].

The exact solution for the dynamics in the large $N$ limit, being available both in the condensed matter TDGL description of phase ordering kinetics and in Quantum Field Theory in Minkowsky and Cosmological space times, allow us to compare *directly* the physics of phase ordering in these situations. Thus we begin by implementing this scheme in the NCOP case for the TDGL description.

What is the $\langle \cdots \rangle$ in the equations of the previous section?: consider that *before* the quench the system in in equilibrium in the *disordered* phase at $T >> T_c$ and with a very short correlation length $(\xi(T) \approx 1/T)$. The ensemble average in this initial state is therefore

$$\begin{aligned} \langle M^i(\vec{r},0) M^j(\vec{r}',0) \rangle &= \Delta \, \delta^{ij} \delta^3(\vec{r}-\vec{r}') \\ \langle M^i(\vec{r},0) \rangle &= 0 \end{aligned} \tag{19}$$

where $\Delta$ specifies the initial correlation. Now consider a critical quench where the system is rapidly cooled through the phase transition to almost zero temperature but in the *absence* of explicit symmetry breaking fields (for example a magnetic field). The average of the order parameter will remain zero through the process of spinodal decomposition and phase ordering. During the initial stages, linear instabilities will grow exponentially with $m^i_k(t) \approx m^i_k(0) \, e^{\omega(k)t} \;\; ; \;\; \omega(k) = k^2 - r(0)$ for $k^2 < r(0)$ and at early times

$$\langle m^i_{\vec{k}}(t) m^j_{-\vec{k}}(t) \rangle \approx \Delta \, e^{2\omega(k)t} \tag{20}$$

hence fluctuations begin to grow exponentially and eventually will sample the broken symmetry states and the exponential growth must shut-off. The large $N$ limit is implemented by writing the potential term in the free energy as

$$V[\vec{M}] = -\frac{r(T)}{2} \, \vec{M}^2 + \frac{\lambda}{4N} (\vec{M}^2)^2 \; ; \;\; \vec{M}^2 = \vec{M} \cdot \vec{M} \tag{21}$$

where $\lambda$ is kept finite in the large $N$ limit. We will focus on the NCOP case with a quench to zero temperature and rescale the order parameter, time and space as

$$\vec{M} = \sqrt{\frac{r(0)}{\lambda}} \, \vec{\eta} \;\; ; \;\; r(0) \, \Gamma_0 \, t = \tau \;\; ; \;\; \sqrt{r(0)} \, \vec{x} = z \tag{22}$$

after which the evolution equation for the NCOP case becomes

$$\frac{\partial \vec{\eta}}{\partial \tau} = \nabla^2 \vec{\eta} + \left( 1 - \frac{\vec{\eta}^2}{N} \right) \vec{\eta} \tag{23}$$

where derivatives are now with respect to the rescaled variables. The large $N$ limit is solved by implementing a Hartree-like factorization[7]

$$\vec{\eta}^2 \to \langle \vec{\eta}^2 \rangle = N \langle \eta_i^2 \rangle \quad \text{no sum over i} \tag{24}$$

Then for each component the NCOP equation becomes

$$\frac{\partial \eta_i}{\partial \tau} = \left[\nabla^2 + M^2(t)\right] \eta_i \tag{25}$$

$$M^2(t) = 1 - \langle \eta_i^2 \rangle \tag{26}$$

the eq.(26) is a *self-consistent* condition that must be solved simultaneously with the equation of motion for the components. Thus the large $N$ approximation linearizes the problem at the expense of a self-consistent condition. The solution for each component is obviously

$$\eta_{i,\vec{k}}(\tau) = \eta_{i,\vec{k}}(0)\; e^{-k^2\tau + b(\tau)} \;;\;\; b(\tau) = \int_0^\tau M^2(\tau')d\tau' \tag{27}$$

Consider for a moment that the $\vec{k} = 0$ mode is slightly displaced at the initial time, then it will roll down the potential hill to a final equilibrium position for which $M^2(\infty)\ \eta_i(\infty) = 0$ (so the time derivative vanishes in equilibrium). If $\eta_i(\infty) \neq 0$ is a broken symmetry minimum of the free energy, then $M^2(\tau) \to 0$ when $\tau \to \infty$. This is the statement of Goldstone's theorem that guarantees that the perpendicular fluctuations are soft modes. This asymptotic limit allows the solution of the self-consistent condition

$$M^2(\tau) = 1 - \langle \eta_i^2(\tau) \rangle = 1 - \Delta\; e^{2b(\tau)} \int \frac{d^dk}{(2\pi)^d}\; e^{-k^2\tau} = 1 - \Delta\; e^{2b(\tau)}\; (8\pi t)^{-\frac{d}{2}} \tag{28}$$

The vanishing of the right hand side in the asymptotic time regime leads to the self-consistent solution

$$b(\tau) \to \frac{d}{4} \ln\left[\frac{\tau}{\tau_0}\right] \quad \Rightarrow M^2(\tau) \to \frac{d}{4\tau} \tag{29}$$

where $\tau_0$ is a constant related to $\Delta$. This self-consistent solution results in the following asymptotic behavior

$$\eta_{i,\vec{k}}(\tau) \to \eta_{i,\vec{k}}(0) \left(\frac{\tau}{\tau_0}\right)^{\frac{d}{4}} e^{-k^2\tau} \tag{30}$$

Introducing the *dynamical length scale* $L(\tau) = \tau^{\frac{1}{2}}$ it is straightforward to find the structure factor and the pair correlation function

$$S(\vec{k}, t) \propto L^d(t)\; e^{-2(kL(t))^2} \tag{31}$$

$$C(\vec{r}, t) \propto e^{-\frac{r^2}{8L^2(t)}} \;\;;\;\; L(t) = t^{\frac{1}{2}} \tag{32}$$

This behavior *should not* be interpreted as diffusion, because of the $L^d(t)$ in eqn. (31) which is a result of the self-consistent condition.

**Important Features:**

- The 'effective squared mass' $M^2(t) \overset{t\to\infty}{\to} 0$: asymptotically there are massless excitations identified as Goldstone bosons.
- Since $M^2(t) \to 0$ asymptotically, the self-consistent condition results in that $\langle \vec{M}^2 \rangle \to Nr(0)/\sqrt{\lambda}$, i.e. the fluctuations sample the broken symmetry states, which are equilibrium minima of the free energy. These fluctuations begin to grow exponentially at early times due to spinodal instabilities.
- A dynamical correlation length emerges $L(t) = t^{1/2}$ which determines the size of the correlated regions or 'domains'. A scaling solution emerges asymptotically with the natural scale determined by the size of the ordered regions. These regions grow with this law until they become macroscopically large. Although this a result obtained in the large $N$ limit, similar growth laws had been found for NCOP both analytically and numerically for $N = 1$ etc.[7]
- **Coarsening:** The expression for the structure factor (31) shows that at large times only the very small wavevectors contribute to $S(\vec{k}, t)$, however the self-consistency condition forces the $\int k^{d-1}\, dk\, S(k,t) \to$ constant thus asymptotically the structure factor is peaked at wavevectors $k \approx L^{-1}(t)$ with an amplitude $L^d(t)$ thus becoming a *delta function* $S(\vec{k}, t) \overset{t\to\infty}{\to} \delta^d(\vec{k})$. The position of the peak in $S(\vec{k}, t)$ moving towards longer wavelength is the phenomenon of coarsening and is observed via light scattering. At long times a zero momentum condensate is formed[10] and a Bragg peak develops at zero momentum, this condensate however grows as a power of time and only becomes macroscopic at asymptotically large times. Coarsening is one of the experimental hallmarks of the process of phase ordering, revealed for example in light scattering[12] and is found numerically in many systems[7]. Thus the large $N$ limit, although not being able to describe topological defects offers a very good description of the ordering dynamics.

## 3. Phase ordering in Quantum Field Theory I: Minkowski space-time

### 3.1. A QUENCH IN Q.F.T.

The dynamics is completely determined by the microscopic field theoretical Hamiltonian. For a simple scalar theory the Hamiltonian operator is given by

$$\hat{H} = \int d^3x \left\{ \frac{1}{2}\Pi^2(\vec{x},t) + \frac{1}{2}[\vec{\nabla}\Phi(\vec{x},t)]^2 + V[\Phi(\vec{x},t)] \right\} \tag{33}$$

where $\Phi$ is the quantum mechanical field and $\Pi$ its canonical momentum. We want to describe a quenched scenario where the initial state of the system for $t < 0$ is the ground state (or density matrix, see later) of a Hamiltonian for which the potential is convex for all values of the field, for example that of an harmonic oscillator, in which case the wave function(al) $\Psi[\Phi]$ is a Gaussian centered at the origin. At $t = 0$ the potential is changed so that for $t > 0$ it allows for broken symmetry states. This can be achieved for example by the following form

$$V[\Phi] = \frac{1}{2}m^2(t)\Phi^2 + \frac{\lambda}{4}\Phi^4 \tag{34}$$

$$m^2(t) = \begin{cases} +m_0^2 > 0 & \text{for } t < 0 \\ -m_0^2 < 0 & \text{for } t > 0 \end{cases} \tag{35}$$

Although in Minkowski space-time this is an *ad-hoc* choice of a time dependent potential that mimics the quench[21], we will see in the next section that in a cosmological setting the mass term naturally depends on time through the temperature dependence and that it changes sign below the critical temperature as the Universe cools off. Most of the results obtained in Minkowski space-time will translate onto analogous results in a Friedmann-Robertson-Walker cosmology. Unlike the phenomenological (but succesful) description of the dynamics in condensed matter systems, in a microscopic quantum theory the dynamics is completely determined by the Schrödinger equation for the time evolution of the wave function or alternatively the Liouville equation for the evolution of the density matrix in the case of mixed states. We will cast our study in terms of a density matrix in general, such a density matrix could describe pure or mixed states and obeys the quantum Liouville equation

$$i\frac{\partial \hat{\rho}(t)}{\partial t} = \left[\hat{H}(t), \hat{\rho}(t)\right] \tag{36}$$

**Question:** How does the wave function(al) or the density matrix evolve after a quench?

### 3.2. A SIMPLE QUANTUM MECHANICAL PICTURE

In order to gain insight into the above question, let us consider a simple case of one quantum mechanical degree of freedom $q$ and the quench is described in terms of an harmonic oscillator with a time dependent frequency $\omega^2(t) = -\epsilon(t)\ \omega_0^2\ ;\ \omega_0^2 > 0$ with $\epsilon(t)$ the sign function, so that

$\omega^2(t<0)>0$ ; $\omega^2(t>0)<0$. Furthermore let us focus on the evolution of a pure state (the density matrix is simple the product of the wave function and its complex conjugate). Consider that at $t<0$ the wave function corresponds to the ground state of the (upright) harmonic oscillator. For $t>0$ the wave function obeys

$$i\frac{\partial\Psi[q,t]}{\partial t}=\left[-\frac{1}{2}\frac{d^2}{dq^2}-\frac{1}{2}\omega_0^2\,q^2\right]\Psi[q,t] \tag{37}$$

Since the initial wave function is a gaussian and under time evolution with a quadratic Hamiltonian Gaussians remain Gaussians, the solution of this Schrödinger equation is given by

$$\Psi[q,t] = N(t)\,e^{-\frac{A(t)}{2}q^2} \tag{38}$$

$$\frac{dlnN(t)}{dt} = -\frac{i}{2}A(t) \tag{39}$$

$$i\frac{dA}{dt} = A^2+\omega_0^2$$

Separating the real and imaginary parts of $A(t)$ it is straightforward to find that $|N(t)|^4/\mathrm{Re}[A(t)]$ is constant, a consequence of unitary time evolution. Eq.(40) can be cast in a more familiar form by a simple substitution

$$A(t)=-i\frac{\dot{\phi}(t)}{\phi(t)}\Rightarrow\ddot{\phi}(t)-\omega_0^2\,\phi(t)=0 \tag{40}$$

where the equation for $\phi$ was obtained by inserting the above expression for $A(t)$ in (40). The solution is $\phi(t)=a\,e^{\omega_0\,t}+b\,e^{-\omega_0\,t}$ featuring exponential growth. This is the quantum mechanical analog of the spinodal instabilities described in the previous section. The equal time two-point function is given by

$$\langle q^2\rangle(t)=A_R^{-1}(t)=|\phi(t)|^2\approx e^{2\omega_0\,t} \tag{41}$$

The width of the Gaussian state increases in time (while the amplitude decreases to maintain a constant norm) and the quantum fluctuations grow exponentially. As the Gaussian wave function spreads out the probability for finding configurations with large amplitude of the coordinates increases. These is the quantum mechanical translation of the linear spinodal instabilities. When the non-linear contributions to the quantum mechanical potential are included the single particle quantum mechanical wave function will simply develop two peaks and eventually re-collapse by focusing near the origin undergoing oscillatory motion between 'collapses' and 'revivals'. In the case of a full quantum field theory there are infinitely many degrees of freedom and the energy is transferred between many modes. This simple quantum mechanical example paves the way for understanding in a simple manner the main features of a quench in the large $N$ limit in quantum field theory, to which we now turn our attention.

### 3.3. BACK TO THE ORIGINAL QUESTION: LARGE $N$ IN Q.F.T.

We now consider the large $N$ limit of a full Q.F.T. in which

$$\vec{\Phi}(\vec{x},t) = (\Phi_1(\vec{x},t), \Phi_2(\vec{x},t), \cdots, \Phi_N(\vec{x},t)) \tag{42}$$

and similarly for the canonical momenta $\vec{\Pi}$. The Hamiltonian operator is of the form (33) with

$$V[\vec{\Phi}] = \frac{1}{2}m^2(t)\, \vec{\Phi}\cdot\vec{\Phi} + \frac{\lambda}{8N}[\vec{\Phi}\cdot\vec{\Phi}]^2 \tag{43}$$

with $m^2(t)$ given by (35). Let us focus on the case in which the initial state pure and symmetric, i.e. $\langle\Phi\rangle = 0$, with $< \cdots >$ being the expectation value in this initial state. The more complicated case of a mixed state, described by a density matrix is studied in detail in[16, 17, 18] and the main features are the same as those revealed by the simpler scenario of a pure state. The large $N$ limit is implemented in a similar manner as in the TDGL example, via a Hartree like factorization

$$(\vec{\Phi}\cdot\vec{\Phi})^2 \rightarrow 2\,\langle\vec{\Phi}\cdot\vec{\Phi}\rangle\, \vec{\Phi}\cdot\vec{\Phi} \tag{44}$$

where the expectation value is in the time evolved quantum state (in the Schrödinger picture) or in the initial state of the Heisenberg operators (in the Heisenberg picture). Via this factorization the Hamiltonian becomes quadratic at the expense of a self-consistent condition as it will be seen below. It is convenient to introduce the spatial Fourier transform of the fields as

$$\vec{\Phi}(\vec{x},t) = \frac{1}{\sqrt{\Omega}}\sum_{\vec{k}} \vec{\Phi}_{\vec{k}}(t)\; e^{i\vec{k}\cdot\vec{x}} \tag{45}$$

with $\Omega$ the spatial volume, and a similar expansion for the canonical momentum $\Pi(\vec{x},t)$. The Hamiltonian becomes

$$H = \sum_{\vec{k}}\left\{\frac{1}{2}\vec{\Pi}_{\vec{k}}\cdot\vec{\Pi}_{-\vec{k}} + \frac{1}{2}W^2_{\vec{k}}(t)\;\vec{\Phi}_{\vec{k}}\cdot\vec{\Phi}_{-\vec{k}}\right\} \tag{46}$$

$$W^2_{\vec{k}}(t) = m^2(t) + k^2 + \frac{\lambda}{2N}\int\frac{d^3k}{(2\pi)^3}\,\langle\vec{\Phi}_{\vec{k}}\cdot\vec{\Phi}_{-\vec{k}}\rangle(t) \tag{47}$$

The problem now has decoupled in a set of infinitely many harmonic oscillators, that are only coupled through the self-consistent condition in the frequencies (47). To induce a quench, the time dependent mass term has the form proposed in eq. (35).

Just as in the simple quantum mechanical case, we consider the initial state to be a Gaussian centered at the origin in field space, which is the ground state of the (upright) harmonic oscillators for $t < 0$. Since a Gaussian is always a Gaussian under time evolution with a quadratic Hamiltonian, we propose the wave function(al) that describes the (pure) quantum mechanical state to be given by

$$\Psi[\vec{\Phi}, t] = \Pi_k \left\{ N_k(t)\, e^{-\frac{A_k(t)}{2}\vec{\Phi}_{\vec{k}}\cdot\vec{\Phi}_{-\vec{k}}} \right\} \;\; ; \;\; A_k(t=0) = W_k(t<0) \tag{48}$$

Time evolution of this wavefunction(al) is determined by the Schrödinger equation: in the Schrödinger representation the canonical momentum becomes a differential (functional) operator, $\vec{\Pi}_{\vec{k}} \rightarrow -i\delta/\delta\vec{\Phi}_{-\vec{k}}$ and the Schrödinger equation becomes a functional differential equation. Comparing the powers of $\Phi_{\vec{k}}$ in this differential equation, one obtains the following evolution equations for $N_k(t)$ and $A_k(t)$

$$\frac{d}{dt}\ln N_k(t) = -\frac{i}{2}A_k(t) \tag{49}$$

$$i\frac{dA_k(t)}{dt} = A_k^2(t) - W_k^2(t) \tag{50}$$

As in the single particle case, the constancy of $|N_k(t)|^4/\mathrm{Re}[A_k(t)]$ is a consequence of unitary time evolution. The non-linear equation for the kernel $A_k(t)$ can be simplified just as in the single particle case by writing

$$A_k(t) = -i\frac{\dot{\phi}_k(t)}{\phi_k(t)} \Rightarrow \ddot{\phi}_k(t) + W_k^2(t)\,\phi_k(t) = 0 \tag{51}$$

and taking the expectation value of $\Phi^2$ in this state we obtain

$$\langle\vec{\Phi}_{\vec{k}}\cdot\vec{\Phi}_{-\vec{k}}\rangle(t) = N\,|\phi_k(t)|^2 \tag{52}$$

Hence we find a self-consistent condition much like the one obtained in the large $N$ limit for TDGL. The equations for the mode functions and the self-consistent condition for $t > 0$ are therefore given by

$$\ddot{\phi}_k(t) + [k^2 + M^2(t)]\,\phi_k(t) = 0 \tag{53}$$

$$M^2(t) = -m_0^2 + \frac{\lambda}{2}\int\frac{d^3k}{(2\pi)^3}|\phi_k(t)|^2 \tag{54}$$

where the integral in the self-consistent term in (54) is simply $\langle\Phi_i^2\rangle$. There are two fundamental *differences* between the quantum dynamics determined by the equations of motion and the classical dissipative dynamics of the TDGL phenomenological description given in sec. II:

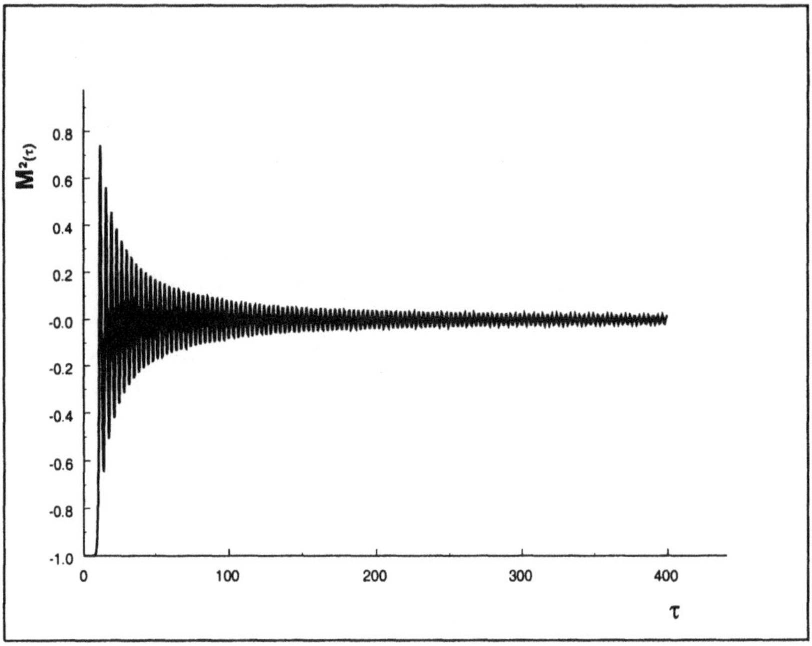


*Figure 1.* $\mathcal{M}^2(\tau)$ vs. $\tau$, $g = 10^{-7}$

- The equations of motion and the self-consistency condition equations (53)-(54) lead immediately to the conservation of energy[15, 16].
- The evolution equations are *time reversal invariant.*

These properties must be contrasted to the purely dissipative evolution dictated by the TDGL equations as is clear from eq. (9). Consider a very weakly coupled theory $\lambda << 1$ and very early times, then the self-consistent term can be neglected and we see that for $k^2 < m_0^2$ the modes grow exponentially. This instability again is the manifestation of spinodal growth[22, 23, 24, 16, 17]. Since the mode functions grow exponentially, fairly soon, at a time scale $t_s \approx m_0^{-1} \ln(1/\lambda)$ the self-consistent term begins to cancel the negative mass squared and $M^2(t)$ becomes smaller. We find numerically that this effective mass vanishes asymptotically, as shown in Fig. 1.

### 3.4. EMERGENCE OF CONDENSATES AND CLASSICALITY

The physical mechanism here is similar to that in the classical TDGL, but in terms of quantum fluctuations. The quantum fluctuations with wave vectors inside the spinodally unstable band grow exponentially, these make the $\langle \Phi^2 \rangle$

self-consistent field to grow non-perturbatively large until when $\langle \Phi^2 \rangle \approx m_0^2/\lambda$ when the self-consistent (mean) field begins to be of the same order as $m_0^2$ (the tree level mass term). At this point the *quantum* fluctuations become non-perturbatively large and sample field configurations near the equilibrium minima of the potential. The spinodal instabilities are shutting off since the effective squared mass $M^2(t)$ is vanishing.

When $M^2(t)$ vanishes, the equations for the mode functions become those of a free massless field, with solutions of the form $\phi_k(t) = A_k\, e^{ikt} + B_k\, e^{-ikt}$, whereas for the $k = 0$ mode the solution must be of the form $\phi_0(t) = a + bt$ with $a; b \neq 0$ since the Wronskian of the mode function and its complex conjugate is a constant. This in turn determines that the low $k$ (long wavelength) behavior of the mode functions is given by

$$\phi_k(t) = a \cos kt + b\, \frac{\sin kt}{k} \tag{55}$$

This behavior at long wavelength has a remarkable consequence: at very long time the power spectrum $|\phi_k(t)|^2$, which is the equivalent of $S(k,t)$ for TDGL (see eq. (17)) is dominated by the small $k$-region, in particular $k << 1/t$, with an amplitude that grows quadratically with time. Then the structure factor $S(\vec{k},t) = |\phi_k(t)|^2$ features a peak that moves towards longer wavelengths at longer times and whose amplitude grows with time in such a way that asymptotically $\int_0^\infty k^2 S(\vec{k},t) dk/2\pi^2 \rightarrow m_0^2/\lambda$ and the integral is dominated by a very small region in $k$ that gets narrower at longer times. This is the equivalent of *coarsening* in the TDGL solution in the large $N$ limit, where the asymptotic time regime was dominated by the formation of a long-wavelength condensate. Fig. 2 shows the power spectrum at two (large) times displaying clearly the phenomenon of coarsening and the formation of a non-perturbative condensate.

The pair correlation function can now be calculated using this power spectrum[17]

$$C(\vec{r},t) = \frac{1}{2\,\pi^2 r} \int_0^\infty k \sin kr \, |\phi_k^2(t)| \, dk \; . \tag{56}$$

At long times and distances the integral is dominated by the very long wavelength modes, in particular by the term $\propto \sin[kt]/k$ of $\phi_k(t)$, hence the integral can be done analytically and we find

$$C(\vec{r},t) = \frac{A}{r}\, \Theta(2t - r) \tag{57}$$

with $A$ a constant. This is a remarkable result: the correlation falls off as $1/r$ inside domains that grow at the speed of light. This correlation function is shown in Fig. 3 at several different (large) times. This correlation function

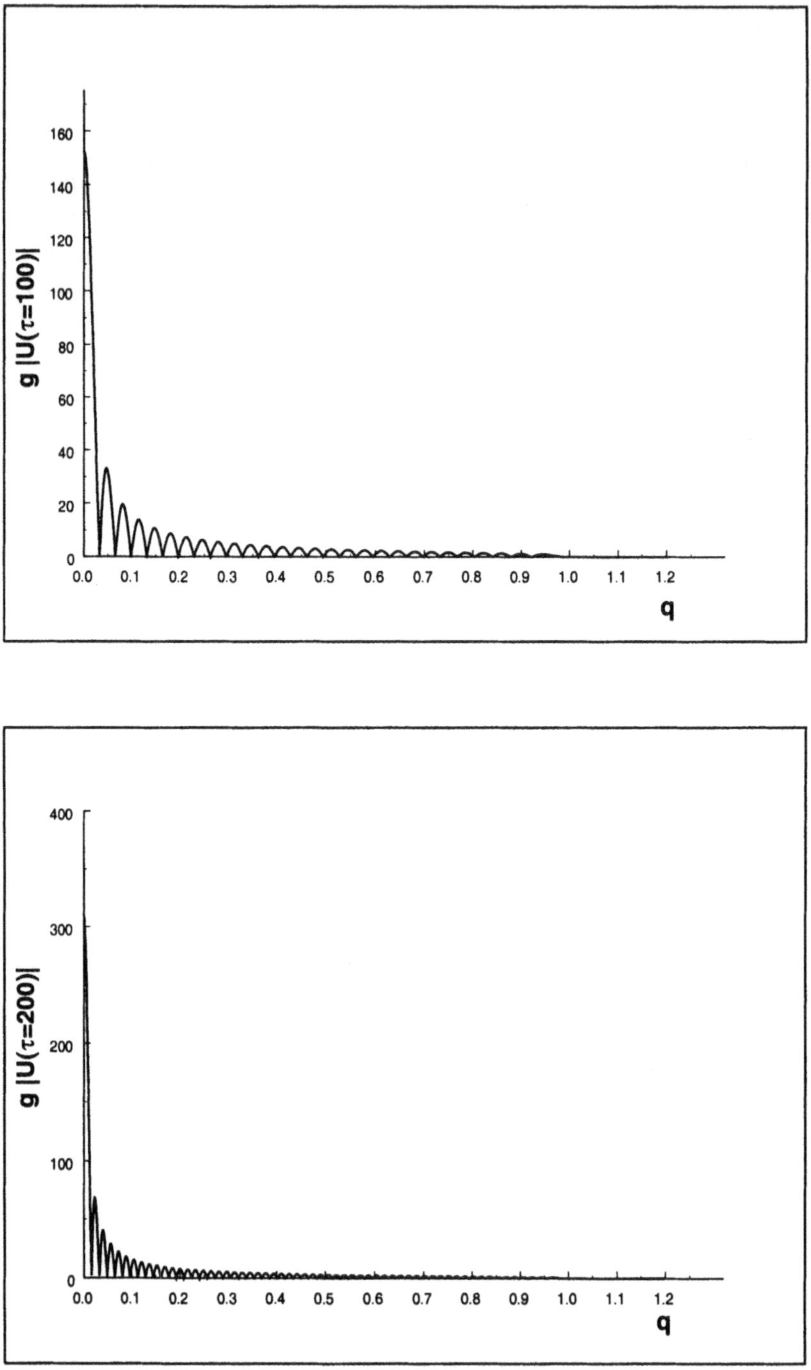


*Figure 2.* $g|\phi_k(\tau = 100, 200)|^2$ vs. $q = k/|m_0|$, $g = 10^{-7}$

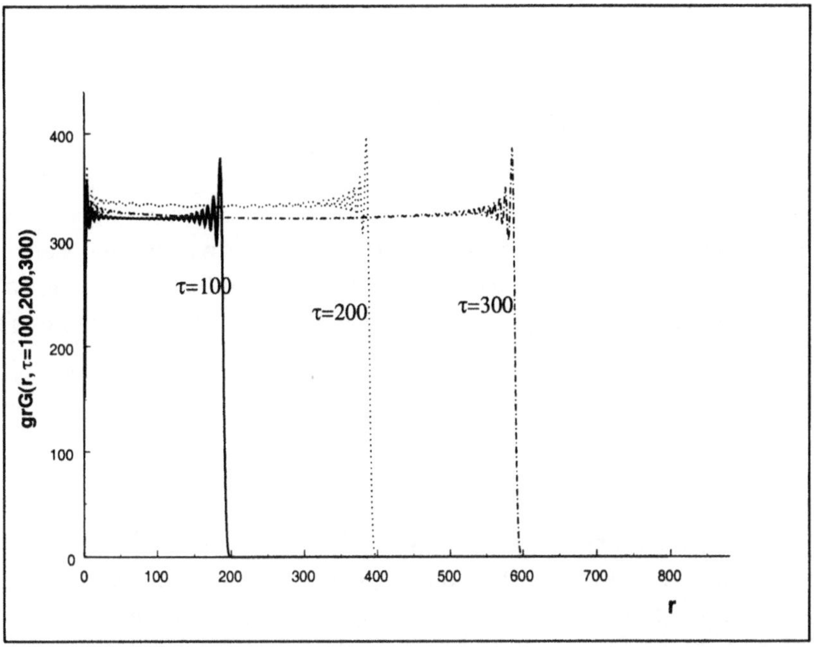


*Figure 3.* : $rC(\vec{r},t)$ vs. $r$ for $t = 100, 200, 300 m_0^{-1}$, $g = 10^{-7}$

is of the *scaling form*: introducing the dynamical length scale $L(t) = t$ it is clear that[17]

$$C(\vec{r},t) \propto L^{-1}(t) f(r/L(t)) \;\; ; \;\; f(s) = \frac{\Theta(2-s)}{s} \tag{58}$$

We interpret these 'domains' as being a non-perturbative condensate of Goldstone bosons, with a non-perturbatively large number of them $\propto 1/\lambda$, such that the mean square root fluctuation of the field samples the (non-perturbative) equilibrium minima of the potential. In particular an important conclusion of this analysis is that the long-wavelength modes acquire very large amplitudes, their phases vary slowly as a function of time (for $k << 1/t$), therefore these fluctuations which began their evolution as being quantum mechanical, now have become *classical.*

### 3.5. O.K...O.K. BUT WHERE ARE THE DEFECTS?

At this point our analysis begs this question. To understand the answer it is convenient to back track the analysis to the beginning. The initial quantum state is given by a the wave-function(al) (48), thus the most probable field configurations found in this ensemble are those whose spatial Fourier transform are given by

$$|\Phi_k| \propto \frac{1}{\sqrt{W_k(t<0)}} \propto \frac{1}{\sqrt{k^2+m_0^2}} \tag{59}$$

(restoring would multiply $\Phi_k$ by $\sqrt{\ }$). Then typical long-wavelength field configurations that are represented in the quantum ensemble described by this initial wave-function(al) are of rather small amplitude. The initial correlations are also rather short ranged on scales $m_0^{-1}$. Under time evolution the probability distribution is given by

$$\mathcal{P}[\Phi,t] = |\Psi[\Phi,t]|^2 = \Pi_{i=1}^{N}\Pi_k \left\{ |N_k(t)|^2 e^{-\frac{|\Phi^i_k(t)|^2}{|\phi_k(t)|^2}} \right\} \tag{60}$$

At times longer than the regime dominated by the exponential growth of the spinodally unstable modes, the power spectrum $|\phi_k^2(t)|^2$ obtains the largest support for long wavelengths $k << m_0^2$ and with amplitudes $\approx m_0^2/\lambda$. Therefore field configurations with typical spatial Fourier transform $\phi_k(t)$ are very likely to be found in the ensemble. These field configurations are primarily made of long-wavelength modes and their amplitudes are non-perturbatively large, of the order of the amplitude of the fields in the broken symmetry minima. A typical such configuration can be written as

$$\Phi^i(\vec{x},t)_{typical} \approx \sum_k |\phi_k(t)| \cos[\vec{k}\cdot\vec{x} + \delta^i_{\vec{k}}] \tag{61}$$

where the phases $\delta^i_{\vec{k}}$ are randomly distributed with a Gaussian probability distribution since the density matrix is gaussian in this approximation. We note that a particular choice of these phases leads to a realization of a likely configuration in the ensemble that *breaks translational invariance.* In fact translations can be absorbed by a change in the phases, thus averaging over these random phases restores translational invariance. Since the quantum state (or density matrix) is translational invariant a particular spatial profile for a field configuration corresponds to a particular representative of the ensemble. Combining all of the above results together we can present the following consistent interpretation of the ordering process and the formation of coherent non-perturbative structures during the dynamics of symmetry breaking in the large $N$ limit[17] :

- The early time evolution occurs via the exponential growth of spinodally unstable long wavelength modes. This unstable growth leads to a rapid growth of fluctuations $\langle\Phi^2\rangle(t)$ which in turn increases the self-consistent contribution and tends to cancel the negative mass squared. The effective mass of the excitations $-m_0^2 + \frac{\lambda}{2N}\langle\Phi^2\rangle(t) \to 0$ and the asymptotic excitations are Goldstone bosons.

- At times larger than the spinodal time $t_s \approx m_0^{-1}\ln(1/\lambda)$, the effective mass vanishes and the power spectrum or structure factor $S(k,t) = |\phi_k(t)|^2$ displays the features of coarsening: a peak that moves towards longer wavelengths and increases in amplitude, resulting in a long-wavelength condensate at asymptotically long times.
- For large time a dynamical correlation length emerges $L(t) = t$ and at long distances the pair correlation function is of the scaling form $C(\vec{r},t) \propto L^{-1}(t)f(r/L(t))$. The length scale $L(t)$ determines the size of the correlated regions and determines that these regions grow at the speed of light. Inside these regions there is a non-perturbative condensate of Goldstone bosons with a typical amplitude of the order of the value of the homogeneous field at the equilibrium broken symmetry minima.

The similarity between these results and those of the more phenomenological TDGL description in condensed matter systems is rather striking. The features that are determined by the structure of the quantum field theory are[17]: i) the scaling variable $s = r/t$ with equal powers of distance and time is a consequence of the Lorentz invariance of the underlying theory, ii) the fact that the pair correlation function vanishes for $r > 2t$ is manifestly a consequence of causality. An analysis of the correlations and defect density during the spinodal time scale has been performed in[25] and related recent studies had been performed in[26].

## 4. Phase ordering in Quantum Field Theory II: FRW Cosmology

### 4.1. COSMOLOGY 101 (THE BASICS):

On large scales > 100 Mpc the Universe appears to be homogeneous and isotropic as revealed by the isotropy and homogeneity of the cosmic microwave background and some of the recent large scale surveys[5]. The cosmological principle leads to a simple form of the metric of space time, the Friedmann-Robertson-Walker (FRW) metric in terms of a scale factor that determines the Hubble flow and the curvature of spatial sections. Observations seem to favor a flat Universe for which the space time metric is rather simple:

$$ds^2 = dt^2 - a^2(t)\, d\vec{x}^2 \tag{62}$$

the time and spatial variables $t, \vec{x}$ in the above metric are called comoving time and spatial distance respectively and have the interpretation of being the time and distance measured by an observer locally at rest with respect to the Hubble flow. At this point we must note that *physical distances* are given by $\vec{l}_{phys}(t) = a(t)\,\vec{x}$. An important concept is that of causal (particle) horizons: events that cannot be connected by a light signal are causally disconnected. Since light travels on null geodesics $ds^2 = 0$ the maximum *physical* distance that can be reached by a light signal at time $t$ is given by

$$d_H(t) = a(t) \int_0^t \frac{dt'}{a(t')} \tag{63}$$

It will prove convenient to change coordinates to *conformal time* by defining a conformal time variable

$$\eta = \int_0^t \frac{dt'}{a(t')} \Rightarrow ds^2 = C^2(\eta)\,(d\eta^2 - d\vec{x}^2) \;\; ; \;\; C(\eta) = a(t(\eta)) \tag{64}$$

in terms of which the causal horizon is simply given by $d_H(\eta) = C(\eta)\,\eta$ and physical distances as $\vec{x}_{phys} = C(\eta)\,\vec{x}$. This metric is of the same form as that of Minkowski space time. For energies well below the Planck scale $M_{Pl} \approx 10^{19}$Gev gravitation is well described by *classical* General Relativity and the Einstein equations:

$$R^{\mu\nu} - \frac{1}{2}\,g^{\mu\nu}R = \frac{8\pi}{M_{Pl}^2}\,T^{\mu\nu} \tag{65}$$

where we have been cavalier and set $c = 1$ (as well as $\quad = 1$). $R^{\mu\nu}$ is the Ricci tensor, $R$ the Ricci scalar and $T^{\mu\nu}$ the matter field energy momentum tensor. The above equation is classical but one seeks to understand the dynamics of the Early Universe in terms of a *quantum field theory* that describes particle physics, thus the question: what is exactly the energy momentum tensor?, in Einstein's equations it is a classical object, but in QFT it is an operator. The answer to this question is: gravity is classical, fields are quantum mechanical, but $T^{\mu\nu} \rightarrow \langle T^{\mu\nu} \rangle$, i.e. it is the expectation value of a *quantum mechanical operator in a quantum mechanical state.* This quantum mechanical state, either pure or mixed is described by a wavefunction(al) or a density matrix whose time evolution is dictated by the quantum equations of motion: the Schrödinger equation for the wave functions or the quantum Liouville equation for a density matrix. Consistency with the postulate of homogeneity and isotropy requires that the expectation value of the energy momentum tensor must have the fluid form and in the rest frame of the fluid takes the form $\langle T^{\mu\nu} \rangle = \text{diagonal}(\rho, p, p, p)$ with $\rho$ the energy density and $p$ the pressure. The time and spatial components of Einstein's equations lead to the Friedman equation

$$\frac{\dot{a}^2(t)}{a^2(t)} = \frac{8\pi}{3M_{Pl}^2}\rho(t) \tag{66}$$

$$2\frac{\ddot{a}(t)}{a(t)} + \frac{\dot{a}^2(t)}{a^2(t)} = -\frac{8\pi}{M_{Pl}^2}p(t) \tag{67}$$

Combining these two equations one arrives at a simple and intuitive equation which is reminiscent of the first law of thermodynamics:

$$\frac{d}{dt}(\rho a^3(t)) = -p\frac{da^3(t)}{dt} \Rightarrow \dot{\rho} + 3\frac{\dot{a}}{a}(\rho + p) = 0 \tag{68}$$

The alternative form shown on the right hand side of (68) is the *covariant conservation of energy*. Since the physical volume of space is $V_0\, a^3(t)$ (with $V_0$ the comoving volume) the above equation is recognized as $dU = -p\, dV$ which is the first law of thermodynamics for *adiabatic* processes. To close the set of equations and obtain the dynamics we need an equation of state $p = p(\rho)$: two very relevant cases are: i) radiation dominated (RD) with $p = \rho/3$ and matter dominated (MD) $p = 0$ (dust) Universes. In our study we will focus on the RD case. The equation of state for RD is that for black-body radiation for which the entropy is $S = CVT^3$ (with C a constant). Since $V(t) = V_0\, a^3(t)$ is the physical volume, the equation (68) which dictates adiabatic (isoentropic) expansion leads to a time dependence of the temperature: $T(t) = T_0/a(t)$. Now the cooling is done by the expansion of the Universe and a phase transition will occur when the Universe cools below the critical temperature for a given theory. For the GUT transition $T_c \approx 10^{16}\,\text{Gev} \approx 10^{29}K$, for the EW transition $T_c \approx 100\,\text{Gev} \approx 10^{15}K$. Returning now back to the large $N$ study of the dynamics of phase transitions, we can include the effect of cooling by the expansion of the Universe by replacing the time dependent mass term $m^2(t)$ in (43) by

$$m^2(t) = m_0^2\left[\frac{T^2(t)}{T_c^2} - 1\right] \quad ; \quad T(t) = \frac{T_i}{a(t)} \tag{69}$$

This form is consistent with the Landau-Ginzburg description including the time dependence of the temperature via the isentropic expansion of the Universe, but perhaps more importantly it can be proven in a detailed manner from the self-consistent renormalization of the mass in an expanding Universe[15]. Thus the large $N$ limit in a RD FRW cosmology will be studied by using the potential (43) but with the time dependent mass given by (69).

### 4.2. LARGE $N$ IN A RD FRW COSMOLOGY

The large $N$ limit is again implemented via the Hartree-like factorization (44) performing the spatial Fourier transforms of the fields and their canonical momenta and including the proper scale factors, the Hamiltonian now becomes[15]

$$H(t) = \sum_k \left\{ \frac{1}{2a^3(t)} \vec{\Pi}_{\vec{k}} \cdot \vec{\Pi}_{-\vec{k}} + W_k^2(t)\, \vec{\Phi}_{\vec{k}} \cdot \vec{\Phi}_{-\vec{k}} \right\} \tag{70}$$

$$W_k^2(t) = \frac{k^2}{a^2} + m^2(t) + \frac{\lambda}{2N} \langle \vec{\Phi}_{\vec{k}} \cdot \vec{\Phi}_{-\vec{k}} \rangle \tag{71}$$

where now the expectation value is in terms of a *density matrix* $\rho[\Phi(\vec{\cdot}), \tilde{\Phi}(\vec{\cdot}); t]$ since we are considering the case of a thermal ensemble as the initial state.

We propose the following Gaussian ansatz for the functional density matrix elements in the Schrödinger representation[15]

$$\rho[\Phi, \tilde{\Phi}, t] = \prod_{\vec{k}} \mathcal{N}_k(t) \exp\left\{ -\frac{A_k(t)}{2}\, \vec{\Phi}_{\vec{k}} \cdot \vec{\Phi}_{-\vec{k}} + \frac{A_k^*(t)}{2}\, \tilde{\vec{\Phi}}_{\vec{k}} \cdot \tilde{\vec{\Phi}}_{-\vec{k}} + B_k(t)\, \vec{\Phi}_{\vec{k}} \cdot \tilde{\vec{\Phi}}_{-\vec{k}} \right\} \tag{72}$$

This form of the density matrix is dictated by the hermiticity condition $\rho^\dagger[\Phi, \tilde{\Phi}, t] = \rho^*[\tilde{\Phi}, \Phi, t]$; as a result of this, $B_k(t)$ is real. The kernel $B_k(t)$ determines the amount of mixing in the density matrix, since if $B_k = 0$, the density matrix corresponds to a pure state because it is a wave functional times its complex conjugate. The kernels $A_k(0)$ ; $B_k(0)$ are chosen such that the initial density matrix is thermal with a temperature $T_i > T_c$[15]. Following the same steps as in Minkowski space time, the time evolution of this density matrix can be found in terms of a set of mode functions $\phi_k(t)$ that obey the following equations of motion and self-consistency condition

$$\ddot{\phi}_k(t) + 3\,\frac{\dot{a}}{a}\,\dot{\phi}_k(t) + \left[\frac{k^2}{a^2(t)} + M^2(t)\right]\phi_k(t) = 0 \tag{73}$$

$$M^2(t) = m_0^2 \left[\frac{T_i^2}{T_c^2 a^2(t)} - 1\right] + \frac{\lambda}{2} \int \frac{d^3k}{(2\pi)^3}\, |\phi_k(t)|^2 \coth \frac{W_k(0)}{2T_i}\,. \tag{74}$$

This equations can be cast in a more familiar form by changing coordinates to conformal time (see eq. (64)) and (conformally) rescaling the mode functions $\phi_k(t) = f_k(\eta)/C(\eta)$ to obtain the following equations for the conformal time mode functions $f_k(\eta)$ in an RD FRW cosmology

$$f_k''(\eta) + \left[k^2 + C^2(\eta) M^2(\eta)\right] f_k(\eta) = 0 \tag{75}$$

where primes now refer to derivatives with respect to conformal time. For RD FRW $C(\eta) = 1 + \eta/2$ (in units of $m_0^{-1}$ which is the only dimensionful variable). The above equations of motion has now an analogous form as those solved in the case of Minkowski space-time.

As the temperature falls below the critical the effective squared mass term becomes negative and spinodal instabilities trigger the process of

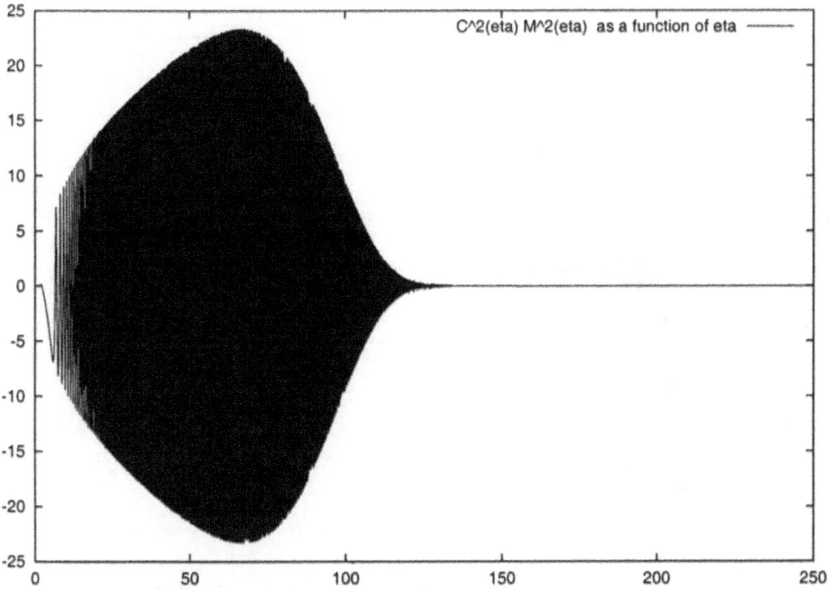


*Figure 4.* $C^2(\eta)M^2(\eta)$ vs. $\eta$(conformal time in units of $m_0^{-1}$) for $\frac{T_i}{T_c} = 3$, $g = 10^{-5}$. R.D. Universe

phase ordering. This results in that the quantum fluctuations quantified by $\langle\vec{\Phi}^2\rangle$ grow exponentially. These spinodal instabilities make the self-consistent field grow at early times and tends to overcome the negative sign of the squared mass, eventually reaching an asymptotic regime in which the total effective mass $M^2(\eta)$ vanishes.

Again this behavior determines that the fluctuations are sampling the equilibrium broken symmetry minima of the initial potential, i.e. $\langle\vec{\Phi}^2\rangle \to \frac{2Nm_0^2}{\lambda}$.

Although, just as in Minkowski space-time the effective mass vanishes asymptotically, the non-equilibrium evolution is rather *different*. We find numerically[19] that asymptotically the effective mass term $C^2(\eta)M^2(\eta)$ vanishes as $-15/4\eta^2$.

Fig. 4 displays $C^2(\eta)M^2(\eta)$ as a function of conformal time for the case of $T_i/T_c = 1.1$ with $T_c \propto m_0/\sqrt{\lambda}$[15, 24].

We see that at very early time the mass is positive, reflecting the fact that the initial state is in equilibrium at an initial temperature larger than the critical. As time evolves the temperature is red-shifted and cools and at some point the phase transition occurs, when the mass vanishes and becomes negative.

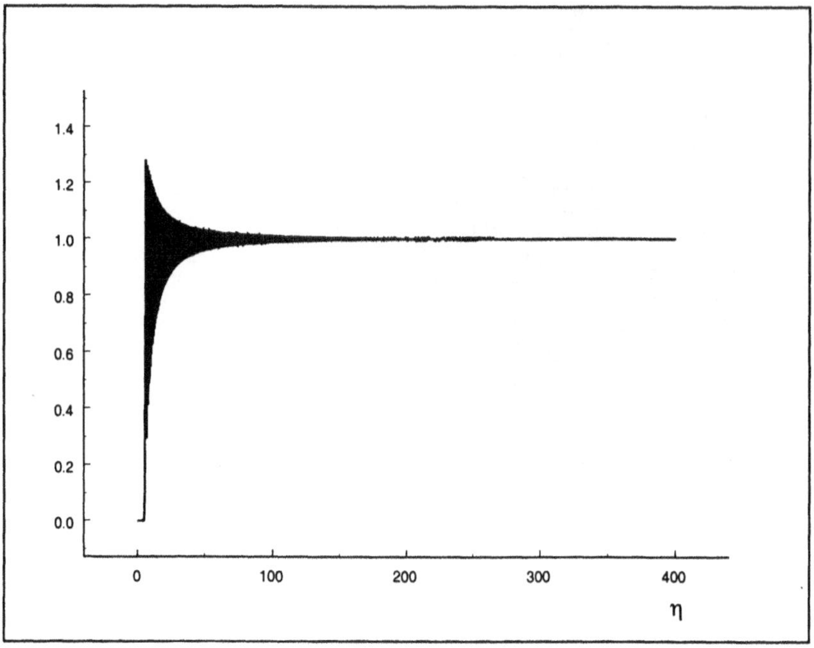


*Figure 5.* $\frac{\lambda}{2Nm_0^2}\langle\vec{\Phi}^2\rangle(\eta)$ vs. $\eta$ (conformal time in units of $m_0^{-1}$) for $\frac{T_i}{T_c}=3$, $g=10^{-5}$. R.D. Universe

Figure 5 displays $\frac{\lambda}{2Nm_0^2}\langle\vec{\Phi}^2\rangle(\eta)$ vs. $\eta$ in units of $m_0^{-1}$ for $\frac{T_i}{T_c}=3$, $g=10^{-5}$ for an R.D. Universe. Clearly at large times the non-equilibrium fluctuations probe the broken symmetry states.

This particular asymptotic behavior of the mass determines that the mode functions $f_k(\eta)$ grow as $\eta^{5/2}$ for $k<1/\eta$ and oscillate in the form $e^{\pm ik\eta}$ for $k>1/\eta$. This behavior is confirmed numerically[19]. We find both analytically and numerically that asymptotically the mode functions are of the *scaling form*

$$f_k(\eta) = A\eta^{\frac{5}{2}}\,\frac{J_2(k\eta)}{(k\eta)^2} \tag{76}$$

Where $A$ is a numerical constant and $J_2(x)$ is a Bessel function.

Figure 6 displays $\eta^{-5}|f_k(\eta)|^2$ as a function of the scaling variable $k\eta$ revealing the scaling behavior.

It is remarkable that this is exactly the same scaling solution found in the *classical* non-linear sigma model in the large $N$ limit and that describes

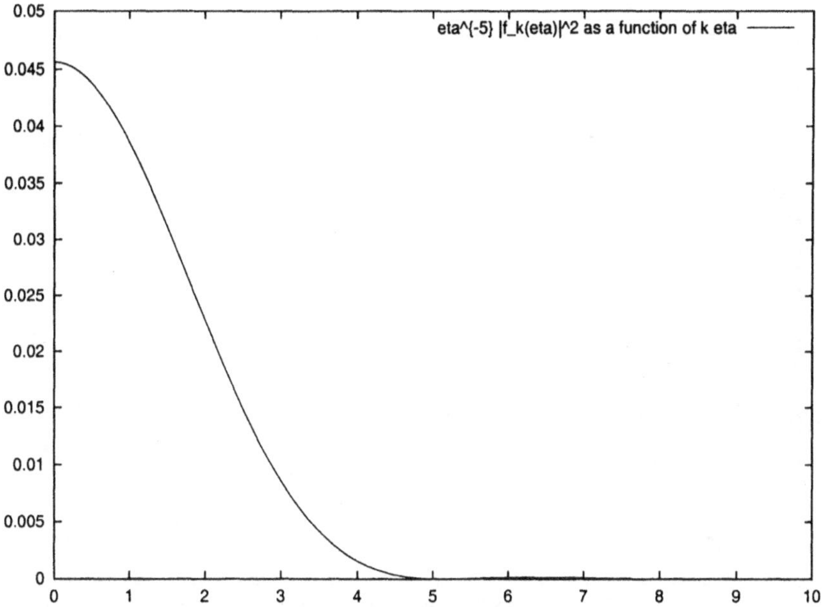


*Figure 6.* $\eta^{-5}|f_k(\eta)|^2$ vs. $k\eta$ for the same case as in Fig. 5

the collapse of textures[13], and also within the context of TDGL equations in the large $N$ limit applied to cosmology[14].

The growth of the long-wavelength modes and the oscillatory behavior of the short wavelength modes again results in that the peak of the structure factor $S(k,\eta) = |f_k(\eta)|^2$ moves towards longer wavelengths and the maximum amplitude increases. This is the equivalent of coarsening and the onset of a condensate.

Although quantitatively different from Minkowsky space time, the qualitative features are similar. Asymptotically the non-equilibrium dynamics results in the formation of a non-perturbative condensate of long-wavelength Goldstone bosons. We can now compute the pair correlation function $C(r,\eta)$ from the mode functions (76) and find that it is cutoff by causality at $r = 2\eta$. The correlation function is depicted in Fig. 4 for two different (conformal) times.

The scaling form of the pair correlation function is

$$C(r,\eta) = \eta^2 \, \chi(r/\eta)$$

where $\chi(x)$ is a hump-shaped function as shown in Fig. 7.

Clearly a *dynamical* length scale $L(\eta) = \eta$ emerges as a consequence of causality, much in the same manner as in Minkowsky space time. The *physical* dynamical correlation length is therefore given by $\xi_{phys}(\eta) = C(\eta)L(\eta) =$

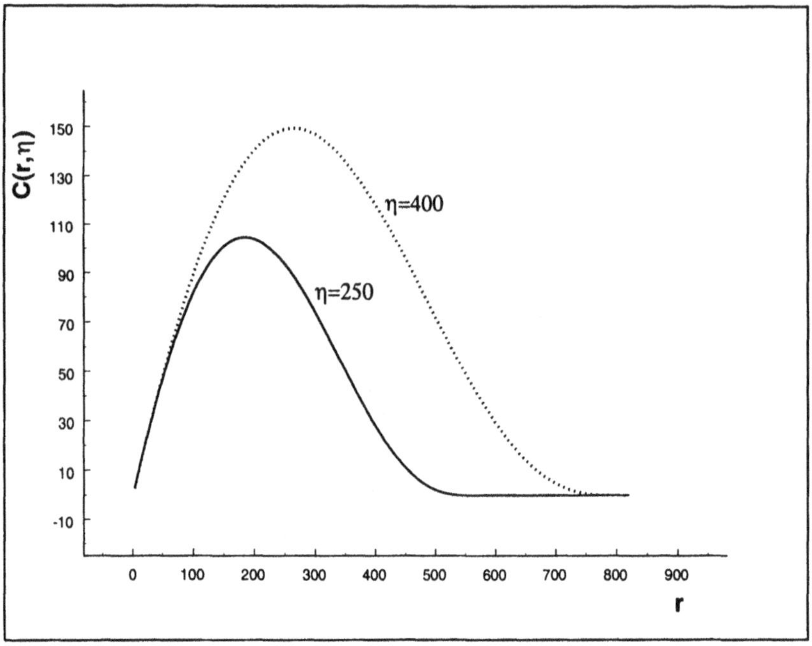


*Figure 7.* $C(r,\eta)$ vs. $r$ for $\eta = 250, 400$ (in units of $m_0^{-1}$) for $\frac{T_i}{T_c} = 3$, $g = 10^{-5}$. R.D. FRW Universe

$d_H(t)$, that is the correlated domains grow again at the speed of light and their size is given by the causal horizon. The interpretation of this phenomenon is that within one causal horizon there is one correlated domain, inside which the mean square root fluctuation of the field is approximately the value of the equilibrium minima of the tree level potential, this is clearly consistent with Kibble's original observation[1, 2]. Inside this domain there is a non-perturbative condensate of Goldstone bosons[19].

## 5. Conclusions and looking ahead

In these lectures we have discussed the multidisciplinary nature of the problem of phase ordering kinetics and non-equilibrium aspects of symmetry breaking. Main ideas from condensed matter were discussed and presented in a simple but hopefully illuminating framework and applied to the rather different realm of phase transitions in quantum field theory as needed to understand cosmology and particle physics. The large $N$ approximation has provided a bridge that allows to cross from one field to another and borrow many of the ideas that had been tested both theoretically and experimentally in condensed matter physics. There are, however, major differences between the condensed matter and particle physics-cosmology applications that require a very careful treatment of the quantum field theory that cannot be replaced by simple arguments. The large $N$ approximation in field theory provides a robust, consistent non-perturbative framework that allows the study of phase ordering kinetics and dynamics of symmetry breaking in a controlled and consistently implementable framework, it is renormalizable, respects all symmetries and can be improved in a well defined manner. This scheme extracts cleanly the non-perturbative behavior, the quantum to classical transition and allows to quantify in a well defined manner the emergence of classical stochastic behavior arising from non-perturbative physics. The emergence of scaling and a dynamical correlation length are robust features of the dynamics and the Kibble-Zurek scenario describes fairly well the general features of the dynamics, albeit the details require careful study, both analytically and numerically.

Of course this is just the beginning, we expect a wealth of important phenomena to be revealed beyond the large $N$, such as the approach to equilibrium, the emergence of other time scales associated with a hydrodynamic description of the evolution at late times and a more careful understanding of the reheating process and its influence on cosmological observables. Although within very few years the wealth of observational data will provide a more clear picture of the cosmological fluctuations, it is clear that the program that pursues a fundamental understanding of the underlying physical mechanisms will continue seeking to provide a consistent microscopic description of the dynamics of cosmological phase transitions.

## 6. Acknowledgements

D. B. thanks the organizors of the school, Henri ('Quique') Godfrin and Yuri Bunkov for a very stimulating school and for their warm hospitality and Tom Kibble and Ana Achucarro for their kind invitation and patience. D. B. thanks the N.S.F for partial support through grant awards: PHY-9605186 and INT-9815064 and LPTHE for warm hospitality, H. J. de Vega thanks the Dept. of Physics at the Univ. of Pittsburgh for hospitality. R. H., is supported by DOE grant DE-FG02-91-ER40682. We thank NATO for partial support.

## References

1. Kibble, T. W. B., *Topology of Cosmic Domains and Strings*, J. Phys. **A 9**, 1387 (1976), and contribution to these proceedings.
2. Hindmarsh, M. B. and Kibble, T.W.B., *Cosmic Strings*, Rep. Prog. Phys. **58**:477

(1995).
3. For a classification of topological defects in terms of the underlying group structure see T. Kibble's contribution to these proceedings.
4. Vilenkin, A. and Shellard, E.P.S, *Cosmic Strings and other Topological Defects*, Cambridge Monographs on Math. Phys. (Cambridge Univ. Press, 1994).
5. For a comprehensive review of the status of theory and experiment see: Proceedings of the 'D. Chalonge' School in Astrofundamental Physics at Erice, edited by N. Sánchez and A. Zichichi, 1996 World Scientific publisher and 1997, Kluwer Academic publishers. In particular the contributions by G. Smoot, A. N. Lasenby and A. Szalay., Durrer, R. , Kunz, M. and Melchiorri, A. *Cosmic Microwave Background Anisotropies from Scaling Seeds: Global Defect Models* Phys.Rev. **D59** (1999) 123005, Kunz M. and Durrer R. *Microwave Background Anisotropies Induced by Global Scalar Fields: The Large N Limit* , Phys. Rev. **D55** , 4516 (1997) and R. Durrer's lectures in these proceedings.
6. See for example: Rajagopal, K. in 'Quark Gluon Plasma 2', (Ed. R. C. Hwa, World Scientific, 1995).
7. Bray, A. J., *Theory of Phase Ordering Kinetics* Adv. Phys. **43**, 357 (1994).
8. Langer, J. S. in 'Solids far from Equilibrium', Ed. C. Godrèche, (Cambridge Univ. Press 1992); Langer, J. S. in 'Far from Equilibrium Phase Transitions', Ed. L. Garrido, (Springer-Verlag, 1988); Langer, J. S. in 'Fluctuations, Instabilities and Phase Transitions', Ed. T. Riste, Nato Advanced Study Institute, Geilo Norway, 1975 (Plenum, 1975).
9. Mazenko, G. in 'Far from Equilibrium Phase Transitions', Ed. L. Garrido, (Springer-Verlag, 1988).
10. Castellano, C. and Zannetti, M., *Overall time evolution in phase-ordering kinetics* cond-mat/9807242; Castellano, C., Corberi, F. and Zannetti, M. *Condensation vs. phase-ordering in the dynamics of first order transitions* Phys. Rev. **E56**, 4973 (1997); F. Corberi, A. Coniglio and M. Zannetti, *Early stage scaling in phase ordering kinetics* Phys. Rev. **E51**, 5469 (1995).
11. Zurek, W. H., Nature **317**, 505 (1985); Acta Physica Polonica **B24**, 1301 1993); *Cosmological Experiments in Condensed Matter Systems* , Phys. Rep. 276, (1996), see also W. H. Zurek's contribution to these proceedings.
12. Goldburg, W. I. and Huang, J. S., in 'Fluctuations, Instabilities and Phase Transitions', Ed. Riste, T., Nato Advanced Study Institute, Geilo Norway, 1975 (Plenum, 1975); Huang, J. S., Goldburg, W. I. and Moldover, M. R., Phys. Rev. Lett. **34**, 639 (1975).
13. Turok, N. and Spergel, D. N., *Scaling Solution for Cosmological $\sigma$ models at Large N*, Phys. Rev. Lett. **66**, 3093 (1991); Spergel, D.N , Turok, N., Press, W. H. and Ryden, B. S., *Global texture as the origin of large-scale structure: Numerical simulations of evolution* Phys. Rev. **D43**, 1038 (1991).
14. Filipe J. A. N. and Bray A. J., *Phase ordering dynamics of cosmological models*, Phys. Rev. **E50**, 2523 (1994); J. A. N. Filipe, (Ph. D. Thesis, 1994, unpublished).
15. D. Boyanovsky, H. J. de Vega and R. Holman, *Nonequilibrium evolution of scalar fields in FRW cosmologies* , Phys. Rev. **D 49**, 2769 (1994); D. Boyanovsky, D. Cormier, H. J. de Vega, R. Holman and S. Prem Kumar, *Nonperturbative quantum dynamics of a new inflation model* Phys. Rev. **D57**, 2166, (1998), (and references therein).
16. Boyanovsky, D., de Vega, H. J., Holman, R., Lee, D.-S., and Singh, A. *Dissipation via particle production in scalar field theories*, Phys. Rev. **D51**, 4419 (1995). Boyanovsky, D. , de Vega H. J. and Holman, R., in the Proceedings of the Second Paris Cosmology Colloquium, Observatoire de Paris, June 1994, pp. 127-215, H. J. de Vega and N. Sánchez, Editors (World Scientific, 1995); Advances in Astrofundamental Physics, Erice Chalonge School, N. Sánchez and A. Zichichi Editors, (World Scientific, 1995). Boyanovsky, D., de Vega, H. J., Holman, R., and Salgado, J., Phys. Rev. **D54**, 7570 (1996); Boyanovsky, D., Cormier, D., de Vega, H. J., Holman, R.,

Singh, A., and Srednicki, M., Phys. Rev. **D56** (1997) 1939. Boyanovsky D. , de Vega, H. J. and Holman, R. in Vth. Erice Chalonge School, Current Topics in Astrofundamental Physics, N. Sánchez and A. Zichichi Editors, World Scientific, 1996, p. 183-270. Boyanovsky, D., D'Attanasio, M., de Vega, H. J., Holman, R. and Lee, D.-S., *Linear versus nonlinear relaxation: Consequences for reheating and thermalization* Phys. Rev. **D52**, 6805 (1995). Boyanovsky, D., Destri, C., de Vega, H. J., Holman, R. and Salgado, J., *Asymptotic dynamics in scalar field theory: Anomalous relaxation* Phys. Rev. **D57**, 7388 (1998).

17. Boyanovsky, D., de Vega, H. J., Holman R. and Salgado, J., *Non-Equilibrium Bose-Einstein Condensates, Dynamical Scaling and Symmetric Evolution in large N $Phi^4$ theory*, Phys. Rev. **D59** 125009 (1999).
18. Cooper, F., Habib, S., Kluger, Y., and Mottola, E. *Nonequilibrium dynamics of symmetry breaking in lambda Phi 4 theory*, Phys.Rev. **D55** (1997), 6471; Cooper F., Habib, S., Kluger, Y., Mottola, E., Paz, J.P., and Anderson, P., *Nonequilibrium quantum fields in the large-N expansion* Phys. Rev. **D50**, 2848 (1994). Cooper F., Kluger, Y., Mottola, E. and Paz, J. P., *Quantum evolution of disoriented chiral condensates*, Phys. Rev. **D51**, 2377 (1995); Cooper, F., and Mottola, E., *Initial-value problems in quantum field theory in the large-N approximation* , Phys. Rev. **D36**, 3114 (1987); Cooper, F., Pi, S.-Y., and Stancioff, P.N., *Quantum dynamics in a time-dependent variational approximation* Phys. Rev. **D34**, 3831 (1986).
19. Boyanovsky, D., de Vega, H.J. *Dynamics of Symmetry Breaking Phase Transitions in FRW Cosmologies* in preparation.
20. Cugliandolo, L.F., and Dean, D.S., *Full dynamical solution for a spherical spin-glass model* J. Phys. **A28**, 4213 (1995); *On the dynamics of a spherical spin-glass*; *ibid* L453, (1995); Cugliandolo, L.F., Kurchan, J., Parisi, G., *Off equilibrium dynamics and aging in unfrustrated systems*, J. Physique (France) **4**, 1641 (1994).
21. Relaxing the assumption of an instantaneous quench and allowing for a time dependence of the cooling mechanism has been recently studied by Bowick, M., Momen, A., *Domain Formation in Finite-Time Quenches* Phys.Rev. **D58** (1998) 085014.
22. Weinberg, E.J., and Wu, A., *Understanding complex perturbative effective potentials* Phys. Rev. **D36**, 2474 (1987); Guth, A., and Pi, S.-Y., *Quantum mechanics of the scalar field in the new inflationary universe*, Phys. Rev. **D32**, 1899 (1985).
23. Boyanovsky, D. and de Vega, H. J., *Quantum rolling down out of equilibrium*, Phys. Rev. **D47**, 2343 (1993); Boyanovsky, D., *Quantum Spinodal Decomposition*, Phys. Rev. **E48**, 767 (1993).
24. Boyanovsky, D., Lee, D.S., and Singh, A., *Phase transitions out of equilibrium: Domain formation and growth*, Phys. Rev. **D48**, 800 (1993).
25. Karra, G. and Rivers, R.J., *Initial Vortex Densities after a Temperature Quench*, Phys.Lett. **B414** (1997), 28; Rivers, R.J., in 3rd. Colloque Cosmologie, Observatoire de Paris, June 1995, p. 341 in the Proceedings edited by H J de Vega and N. Sánchez, World Scientific. Gill, A.J., and Rivers, R.J., *Dynamics of vortex and monopole production by quench-induced phase separation* , Phys.Rev. **D51** (1995), 6949; Cheetham, G.J., Copeland, E.J., Evans, T.S. and Rivers, R.J., *Role of quantum fluctuations in defect-dominated transitions* Phys.Rev. **D47** (1993),5316.
26. Stephens, G.J., Calzetta, E. A., Hu, B.L., and Ramsey, S. A., *Defect Formation and Critical Dynamics in the Early Universe*, gr-qc/9808059 (1998); Ibaceta, D., and Calzetta, E.A., *Counting Defects in an Instantaneous Quench* hep-ph/9810301 (1998).

# DEFECT FORMATION THROUGH BOSON CONDENSATION IN QUANTUM FIELD THEORY

GIUSEPPE VITIELLO
*Dipartimento di Fisica and INFM Unita' di Salerno*
*Università di Salerno, 84100 Salerno, Italy*

## 1. Introduction

The study of many body physics as well as the study of elementary particle physics has convinced us that at a very basic level Nature is ruled by quantum dynamical laws. On the other hand, we also know and observe several systems, such as superconductors, superfluids, crystals and ferromagnets which behave as *macroscopic* quantum systems.

The question then arises of how the quantum dynamics may generate the observed macroscopic properties. In other words, how it happens that the macroscopic scale characterizing those systems is dynamically generated out of the microscopic scale of the quantum elementary components[1].

Moreover, we also observe a variety of phenomena where quantum objects coexist and interact with extended macroscopic objects which show classical behavior, e.g. vortices in superconductors and superfluids, magnetic domains in ferromagnets, dislocations and other *defects* in crystals.

Thus, we are faced also with the question of the quantum origin of the macroscopically behaving extended objects and of their interaction with quanta[2]. Even for structures at cosmological scale, the question of their dynamical origin from elementary components asks for an answer consistent with quantum dynamical laws[3].

Macroscopic quantum systems are quantum systems not, of course, in the rather trivial sense that they are made by quantum components, but in the sense that, although they behave classically, nevertheless some of their macroscopic features cannot be understood without recurse to quantum



*Y.M. Bunkov and H. Godfrin (eds.),*
*Topological Defects and the Non-Equilibrium Dynamics of Symmetry Breaking Phase Transitions*, 171–191.

theory. Quantum theory thus appears not confined to microscopic phenomena.

In this respect it is remarkable that these "classical" systems present observable ordered patterns, e.g. crystal ordering, phase coherence, ferromagnetic ordering, etc.. Moreover, most extended objects present some topological singularity, and interesting enough, these topologically non-trivial defects only appear in systems presenting an ordered state.

The formation of defects in the course of phase transitions provides a further source of questions which are attracting much attention since it appears that defect formation during phase transitions may reveal unifying understanding of phenomena belonging to a wide energy scale[4].

The task of this paper is to review some of the main aspects in the Quantum Field Theory (QFT) description of topological defect formation, which also illustrates how to get the macroscopic scale out of the quantum dynamics. I will further mention some recent developments dealing with temperature effects on defect formation[5].

The paper is organized as follows. I will consider the problem of dynamical generation of order in quantum systems in Section 2. The key ingredients are the mechanism of spontaneous breakdown of symmetry (SBS) and the consequent appearance of Nambu-Goldstone (NG) boson particles[6, 7] (such as phonons in crystals). In order to present a general, model independent discussion, I will use functional integration techniques. As we will see, NG modes manifest as long range correlations and thus they are responsible of the above mentioned change of scale, from microscopic to macroscopic. The coherent boson condensation of NG modes turns out to be the mechanism by which order is generated. From the point of view of the invariance properties of the theory, the mathematical structure of the contraction[8] of the symmetry group is the one controlling the SBS mechanism[9].

I will show how topologically non-trivial defects are generated in quantum systems by non-homogeneous boson condensation in Section 3. Here the so called boson transformation revealed to be the crucial tool. I will prove that topological defects only can be formed in the presence of NG modes, i.e. in the presence of ordering. This sheds some light on the mechanism by which defect formation occurs in phase transitions, i.e. in the presence of gradients of the order parameter. Interaction of defects with quanta is also very briefly considered in this Section. Explicit vortex solutions in terms of boson condensation are presented in Section 4.

Temperature effects and volume effects on SBS, on defects formation and on symmetry restoration are considered in Section 5. Contact with the problem of defect formation in phase transition processes are also made in this Section.

A preliminary remark to my subsequent discussion is the following.

The von Neumann theorem in Quantum Mechanics (QM) [10] states that for systems with a finite number of degrees of freedom all the representations of the canonical commutation relations are unitarily equivalent. This theorem actually states that in QM the physical system can only live in one single phase: unitary equivalence means indeed physical equivalence and thus there is no room (no representations) to represent different physical phases. Fortunately, such a situation drastically changes in QFT where systems with infinitely many degrees of freedom are studied. In such a case the von Neumann theorem does not hold and infinitely many unitarily inequivalent representations of the canonical commutation relations do in fact exist[11]. It is such a richness of QFT which allows the description of different physical phases. The occurrence of spontaneous breakdown of symmetry and of the related NG boson condensation thus becomes possible in QFT.

Although one can set up many formal devices based on more or less sophisticated approximations, or even on semi-classical methods, which may nevertheless lead to phenomenologically successful results, it should be always understood that the proper theoretical framework dealing with phase transitions, defect formation and all that is in the larger manifold of unitarily inequivalent representations provided by QFT.

## 2. Spontaneous breakdown of symmetry and group contraction

In QFT the dynamics is described by a set of field equations for the interacting operator fields, say $\psi(x)$, also called the Heisenberg fields. These are the basic fields of the theory satisfying equal-time canonical commutation relations and the Heisenberg field equations

$$\Lambda(\partial)\,\psi(x) = j[\psi](x) \ , \tag{1}$$

where $x \equiv (t, \mathbf{x})$. $j$ is a functional of the $\psi$ field describing the interaction.

Observable phenomena are on the other hand, described by observable (physical) operator fields (such as phonons), say $\varphi(x)$. They also obey equal-time canonical commutation relations and satisfy free field equations, which, with convenient care in the renormalization procedure, may be written as

$$\Lambda(\partial)\,\varphi(x) = 0 \,. \tag{2}$$

The Hilbert space, say $\mathcal{H}$, for the physical states is the Fock space for the fields $\varphi$. Solving the dynamical problem thus means computing by means of eq. (1) the matrix elements of $\psi$ in the space $\mathcal{H}$: this will relate the basic dynamics to the observable properties of the physical states. In this way we obtain the *dynamical map* between interacting fields and physical fields[1, 2]:

$$\psi(x) = F\,[x;\varphi] \;. \tag{3}$$

Eq. (3) is also called the Haag expansion in the LSZ formalism[12]. I have to stress that the equality in (3) is a "weak" equality: it must be understood as an equality among matrix elements computed in $\mathcal{H}$.

This a crucial point and a couple of remarks need to be made. First, I observe that the set of $\varphi$ fields must be an irreducible set; however, it may happen that not all the elements of the set are known since the beginning. For example there might be composite (bound states) fields or even elementary quanta whose existence is ignored in a first recognition. Then the computation of the matrix elements in physical states will lead to the detection of unexpected poles in the Green's functions, which signal the existence of the ignored quanta. One thus introduces the fields corresponding to these quanta and repeats the computation. This way of proceeding is called the self-consistent method[2]. In this connection, I note that it is not necessary to have a one-to-one correspondence between the sets $\{\psi_i\}$ and $\{\varphi_i\}$. This happens in fact when the set $\{\varphi_i\}$ includes composite particles.

Another remark is that, as already mentioned, in QFT the Fock space for the physical states is not unique: indeed one may have several physical phases, e.g. for a metal the normal phase and the superconducting phase, and so on. Fock spaces describing different phases are unitarily inequivalent spaces and correspondingly we have different expectation values for certain observables and even different irreducible sets of physical quanta; for example, in ferromagnets this set includes magnon fields which do not exist in the non-magnetic phase, etc.. Thus, finding the dynamical map involves the "choice" of the Fock space where the dynamics has to be realized: in other words the same dynamics (i.e. same Heisenberg fields and same Heisenberg field equations) may generate different physical phases.

Suppose now that the dynamics is invariant under some group $G$ of transformations of $\psi$:

$$\psi(x) \to \psi'(x) = g\left[\psi(x)\right] , \tag{4}$$

with $g \in G$. Invariance of the dynamics means that the Heisenberg equations (or the Lagrangian from which they may be derived) are invariant (in form) under the transformations of $G$.

One says that symmetry is spontaneously broken when the vacuum state in the Fock space $\mathcal{H}$ is not invariant under the group $G$ but only under one of its subgroups[1, 2, 12].

Eq. (3) implies that when $\psi$ is transformed as in (4), then

$$\varphi(x) \to \varphi'(x) = g'\left[\varphi(x)\right] , \tag{5}$$

such that

$$g\left[\psi(x)\right] = F\left[g'\left[\varphi(x)\right]\right] , \tag{6}$$

with $g'$ belonging to some group of transformations $G'$.

Now it happens, as we will see below, that $G \neq G'$ when symmetry is spontaneously broken, with $G'$ the group contraction of $G$[9]; when symmetry is not broken $G = G'$.

Since $G$ is the invariance group of the dynamics, eq. (3) requires that $G'$ is the group under which free fields equations are invariant, i.e. $\varphi'$ also is a solution of (2). Since eq. (3) is a weak equality, $G'$ depends on the choice of the Fock space $\mathcal{H}$ (among the physically realizable unitarily inequivalent state spaces). Thus we see that the original (same) invariance of the dynamics may manifest itself in different symmetry groups for the $\varphi$ fields according to different choices of the physical state space. Since this process is constrained by the dynamical equations (1), it is called the *dynamical rearrangement of symmetry*[1, 2, 13] .

To be specific, let me consider, in the path-integral formalism, a complex scalar field $\phi(x)$ interacting with a gauge field $A_\mu(x)$ (Anderson-Higgs-Kibble type model)[14, 15, 16]. The lagrangian density $\mathcal{L}[\phi(x), \phi^*(x), A_\mu(x)]$ is invariant under the global and the local gauge transformations:

$$\phi(x) \to e^{i\theta}\phi(x) , \qquad A_\mu(x) \to A_\mu(x) , \tag{7}$$

$$\phi(x) \to e^{ie_0\lambda(x)}\phi(x) , \qquad A_\mu(x) \to A_\mu(x) + \partial_\mu\lambda(x) , \tag{8}$$

respectively, where $\lambda(x) \to 0$ for $|x_0| \to \infty$ and/or $|\mathbf{x}| \to \infty$. The Lorentz gauge $\partial^\mu A_{\mu H}(x) = 0$ is used. I put $\phi(x) = \frac{1}{\sqrt{2}}\left[\psi(x) + i\chi(x)\right]$.

Spontaneous breakdown of symmetry is introduced through the condition $\langle 0|\phi_H(x)|0\rangle \equiv \tilde{v} \neq 0$, with $\tilde{v}$ constant and I put $\rho(x) \equiv \psi(x) - \tilde{v}$. Here $\phi_H(x)$ denotes the Heisenberg field. $\phi(x)$ is the c-number field entering the functional integral. The generating functional, including the gauge constraint, is[17]

$$W[J,K] = \frac{1}{N}\int [dA_\mu][d\phi][d\phi^*][dB] \exp\left[i\int d^4x \left(\mathcal{L}(x) + B(x)\partial^\mu A_\mu(x) + \right.\right.$$
$$\left.\left. K^*\phi + K\phi^* + J^\mu(x)A_\mu(x) + i\epsilon|\phi(x) - v|^2\right)\right] , \tag{9}$$

with $N$ a convenient normalization. $B(x)$ is an auxiliary field which guarantees the gauge condition. The rôle of the $\epsilon-$term is to specify the condition of symmetry breakdown under which we want to compute the path-integral[18, 19]. It may be given the physical meaning of the small external field triggering the symmetry breakdown. The limit $\epsilon \to 0$ must be made at the end of the computations.

As customary, I will use the notation $\langle F[\phi]\rangle_{K,\epsilon}$ to denote functional average and $\langle F[\phi]\rangle_\epsilon \equiv \langle F[\phi]\rangle_{\epsilon,K=0}$ , $\langle F[\phi]\rangle \equiv \lim_{\epsilon\to 0}\langle F[\phi]\rangle_\epsilon$. Note that $\langle\chi(x)\rangle_\epsilon = 0$ because of the invariance under $\chi \to -\chi$.

Invariance of the path-integral under the change of variables (7) (and/or (8) ) leads to

$$\langle\psi(x)\rangle_\epsilon = \sqrt{2}\,\epsilon v\int d^4y\langle\chi(x)\chi(y)\rangle_\epsilon = \sqrt{2}\,\epsilon v\,\Delta_\chi(\epsilon,0)\ . \tag{10}$$

This is one of the Ward-Takahashi identities. Such identities carry the symmetry content of the theory. In momentum space the propagator for the Heisenberg field $\chi$ has the general form

$$\Delta_\chi(0,p) = \lim_{\epsilon\to 0}\left[\frac{Z_\chi}{p^2 - m_\chi^2 + i\epsilon a_\chi} + (continuum\ contributions)\right] . \tag{11}$$

$Z_\chi$ and $a_\chi$ are renormalization constants. The integration in eq.(10) picks up the pole contribution at $p^2 = 0$, and leads to

$$\tilde{v} = \sqrt{2}\frac{Z_\chi}{a_\chi}v \Leftrightarrow m_\chi = 0\ , \quad or \quad \tilde{v} = 0 \Leftrightarrow m_\chi \neq 0\ . \tag{12}$$

The Goldstone theorem[6] is thus proved[18, 19]: if the symmetry is spontaneously broken ($\tilde{v} \neq 0$), a massless mode exists, whose interpolating Heisenberg field is $\chi(x)$. It is the NG boson mode. Since it is massless it manifests as a long range correlation mode. Notice that the NG mode is an elementary field. In other models it may appear as a bound state, e.g. the magnon in ferromagnets[20]. Note that[18, 19] $\tilde{v}$ is independent of $|v|$, although the phase of $|v|$ determines the one of $\tilde{v}$: as in ferromagnets, once an external magnetic field is switched on, the system is magnetized independently of the strength of the external field.

The analysis of the two-point functions of the theory shows[17] that the model contains a massless negative norm state (ghost), besides the NG massless mode $\chi$, and a massive vector field $U^{\mu}$. The dynamical maps are:

$$S =: S[\rho_{in}, U^{\mu}_{in}, \partial(\chi_{in} - b_{in})] : \ , \tag{13}$$

$$\phi_H(x) =: \exp\left\{ i\frac{Z_{\chi}^{\frac{1}{2}}}{\tilde{v}}\chi_{in}(x)\right\} \left[\tilde{v} + Z_{\rho}^{\frac{1}{2}}\rho_{in}(x) + F[\rho_{in}, U^{\mu}_{in}, \partial(\chi_{in} - b_{in})]\right] : \ , \tag{14}$$

$$A^{\mu}_H(x) = Z_3^{\frac{1}{2}} U^{\mu}_{in}(x) + \frac{Z_{\chi}^{\frac{1}{2}}}{e_0\tilde{v}}\partial^{\mu} b_{in}(x) + : F^{\mu}[\rho_{in}, U^{\mu}_{in}, \partial(\chi_{in} - b_{in})] : \ , \tag{15}$$

where the functionals $F$ and $F^{\mu}$ are to be determined within a particular model. These relations are weak equalities and are equivalent to the familiar LSZ reduction formula[12]. It will be also used $A^{0\mu}_H(x) \equiv A^{\mu}_H(x) - e_0\tilde{v} : \partial^{\mu} b_{in}(x) :$. In eqs. (13)-(15) $\chi_{in}$ denotes the NG mode, $b_{in}$ the ghost mode, $U^{\mu}_{in}$ the massive vector field and $\rho_{in}$ the massive matter field. Their field equations are

$$\partial^2\chi_{in}(x) = 0 \ , \quad \partial^2 b_{in}(x) = 0 \ , \quad (\partial^2 + m_{\rho}^2)\rho_{in}(x) = 0 \ , \tag{16}$$

$$(\partial^2 + m_V{}^2) U_{\mu in}(x) = 0 \quad , \quad \partial^{\mu} U_{\mu in}(x) = 0 \ . \tag{17}$$

with $m_V{}^2 = \frac{Z_3}{Z_{\chi}}(e_0\tilde{v})^2$. We also have

$$B_H(x) = \frac{e_0\tilde{v}}{Z_{\chi}^{\frac{1}{2}}}[b_{in}(x) - \chi_{in}(x)] \ . \tag{18}$$

The field equations for $B_H$ and $A_{H\mu}$ are

$$\partial^2 B_H(x) = 0 \ , \quad -\partial^2 A_{H\mu}(x) = j_{H\mu}(x) - \partial_{\mu} B_H(x) \ , \tag{19}$$

with $j_{H\mu}(x) = \delta\mathcal{L}(x)/\delta A^{\mu}_H(x)$. One may then require that the current $j_{H\mu}$ is the only source of the gauge field $A_{H\mu}$ in any observable process. This amounts to imposing the condition: ${}_p\langle b|\partial_{\mu} B_H(x)|a\rangle_p = 0$, i.e.

$$(-\partial^2)\, {}_p\langle b|A^0{}_{H\mu}(x)|a\rangle_p = {}_p\langle b|j_{H\mu}(x)|a\rangle_p \ , \tag{20}$$

where $|a\rangle_p$ and $|b\rangle_p$ denotes two generic physical states. Eq.(20) are the classical Maxwell equations. The condition ${}_p\langle b|\partial_\mu B_H(x)|a\rangle_p = 0$ leads to the Gupta-Bleuler-like condition

$$[\chi_{in}^{(-)}(x) - b_{in}^{(-)}(x)]|a\rangle_p = 0 \,, \tag{21}$$

where $\chi_{in}^{(-)}$ and $b_{in}^{(-)}$ are the positive-frequency parts of the corresponding fields. Thus we see that $\chi_{in}$ and $b_{in}$ do not participate to any observable reaction. Note that, in fact, they are present in the $S$ matrix in the combination $(\chi_{in} - b_{in})$. It is to be remarked that the NG bosons do not disappear from the theory: we will see that in situations in which the vacuum is not translationally invariant, the NG fields can have observable effects.

The study of the dynamical rearrangement of the symmetry shows that local gauge transformations of the Heisenberg fields

$$\phi_H(x) \to e^{ie_0\lambda(x)}\phi_H(x) \,, \quad A_H^\mu(x) \to A_H^\mu(x) + \partial^\mu\lambda(x) \,, \quad B_H(x) \to B_H(x) \tag{22}$$

are induced by the in-field transformations

$$\chi_{in}(x) \to \chi_{in}(x) + \frac{e_0\tilde{v}}{Z_\chi^{\frac{1}{2}}}\lambda(x) \,, \quad b_{in}(x) \to b_{in}(x) + \frac{e_0\tilde{v}}{Z_\chi^{\frac{1}{2}}}\lambda(x) \,, \tag{23}$$

$$\rho_{in}(x) \to \rho_{in}(x) \,, \quad U_{in}^\mu(x) \to U_{in}^\mu(x) \,. \tag{24}$$

The global transformation $\phi_H(x) \to e^{i\theta}\phi_H(x)$ is induced by

$$\chi_{in}(x) \to \chi_{in}(x) + \frac{\tilde{v}}{Z_\chi^{\frac{1}{2}}}\theta \,, \tag{25}$$

$$b_{in}(x) \to b_{in}(x) \,, \; \rho_{in}(x) \to \rho_{in}(x) \,, \; U_{in}^\mu(x) \to U_{in}^\mu(x) \,. \tag{26}$$

Note that under the above in-field transformations the in-field equations and the $S$ matrix are invariant and that $B_H$ is changed by an irrelevant c-number. Thus we see that the original invariance cannot be lost even at the level of the physical fields, although it can manifest there in a different symmetry group structure.

Eq. (25) shows that the physical field $\chi_{in}$ translates by a constant when the Heisenberg field $\phi$ undergoes the global phase transformation (and vice versa): The global $U(1)$ invariance group is dynamically rearranged into the one-parameter constant translation group. This last one is the group contraction of global $U(1)$.

Notice that translation by a constant is an invariant transformation for the $\chi_{in}$ field equation if and only if $\chi_{in}$ is a massless field. Thus the rearrangement into the contraction of global $U(1)$ group has the same content as the Goldstone theorem[9]. It is also interesting to note that while $U(1)$ is a compact group its contraction is not compact. Since the number operator of $\chi_{in}$ field changes under the translation (25) we say that we have $\chi_{in}$ coherent boson condensation.

It must be stressed that the translation by a constant (25) must be actually understood as the limit for $f(x) \rightarrow 1$ of the transformation

$$\chi_{in}(x,\theta) = \chi_{in}(x) + \frac{\tilde{v}}{Z_\chi^{\frac{1}{2}}}\theta f(x) \ , \tag{27}$$

with $f(x)$ a normalizable solution of the $\chi_{in}$ field equation: $\partial^2 f(x) = 0$. Eq. (25) induces infrared singularities in Feynman diagrams with many momentumless and energyless lines. These are smeared out by use of (27). However, notice that matrix elements are well defined even when (25) is used. Also, the function $f(x)$ appearing in the generator of (27) makes it well defined[19, 20, 21].

Since different physical phases (unitarily inequivalent representations ) are associated to different NG boson condensation densities, we see that boson translations, by inducing variation of NG boson condensation, represent transitions through physically different phases. In particular, I will discuss non-homogeneous boson condensation induced by transformations such as (27).

The dynamical rearrangement of symmetry has been studied in many models of physical interest. In the case of global $SU(2)$ invariance group, for example in ferromagnets[20], or in systems with isospin vector fields[21], the NG boson condensation is controlled by the $E(2)$ group, which is the group contraction of $SU(2)$[9]. Unfortunately, for lack of space I cannot report about these and some other interesting cases.

## 3. Observable effects of non-homogeneous boson transformations

Translations of bosonic physical fields (not necessarily massless) by space-time dependent functions, say $\alpha(x)$, satisfying the same field equation of the translated physical field, are called *boson transformations* [1, 2, 19]. Eq. (27) is thus an example of boson transformation.

Let $\phi'_H$ denote the Heisenberg field obtained through the dynamical mapping when the physical field undergoes the boson transformation . The *boson transformation theorem* can be then proved, which states that $\phi'_H$ is also a solution of the Heisenberg field equation[1, 2, 19].

The proof of the theorem rests in showing that the boson-transformed fields, say $\phi_H[x;\chi_{in}(x)+\alpha(x)]$, differ from $\phi_H[\chi_{in}(x)]$ only by an $\epsilon-$dependent factor and therefore are solutions of the same field equations.

In the absence of a gauge field, under boson transformation the order parameter gets space-time dependence given by

$$\tilde{v}(x) = \exp\left\{i\frac{Z_\chi^{\frac{1}{2}}}{\tilde{v}}\alpha(x)\right\}\left[\tilde{v} + V\left(i\frac{Z_\chi^{\frac{1}{2}}}{\tilde{v}}\partial_\mu\alpha(x)\right)\right] , \tag{28}$$

where the expansion of $\tilde{v}(x)$ around $\partial_\mu\alpha(x) = 0$ is used $(V\,(i\partial_\mu\lambda(x)) \to 0$ when $\partial_\mu\alpha(x) \to 0)$. Note that the modulus of $\tilde{v}$ changes and in the limit $\alpha \to$ *const.* only its phase is changed.

In the case that a gauge field is present, it can be shown that any space-time dependence of the $\epsilon-$term can be eliminated by a gauge transformation when $\alpha(x)$ is a regular (i.e. Fourier transformable) function and the only effect is the appearance of a phase factor in the order parameter: $\tilde{v}(x) = e^{ic\alpha(x)}\tilde{v}$, with $c$ a constant.

The conclusion is that when a gauge field is present, the boson transformation with regular $\alpha(x)$ is equivalent to a gauge transformation. On the contrary, in a theory with global invariance only, non-singular boson transformations of the NG fields can produce non-trivial physical effects (like linear flow in superfluidity).

I want to stress that, in the case of global phase transformations as well as in the case of local gauge transformations, the proof of the boson transformation theorem relies on the fact that $\alpha(x)$ is a regular function. If one wants to consider functions with some singularities (divergence and topological singularities) one has to carefully exclude the singularity regions when integrating on space and/or time. For example, if $\alpha(x)$ is singular on the axis of a cylinder (at $r = 0$) one must exclude the singular line $r = 0$ by a cylindrical surface of infinitesimal radius. The phase of the order parameter will be singular on that line. This means that SBS does not occur in that region (the core): there we have the "normal" state rather than the ordered one. Provided one uses such care, the boson transformation can be safely (and advantageously) used also in the case of singular $\alpha(x)$.

The boson theorem has relevant physical meaning since it shows that the same dynamics may describe homogeneous and non-homogeneous phenomena. When a theory allows SBS, there always exist solutions of the field equations with space and/or time-dependent vacuum. These solutions are obtained from the translationally invariant ones by the boson transformation of the NG field: they results from a local Bose condensation of the $\chi_{in}$ particles. This directly leads us to the mechanism of formation of extended objects (defect formation) .

Notice that in local gauge theories the boson transformation must be compatible with Heisenberg field equations but also with the physical state condition (21). Under the boson transformation with $\alpha(x) = \frac{\tilde{v}}{Z_\chi^{\frac{1}{2}}} f(x)$ and $\partial^2 f(x) = 0$, $B_H$ changes as

$$B_H(x) \to B_H(x) - \frac{e_0 \tilde{v}^2}{Z_\chi} f(x) \,, \tag{29}$$

Eq. (20) is thus violated when the Gupta-Bleuler-like condition is imposed. In order to restore it, the shift in $B_H$ must be compensated by means of the transformation on $U_{in}$:

$$U^\mu_{in}(x) \to U^\mu_{in}(x) + Z_3{}^{-\frac{1}{2}} a^\mu(x) \quad , \quad \partial_\mu a^\mu(x) = 0 \,, \tag{30}$$

with a convenient c-number function $a^\mu(x)$. The dynamical maps of the various Heisenberg operators are not affected by (30) since they contain $U^\mu_{in}$ and $B_H(x)$ in a combination such that the changes of $B_H$ and of $U^\mu_{in}$ compensate each other provided

$$(\partial^2 + m_V^2) a_\mu(x) = \frac{m_V^2}{e_0} \partial_\mu f(x) \,. \tag{31}$$

Eq. (31) thus obtained is the Maxwell equation for the massive potential vector $a_\mu$[17, 22]. The classical ground state current $j_\mu$ turns out to be

$$j_\mu(x) \equiv \langle 0 | j_{H\mu}(x) | 0 \rangle = m_V^2 \left[ a_\mu(x) - \frac{1}{e_0} \partial_\mu f(x) \right] \,. \tag{32}$$

The term $m_V^2 a_\mu(x)$ is the *Meissner current* , while $\frac{m_V^2}{e_0} \partial_\mu f(x)$ is the *boson current . The macroscopic field and current are thus given in terms of the boson condensation function.*

The (classical) boson current is given by $\partial_\mu f$, i.e. by variations in the non-homogeneous boson condensate: boson condensation functions must play a role in phase transitions where the boson condensate indeed changes.

*Let me now show that boson transformation functions carrying topological singularities are only allowed for massless bosons*[2, 23, 24]. Suppose the function $f(x)$ for the boson transformation of the field $\chi_{in}$ carries a topological singularity and is thus path-dependent:

$$G^{\dagger}_{\mu\nu}(x) \equiv [\partial_\mu, \partial_\nu]\, f(x) \neq 0 \;, \qquad \textit{for certain}\ \mu\,,\ \nu\,,\ x\;. \tag{33}$$

On the other hand, $\partial_\mu f$, which is related with observables, is single-valued, i.e. $[\partial_\rho, \partial_\nu]\, \partial_\mu f(x) = 0$. Recall that $f(x)$ is solution of the $\chi_{in}$ equation and suppose there is a non-zero mass term:

$$(\partial^2 + m^2) f(x) = 0\;. \tag{34}$$

¿From the definition of $G^{\dagger}_{\mu\nu}$ and the regularity of $\partial_\mu f(x)$ it follows that

$$\partial_\mu f(x) = \frac{1}{\partial^2 + m^2} \partial^\lambda G^{\dagger}_{\lambda\mu}(x)\;, \tag{35}$$

which leads to $\partial^2 f(x) = 0$ which implies $m = 0$.

Thus (33) is only compatible with the massless equation for $\chi_{in}$.

The quantity $\partial_\lambda f(x)$ is given by (35) with $m = 0$. ¿From this equation, $f(x)$ can be determined. The topological charge is defined as

$$N_T = \int_C dl^\mu\, \partial_\mu\, f = \int_S dS_\mu \epsilon^{\mu\nu\sigma}\, \partial_\nu \partial_\sigma\, f = \frac{1}{2}\int_S dS^{\mu\nu}\, G^{\dagger}_{\mu\nu}\;. \tag{36}$$

Here $C$ is a contour enclosing the singularity and $S$ a surface with $C$ as boundary. $N_T$ does not depend on the path $C$ provided this does not cross the singularity. The dual tensor $G^{\mu\nu}$ is

$$G^{\mu\nu}(x) \equiv -\frac{1}{2}\epsilon^{\mu\nu\lambda\rho} G^{\dagger}_{\lambda\rho}(x) \tag{37}$$

and satisfies the continuity equation

$$\partial_\mu G^{\mu\nu}(x) = 0 \qquad \Leftrightarrow \qquad \partial_\mu G^{\dagger}_{\lambda\rho} + \partial_\rho G^{\dagger}_{\mu\lambda} + \partial_\lambda G^{\dagger}_{\rho\mu} = 0\;. \tag{38}$$

This equation completely characterizes the topological singularity [2, 24].

Let me now observe that all the macroscopic ground state effects do not occur for regular $f(x)$ ($G^{\dagger}_{\mu\nu} = 0$). In fact, from (31) we obtain $a_\mu(x) = \frac{1}{e_0}\partial_\mu f(x)$ for regular $f$ which implies zero classical current ($j_\mu = 0$) and zero classical field ($F_{\mu\nu} = \partial_\mu a_\nu - \partial_\nu a_\mu$), since the Meissner and the boson current cancel each other.

In conclusion, the vacuum current appears only when $f(x)$ has topological singularities and these can be created only by condensation of massless bosons, e.g. by condensation of NG bosons when SBS occurs. This explains why topological defects appear in the process of phase transitions, where NG modes are present and gradients in their condensate densities are nonzero.

On the other hand, the appearance of space-time order parameter is no guarantee that persistent ground state currents (and fields) will exist: if $f$ is a regular function, the space-time dependence of $\tilde{v}$ can be gauged away by an appropriate gauge transformation.

Since the boson transformation with regular $f$ does not affect observable quantities, the $S$ matrix (13) is actually given by

$$S =: S[\rho_{in}, U^{\mu}_{in} - \frac{1}{m_V}\partial(\chi_{in} - b_{in})] : \ . \tag{39}$$

This is in fact independent of the boson transformation with regular $f$:

$$S \rightarrow S' =: S[\rho_{in}, U^{\mu}_{in} - \frac{1}{m_V}\partial(\chi_{in} - b_{in}) + Z_3^{-\frac{1}{2}}(a^{\mu} - \frac{1}{e_0}\partial^{\mu} f)] : \tag{40}$$

since $a_{\mu}(x) = \frac{1}{e_0}\partial_{\mu} f(x)$ for regular $f$. However, $S' \neq S$ for singular $f$: $S'$ includes the interaction of the quanta $U^{\mu}_{in}$ and $\phi_{in}$ with the defect classical field and current.

Thus we see how quantum fluctuations may interact and have effects on classically behaving macroscopic defects: our picture includes interaction of quanta with macroscopic objects. Much more can be said on the interaction of extended objects with quanta; however, for shortness I will not discuss more on that.

## 4. The vortex solution

The meaning of Eq. (35) with $m = 0$ is of course[2, 22, 24, 25]

$$\partial_{\mu} f(x) = 2\pi \int d^4x' \, G^{\dagger}_{\mu\nu}(x') \partial^{\nu}_{x} K(x - x') \ , \tag{41}$$

$$K(x - x') = -\frac{1}{(2\pi)^4} \int d^4p \, e^{-ip(x-x')} \frac{1}{p^2 + i\epsilon} \ , \tag{42}$$

with the Green's function satisfying $\partial^2 K(x - x') = \delta(x - x')$. Eq.(41) gives upon path integration

$$f(x) = 2\pi \int^{x} dx^{\mu} \int d^4x' \, G^{\dagger}_{\mu\nu}(x') \partial^{\nu}_{x} K(x - x') \ , \tag{43}$$

which is indeed solution of $\partial^2 f(x) = 0$. The classical vector potential is

$$a_\mu(x) = -\frac{m_V^2}{e}\int d^4x'\,\Delta_c(x-x')\partial'_\mu f(x')\ , \tag{44}$$

$$\Delta_c(x-x') = \frac{1}{(2\pi)^4}\int d^4p\, e^{-ip(x-x')}\frac{1}{p^2-m_V^2+i\epsilon}\ . \tag{45}$$

The electromagnetic tensor and the vacuum current are[2, 22, 24]

$$F_{\mu\nu}(x) = \partial_\mu a_\nu(x) - \partial_\nu a_\mu(x) = 2\pi\frac{m_V^2}{e}\int d^4x'\,\Delta_c(x-x')G^\dagger_{\mu\nu}(x')\ , \tag{46}$$

$$j_\mu(x) == -2\pi\frac{m_V^2}{e}\int d^4x'\,\Delta_c(x-x')\partial^\nu_{x'}G^\dagger_{\nu\mu}(x')\ , \tag{47}$$

respectively, and satisfy $\partial^\mu F_{\mu\nu}(x) = -j_\nu(x)$.

The line singularity for the vortex solution can be parameterized by a single line parameter $\sigma$ and by the time parameter $\tau$. The *static vortex* solution is then obtained by setting $y_0(\tau,\sigma) = \tau$ and $\mathbf{y}(\tau,\sigma) = \mathbf{y}(\sigma)$, with $y$ denoting the line coordinate. $G^\dagger_{\mu\nu}$ is non-zero only on the line at $y$ (we can consider more lines but here I limit myself to one line, for simplicity). Thus,

$$G_{0i}(x) = \int d\sigma\,\frac{dy_i(\sigma)}{d\sigma}\,\delta^3[\mathbf{x}-\mathbf{y}(\sigma)]\ ,\quad G_{ij}(x) = 0\ , \tag{48}$$

$$G^\dagger_{ij}(x) = -\epsilon_{ijk}G_{0k}(x)\quad ,\qquad G^\dagger_{0i}(x) = 0\ . \tag{49}$$

Eq.(46) shows that these vortices are purely *magnetic*. Eq.(41) gives

$$\partial_i f(x) = \frac{1}{(2\pi)^2}\int d\sigma\,\epsilon_{ijk}\frac{dy_k(\sigma)}{d\sigma}\partial^x_j\int d^3p\frac{e^{i\mathbf{p}\cdot(\mathbf{x}-\mathbf{y}(\sigma))}}{\mathbf{p}^2}\ , \tag{50}$$

and $\partial_0 f(x) = 0$, i.e., by using the identity $(2\pi)^{-2}\int d^3p\frac{e^{i\mathbf{p}\cdot\mathbf{x}}}{\mathbf{p}^2} = \frac{1}{2|\mathbf{x}|}$,

$$\nabla f(x) = -\frac{1}{2}\int d\sigma\,\frac{d\mathbf{y}_k(\sigma)}{d\sigma}\wedge\nabla_x\frac{1}{|\mathbf{x}-\mathbf{y}(\sigma)|}\ . \tag{51}$$

Note that $\nabla^2 f(x) = 0$ is satisfied.

*A straight infinitely long vortex* is specified by $y_i(\sigma) = \sigma\,\delta_{i3}$ with $-\infty < \sigma < \infty$. The only non vanishing component of $G^{\mu\nu}(x)$ are $G^{03}(x) = G^\dagger_{12}(x) = \delta(x_1)\delta(x_2)$. Eq.(51) gives[2, 22, 24]

$$\frac{\partial}{\partial x_1}f(x) = \frac{1}{2}\int d\sigma\frac{\partial}{\partial x_2}[x_1^2+x_2^2+(x_3-\sigma)^2]^{-\frac{1}{2}} = -\frac{x_2}{x_1^2+x_2^2}\ , \tag{52}$$

$$\frac{\partial}{\partial x_2} f(x) = \frac{x_1}{x_1^2 + x_2^2} \ , \quad \frac{\partial}{\partial x_3} f(x) = 0 \ , \tag{53}$$

which give

$$f(x) = \tan^{-1}(\frac{x_2}{x_1}) = \theta(x) \ . \tag{54}$$

Use of these results gives the vector potential, $F_{\mu\nu}$ and the vacuum current. The only non-zero components of these fields are $a_1$, $a_2$, $F_{12}$, $j_1$ and $j_2$.

Notice that the condition (38) can be shown to be violated if the line singularity has isolated end points inside the system. Thus consistency with the continuity equation (38) implies that either the string is infinite, or that it forms a closed loop. Also, if there is more than one string, the end points of different strings can be connected in a vertex, eq.(38) resulting then in a condition for the relative string tensions $\nu_\alpha$, with $\alpha$ denoting different strings.

Further simple examples are the following[2, 22, 24].

A *circular loop*: $\mathbf{y}(\sigma) = (a\cos\sigma, a\sin\sigma, 0)$, $0 \leq \sigma \leq 2\pi$. $G^{01}(x) = \delta\left[x_2 - \sqrt{a^2 - x_1^2}\right]\delta(x_3)$, $G^{02}(x) = \delta\left[x_1 - \sqrt{a^2 - x_2^2}\right]\delta(x_3)$, $G^{03}(x) = G^{ij}(x) = 0$.

A straight line along the third axis moving in the $x_1$ direction with velocity $v$ is given by $\mathbf{y}(\sigma,\tau) = (v\tau, 0, \sigma)$, $y_0(\sigma,\tau) = \tau$, from which $G^{03}(x) = \nu\,\delta(x_1 - vt)\,\delta(x_2)$ and $G^{13}(x) = v\,\delta(x_1 - vt)\,\delta(x_2)$.

## 5. Finite temperature and finite volume effects

Consider the $U(1)$ invariant model at finite temperature. The breakdown of symmetry condition in the homogeneous condensation case is[26]

$$\langle 0(\beta)|\phi(x)|0(\beta)\rangle = \frac{1}{\sqrt{2}}\sigma(\beta) \ , \tag{55}$$

where $\beta \equiv \frac{1}{K_B T}$. $|0(\beta)\rangle$ denotes the temperature dependent vacuum state in Thermo Field Dynamics[2]. Note that the statistical average of any operator $A$ is given by $< A >_0 = \langle 0(\beta)|A|0(\beta)\rangle$.

I omit to consider here the presence of other fields (such as the ghost fields) for brevity. The fields $\phi$, $\chi$ and $A_\mu$ may undergo translation transformations by c-number functions, say $\sigma$, $\kappa$ and $\alpha_\mu$, respectively, controlling the respective condensate structures. I write $\phi \equiv \rho + \frac{1}{\sqrt{2}}\sigma(\beta)$. Usual gauge transformations are induced by using $\sigma = 0$, $\kappa = \alpha(x)$ and $\alpha_\mu(x) = \partial_\mu\alpha(x)$.

The homogeneous boson condensation of the Higgs field alone ($\sigma(\beta) = \sigma_0(\beta) = const. \neq 0$, $\kappa = 0$ and $\alpha_\mu = 0$) leads to

$$m^2 = 2\lambda\sigma_0^2 \ , \quad M^2 = e^2({\sigma^2}_0 + <: \bar{\rho}^2 :>_0) \ , \tag{56}$$

$$\sigma_0^2 = v^2 - 3 <: \bar{\rho}^2 :>_0 + \frac{e^2}{\lambda} <: \bar{A}_\mu \bar{A}^\mu :>_0 \ , \tag{57}$$

where $m$ and $\lambda$ denote the Higgs field mass and self-coupling, respectively, $v \equiv \langle 0|\phi|0\rangle$ at $T = 0$ and is assumed to be non-zero, $M$ is the gauge field mass and $e$ is the (electric charge) coupling between $A_\mu$ and $\phi$. $\bar{\rho}$ and $\bar{A}$ denote the physical fields.

Eqs.(56) are actually self-consistent equations since $<: \bar{\rho}^2 :>_0$ also depends on $m^2$. In the discontinuous phase transition case the free energy should be examined[26]. The proper phase transition point is defined by the equality between the ordered and the disordered free energy phase.

As $T \to 0$ eqs. (56) show that $\sigma_0 \to v$ thus recovering the original zero temperature symmetry breaking. We have phase transition to the (disordered) phase $\sigma_0(\beta_C) = 0$ at the critical temperature $T_C$ such that

$$v^2 = 3 <: \bar{\rho}^2 :>_0 - \frac{e^2}{\lambda} <: \bar{A}_\mu \bar{A}^\mu :>_0 \ . \tag{58}$$

Above the phase transition point $T > T_C$, and $\sigma_0 = 0$, we have[26]

$$m^2 = -\lambda v^2 + 3\lambda <: \bar{\rho}^2 :>_0 - e^2 <: \bar{A}_\mu \bar{A}^\mu :>_0 \ . \tag{59}$$

Full symmetry restoration (i.e. $v = 0$) occurs at $T \equiv T_{sym}$ such that thermal contributions in (58) compensate each other, and then also $m = 0$. The gauge field mass $M$ goes to zero not at $T_C$, but at $T$ such that

$$\lambda v^2 - 2\lambda <: \bar{\rho}^2 :>_0 + e^2 <: \bar{A}_\mu \bar{A}^\mu :>_0 = 0 \ . \tag{60}$$

The vortex solution arises in the non-homogeneous condensation case obtained by introducing space dependence in the condensate functions. Introducing the cylindrical coordinates, the asymptotic gauge field configuration is imposed by considering the angle function as gauge function at infinity $k(x) = \frac{n}{e}\theta$, $(k(x) = 0$ at $r = 0)$:

$$\alpha^i_{as} = -\frac{n}{er}\mathbf{e}^i_\theta \ . \tag{61}$$

Here $n$ is the winding number and we see that, although, as already observed, the NG bosons do not enter the physical spectrum, nevertheless their condensation is directly related to the topological charge. For $r < \infty$ we assume (the vortex ansatz)[27]

$$\alpha^i = -\frac{n}{er}(1 - K(r))\,\mathbf{e}^i_\theta \ , \quad \sigma(x) = \sigma_0 f(x) \ , \tag{62}$$

where $\sigma_0$ is the Higgs field shift for the homogeneous condensation and

$$K(r) = 1 - \frac{r}{n}A(r), \qquad A(r) \to \frac{n}{r} \;\; for \;\; r \to \infty \ . \tag{63}$$

The vortex ansatz leads to the temperature dependent vortex equations

$$\begin{aligned} &\frac{d}{dr}\left(\frac{1}{r}\frac{d}{dr}(rA)\right) = e^2\left(A - \frac{n}{r}\right)(\langle :\bar{\rho}^2 :\rangle_0 + f^2\sigma_0^2) \ , \\ &\frac{1}{r}\frac{d}{dr}\left(r\frac{df}{dr}\right) = \left(A - \frac{n}{r}\right)^2 f - \lambda\sigma_0^2 f(1 - f^2) \ . \end{aligned} \tag{64}$$

As $T \to 0$ these equations reduce to the usual vortex equations. One can show[26] that in the vortex case the masses are given by

$$m^2(x) = 2\lambda\sigma_0^2 f^2(x) \ , \quad M^2(x) = e^2(\sigma_0^2 f^2(x) + <:\bar{\rho}^2:>_0) \ . \tag{65}$$

These masses act as potential terms in the field equations and only at spatial infinity ($r \to \infty, f(x) \to 1$) ordinary mass interpretation is recovered. We have in fact the asymptotic behavior

$$K(r) \simeq e^{-Mr} = e^{-\frac{r}{R_0}} \ , \quad f(r) \simeq 1 - f_0 e^{-mr} = 1 - f_0 e^{-\frac{r}{r_0}} \ . \tag{66}$$

$R_0 \equiv \frac{1}{M}$ gives the size of the gauge field core and $r_0 \equiv \frac{1}{m}$ the Higgs field core. As $T \to T_C$ the Higgs field core increases and the gauge field core becomes smaller. At $T = T_C$ one obtains the pure gauge field core. Above $T_C$ symmetry is restored. The discussion on temperature dependence of $\sigma_0$ is similar to the one for the homogeneous case.

The 't Hooft-Polyakov monopole and the sphaleron solutions at finite temperature are discussed in [26].

Let me now discuss the effects of the finite size of the system on the boson condensate and the relation between finite size and temperature effects . This will help to understand how temperature variations near $T_C$ control the defect size (and thus the defect number)[5].

In the case of large but finite volume we expect that the condition of symmetry breakdown is still satisfied "inside the bulk" *far* from the boundaries. However, "*near*" the boundaries, one might expect "distortions" in the order parameter: $\tilde{v} = \tilde{v}(x)$ (or even $\tilde{v} \to 0$). "Near" the system boundaries we may have non-homogeneous order parameter. Non-homogeneities in the boson condensation will "smooth out" in the $V \to \infty$ limit. Suppose the integration in eq. (10) is over the finite (but large) volume $V \equiv \eta^{-3}$ and use

$$\delta_\eta(p) = \frac{1}{2\pi} \int_{-\frac{1}{\eta}}^{\frac{1}{\eta}} dx\, e^{ipx} = \frac{1}{\pi p} \sin\frac{p}{\eta} , \tag{67}$$

which, as well known, approaches $\delta(p)$ as $\eta \to 0$: $\lim_{\eta\to 0} \delta_\eta(p) = \delta(p)$. Now

$$\lim_{\eta\to 0} \int dp\, \delta_\eta(p)\, f(p) = f(0) = \lim_{\eta\to 0} \int dp\, \delta(p-\eta)\, f(p) , \tag{68}$$

then, using $\delta_\eta(p) \simeq \delta(p-\eta)$ for small $\eta$, it is

$$\tilde{v}(\mathbf{x}, \epsilon, \eta) = i\epsilon e^{-i\eta\cdot \mathbf{x}}\, \Delta_\chi(\epsilon, \eta, p_0 = 0) , \tag{69}$$

$$\Delta_\chi(\epsilon, \eta, p_0 = 0) = \lim_{\epsilon\to 0} \left[ \frac{Z_\chi}{-\omega^2_{\mathbf{p}=\eta} + i\epsilon a_\chi} + c.c. \right] . \tag{70}$$

with $-\omega^2{}_{\mathbf{p}=\eta} = \eta^2 + m^2{}_\chi$. Thus, $\lim_{\epsilon\to 0}\lim_{\eta\to 0}\tilde{v}(\mathbf{x}, \epsilon, \eta) \neq 0$ only if $m_\chi = 0$, otherwise $\tilde{v} = 0$. Eq. (12) (the Goldstone theorem ) is thus recovered in the infinite volume limit ($\eta \to 0$).

On the other hand, assume that NG modes are there, i.e. $m_\chi = 0$, then $\omega_\eta \equiv -\omega_{\mathbf{p}=\eta} \neq 0$ for $\eta \neq 0$ and it acts as an "effective mass" for the NG boson. Here I mean that the NG boson is massless, but its lowest (zero) energy is prevented from being reached, the lowest energy being given by the non-zero $\omega_\eta$. The effect of the boundaries ($\eta \neq 0$) is to give an "effective mass" (in the above sense) $m_{eff} \equiv \omega_\eta$ to the NG bosons. These will then propagate over a range of the order of $\xi \equiv \frac{1}{\eta}$, which is the system linear size.

Notice that only if $\epsilon \neq 0$ the order parameter can be kept different from zero, i.e. if $\eta \neq 0$ then $\epsilon$ must be non-zero in order to have $\tilde{v} \neq 0$ (at least locally). In such a case the symmetry breakdown is maintained thanks to the non-zero coupling, $\epsilon \neq 0$, with an external field acting as an external pump providing energy: energy supply is required in order to condensate modes of non-zero lowest energy $\omega_\eta$. Boundary effects are thus in competition with breakdown of symmetry[5, 28]. They may preclude its occurrence or, if symmetry is already broken, they may reduce to zero the order parameter.

We have seen that temperature may have similar effects on the order parameter (at $T_C$ symmetry may be restored, cf. eq. (58) and the discussion following it). Since the order parameter goes to zero when NG modes acquire non-zero effective mass, we may represent the effect of thermalization in terms of finite volume effects and put, e.g., $\eta \propto \sqrt{\frac{|T-T_C|}{T_C}}$. In this way temperature fluctuations around $T_C$ may be discussed as fluctuations in the condensate domain size $\xi$. For example, in the presence of an external driving field ($\epsilon \neq 0$), for $T > T_C$ but near to $T_C$ one may have the formation of ordered domains of size $\xi \propto (\sqrt{\frac{T-T_C}{T_C}})^{-1}$ *before* phase transition to fully ordered phase is achieved (as $T \to T_C$). As far as $\eta \neq 0$, the ordered domains are unstable, they disappear as the external field coupling $\epsilon \to 0$.

Of course, if ordered domains are still present at $T < T_C$, they also disappear as $\epsilon \to 0$. The possibility to maintain such ordered domains below $T_C$ depends on the speed by which $T$ is lowered, compared to the speed by which the system is able to get homogeneously ordered. Notice that the speed by which $T \to T_C$ is related to the speed by which $\eta \to 0$.

In the case of the kink solution it can be shown[29] that the mass $\mu_0 = (2\lambda)^{\frac{1}{2}} v(\beta)$ of the "constituent" fields $\rho^{in}$ fixes the kink size $\xi_\beta \propto \frac{2}{\mu_0} = \frac{\sqrt{2}}{\sqrt{\lambda} v(\beta)}$ which thus increases as $T \to T_C$. It is also interesting to note that in the $T \to 0$ limit the kink size is $\xi_0 \propto \frac{\sqrt{2}}{\sqrt{\lambda} v} < \frac{\sqrt{2}}{\sqrt{\lambda} v(\beta)} = \xi_\beta$, since $v^2(\beta) < v^2$.

As $T$ is different from zero, the thermal Bose condensate $\langle : \rho^2 : \rangle_0$ develops which acts as a potential term for the quantum field $\rho(x)$. Such temperature effects manifest at classical level as potential term for the classical kink field. It is such a potential term which actually controls the "size" (and the number) of the kinks. Notice that $\mu_0^2(x, \beta)$ also acts as a potential for the $\rho_\beta^{in}(x)$ field. Only in the limit $v(x, \beta) \to const$ the $\rho_\beta^{in}(x)$ field may be considered as a free field; e.g. far from the kink core.

The $\rho_\beta^{in}(x)$ condensation, whose macroscopic envelope is represented by the soliton solution, is induced by the boson transformation with $f_\beta(x) = const. \, e^{-\mu_0(\beta) x_1}$ playing the role of "form factor". The number of condensed bosons is thus proportional to $|f_\beta(x)|^2 = e^{-2\mu_0(\beta)(x_1 - a)}$, which is maximal near the kink center $x_1 = a$ and decrease over a size $\xi_\beta = \frac{2}{\mu_0(\beta)}$. The meaning of the boson transformation is that the $f_\beta$-translation breaks the homogeneity of the otherwise constant in space order parameter $v(\beta)$.

In conclusion, phase transitions imply "moving" over unitarily inequivalent representations, and this in general implies non-trivial homotopy mapping between the $(x, \beta)$ variability domain and the group manifold. The order parameters $v(x, \beta)$ and $\sigma(x, \beta)$ introduced above provide a mapping between the variation domains of $(x, \beta)$ and the *space of the unitarily inequivalent representations* of the canonical commutation relations, i.e. the set of Hilbert spaces where the operator field $\phi$ is realized for different values of the order parameter. The invariance under the theory symmetry group then *necessarily* leads to NG boson condensation functions with topological singularities. In other words, since phase transitions imply observable changes in the system physical properties, boson condensation functions relevant to phase transitions necessarily carry topological singularities (non-singular functions have no observable effects, indeed). This explains why we observe defect formation in the process of phase transition.

In the case of the kink there are no NG modes, nevertheless the topologically non-trivial kink solution requires the boson condensation function to carry divergence singularity (at spatial infinity).

This work has been partially supported by INFM, MURST and the ESF Network on Topological defect formation in phase transitions.

## References

1. Umezawa, H. (1993) *Advanced Field Theory: Micro, Macro and Thermal Physics.* American Institute of Physics, New York
2. Umezawa, H., Matsumoto, H. and Tachiki, M. (1982) *Thermo Field Dynamics and Condensed States.* North-Holland Publ.Co., Amsterdam
3. Kibble, T.W.B., (1976) Topology of cosmic domains and strings, *J. Phys. A: Math. Gen.*, **A9**, pp. 1387-1398; (1980) Some implications of a cosmological phase transition, *Phys. Rep.*, **67**, pp. 183-199
   Vilenkin, A., (1985) Cosmic strings and domain walls, *Phys. Rep.*, **121**, pp. 263-367
4. Zurek, W.H. (1997) Cosmological experiments in condensed matter systems, *Phys. Rep.*, **276**, pp. 177-221 and refs. therein quoted
5. Alfinito, E., Romei, O. and Vitiello, G. (1999) in preparation
6. Goldstone, J. (1961) Theories with "superconductor" solutions, *Nuovo Cimento*, **19**, pp. 154-164
   Goldstone, J., Salam, A. and Weinberg, S. (1962) Broken symmetries, *Phys. Rev.*, **127**, pp. 965-970
7. Nambu, Y. and Jona-Lasinio, G. (1961) Dynamical model of elementary particles based on an analogy with superconductivity, *Phys. Rev.*, **124**, pp. 246-254
8. Inönü, E. and Wigner, E.P. (1953) On the contraction of groups and their representations, *Proc. Nat. Acad.Sci. US*, **39**, pp. 510-524
9. De Concini, C. and Vitiello, G. (1976) Spontaneous breakdown of symmetry and group contractions, *Nucl. Phys.*, **B116**, pp. 141-156
10. von Neumann, J. (1955) *Mathematical foundation of Quantum Mechanics.* Princeton University Press, Princeton
11. Bratteli, O. and Robinson, D.W. (1979) *Operator Algebras and Quantum Statistical Mechanics.* Springer, Berlin
12. Itzykson, C. and Zuber, J.B. (1980) *Quantum Field Theory.* MacGraw-Hill Book Co., New York
13. Vitiello, G. (1974) *Dynamical Rearrangement of Symmetry.* PhD Thesis, University of Wisconsin, Milwakee

14. Anderson, M.P. (1958) Coherent excited states in the theory of superconductivity: gauge invariance and the Meissner effect, *Phys. Rev.*, **110**, pp. 827-835; New method in the theory of superconductivity, *Phys. Rev.*, **110** pp. 985-986
15. Higgs, P. (1964) Broken Symmetries and the masses of gauge bosons, *Phys. Rev. Lett.*, **13**, pp. 508-509
16. Kibble, T.W.B. (1967) Symmetry breaking in non Abelian gauge theories, *Phys. Rev.*, **155**, pp. 1554-1561
17. Matsumoto, H., Papastamatiou, N.J., Umezawa H. and Vitiello, G. (1975) Dynamical rearrangement of symmetry in the Anderson-Higgs-Kibble mechanism, *Nucl. Phys.*, **B97**, pp. 61-89
18. Matsumoto, H., Papastamatiou, N.J. and Umezawa, H. (1974) The formulation of spontaneos breakdown of symmetry in the path-integral method, *Nucl. Phys.*, **B68**, pp. 236-254
19. Matsumoto, H., Papastamatiou, N.J. and Umezawa, H. (1974) The Goldstone theorem and dynamical rearrangement of symmetry in the path-integral formalism, *Nucl. Phys.*, **B82**, pp. 45-68
20. Shah, M., Umezawa, H. and Vitiello, G. (1974) Relation among spin operators and magnon, *Phys. Rev.*, **B10**, pp. 4724-4736
21. Matsumoto, H., Umezawa, H., Vitiello, G. and Wyly, J.K. (1974) Spontaneous breakdown of a non-Abelian symmetry, *Phys. Rev.*, **D9**, pp. 2806-2813
22. Matsumoto, H., Papastamatiou, N.J. and Umezawa, H. (1975) The boson transformation and the vortex solutions, *Nucl. Phys.*, **B97**, pp. 90-124
23. Matsumoto, H., Sodano, P. and Umezawa, H. (1979) Extended objects in quantum systems and soliton solutions *Phys. Rev.*, **D19**, pp. 511-516
24. Wadati, M., Matsumoto, H. and Umezawa, H. (1978) Extended objects created by Goldstone bosons, *Phys. Rev.*, **D18**, pp. 520-531
25. Tze, H.C. and Ezawa, Z.F. (1975) Quark confinement and quantum vortex strings in Nambu-Heisenberg model, *Phys. Lett.*, **55B**, pp. 63-74; (1975) Dynamics of vacuum excitations and quark confinement in quantum field theories, *Nucl. Phys.*, **B96**, pp.264-284
26. Manka, R. and Vitiello, G. (1990) Topological solitons and temperature effects in gauge field theory, *Ann. Phys. (N.Y.)*, **199**, pp. 61-83
27. Nielsen, H.B. and Olesen, P. (1973) Vortex line models for dual strings, *Nucl. Phys.*, **B61**, pp. 45-61
28. Koma, T. and Tasaki, H. (1994) Symmetry breaking and finite-size effects in quantum many-body systems, *J. Stat. Phys.*, **76**, pp. 745-803
29. Manka, R., Kuczynski, J. and Vitiello, G. (1986) Vacuum structure and temperature effects *Nucl. Phys.*, **B276**, pp. 533-548

# MAGNETIC MONOPOLES AND VORTICES IN THE STANDARD MODEL OF ELECTROWEAK INTERACTIONS

A. ACHUCARRO
*Department of Theoretical Physics*
*The University of the Basque Country*
*Apt 644, 48080 Bilbao, Spain*
*and*
*Institute for Theoretical Physics*
*University of Groningen*
*Nijenborgh 4, 9747 AG Groningen, The Netherlands*

These lectures start with an elementary introduction to the subject of magnetic monopoles which should be accesible from any physics background. In the Weinberg-Salam model of electroweak interactions, magnetic monopoles appear at the ends of a type of non-topological vortices called *electroweak strings.* These will also be discussed, as well as recent simulations of their formation during a phase transition which indicate that, in the (unphysical) range of parameters in which the strings are classically stable, they can form with a density comparable to topological vortices.

## 1. Introduction

Last year marked the 50th anniversary of one of P.A.M. Dirac's most profound and famous papers - on magnetic monopoles. While his 1931 paper is usually considered the official birthday of magnetic monopoles, his 1948 paper is where he really showed that it was possible to have a consistent quantum theory of magnetic poles in conjunction with electric charges, and described the interaction between them [1].

A standard reference on magnetic monopoles is Preskill's lectures in the 1985 Les Houches school [2]. Here I have tried to present magnetic monopoles and vortices in a way that makes them accessible to physicists who are not so familiar with the language of high energy physics, in particular with *e.g.* non-abelian gauge theories. As a result, these lectures are much less technical.



*Y.M. Bunkov and H. Godfrin (eds.),*
*Topological Defects and the Non-Equilibrium Dynamics of Symmetry Breaking Phase Transitions,* 193–211.

The theme of this school is the use of topological defects as a tool to understand the dynamics of phase transitions out of equilibrium. It turns out that, since magnetic monopoles and electroweak vortices are non-topological in the Weinberg-Salam model, their study can be particularly interesting in order to understand the role of gauge fields in defect formation during phase transitions.

## 2. The elusive monopole

The problem with magnetic monopoles is well known in the context of Maxwell's equations. If the electromagnetic field is described by the vector potential $\vec{A}$, then

$$\vec{\nabla} \cdot \vec{B} = 0 \tag{1}$$

so there can be no sources or sinks for the magnetic field.

It would be fair to say that the experimental evidence for the existence of magnetic monopoles is not good. The 1998 Review of Particle Properties by the Particle Data Group [3] shows the result of monopole searches in particle accelerators: not a monopole in sight. On the other hand, cosmic ray searches have essentially only one event for which there seems to be no alternative explanation, observed by Cabrera in Stanford in 1982 [4]. The Cabrera detector, like many others that failed to find anything before or after it, consisted of a superconducting ring where a persistent current was monitored for a long time; in [4] the loop had an area of 20 $cm^2$ and was monitored for a total of 151 days. During this time a single event was recorded which could be ascribed to a magnetically charged particle with one Dirac unit of magnetic charge $q_m = 2\pi\hbar c/e$. A magnetic monopole.

Under the circumstances, the experimental evidence is neatly summarized by the sentence (whose author is unfortunately unknown to me) *"It is not clear that nobody has ever seen a magnetic monopole; what is clear is that nobody has ever seen two"*. And yet since Dirac's seminal work there have been over three thousand papers in the literature about magnetic monopoles! [5]

The reasons behind this fascination with monopoles have evolved with time, but they are basically three:

- the existence of monopoles would explain the quantization of electric charge (for which there is no alternative explanation to this day). In his 1948 paper Dirac says: *"If one supposes that a particle with a single magnetic pole can exist and that it interacts with charged particles, the laws of quantum mechanics lead to the requirement that the electric charges shall be quantized – all charges must be integral multiples of a unit charge e connected with the pole strength by the formula* $eg = \frac{1}{2}\hbar c$*. Since electric charges are known to be quantized and no reason for this has yet been proposed apart from* *the existence of magnetic poles, we have here a reason for taking magnetic monopoles seriously"*. He then goes on to say that the fact that they have not yet been observed may be ascribed to the large value of the quantum of the pole.

- a large class of theories that include electromagnetism as a subset predict magnetic monopoles as solitons, as was shown by 't Hooft and Polyakov in 1974 [6], and

- if magnetic monopoles exist, Maxwell's equations are symmetric under the exchange of electric and magnetic fields. This *duality* symmetry relates small electric charge to large magnetic charge and viceversa. A generalization of this symmetry to non-abelian theories would mean that the dual theory (of weakly coupled monopoles) could be used to understand strongly coupled non-abelian gauge theories and, in particular, confinement, for which there is no other analytic approach. While the idea is not new, some of the most important developments in this area are fairly recent; but they fall outside the scope of these lectures, and I refer the reader to an excellent review by Harvey [7].

In what follows I will take $c = \hbar = 1$. I will also depart from Dirac's notation and use $q_m$ to refer to the magnetic charge; $g$ will be the $SU(2)$ coupling constant.

One final comment. I think everyone attending this school is aware of the language problems between high energy and condensed matter physicists. A relatively common source of confusion is the use of the word *gauge symmetry*, which can mean different things to the two communities. We all agree that electromagnetism has a gauge symmetry, it is the symmetry that allows *local* (that is, position–dependent) changes in the phase of the wave function and a compensating change of gauge in the vector potential

$$\begin{aligned} \psi(t,\vec{x}) &\to e^{ie\chi(t,\vec{x})}\psi(t,\vec{x}) \\ A_\mu(t,\vec{x}) &\to A_\mu(t,\vec{x}) + \nabla_\mu\chi(t,\vec{x}) \qquad \mu = 0,1,2,3\,. \end{aligned} \tag{2}$$

The electromagnetic tensor $F_{\mu\nu} \equiv \nabla_\mu A_\nu - \nabla_\nu A_\mu$, also known as the *field strength*, is unchanged by this transformation, while the covariant derivative of $\psi$, $\nabla_\mu\psi - ieA_\mu\psi$, transforms in the same way as $\psi$ itself (thus the name *covariant*).

High energy physicists use the term "gauge symmetry" to indicate any symmetry which is local, whether or not it corresponds to a $U(1)$ transformation. Most condensed matter physicists, on the other hand, will talk about gauge symmetries to indicate a change in the phase of the wave function (a $U(1)$ transformation), whether or not there are vector potentials around. Thus, the transformation

$$\Psi \to e^{i\alpha}\Psi \qquad \alpha = \text{const} \tag{3}$$

is a gauge transformation in the condensed matter literature, but not in the high energy literature (where it would be called a *global* $U(1)$ transformation). On the other hand a high energy physicist will talk about *e.g.* a $SU(2)$ or $SO(3)$ gauge transformation, meaning what is best described as a "position-dependent rotation" in internal space. Since there are three degrees of freedom associated with rotations, we need three vector potentials

$$\underline{W}_\mu = (W_\mu^1, W_\mu^2, W_\mu^3) \qquad \mathbf{W}_\mu \equiv \underline{W}_\mu \cdot \underline{\tau} = \frac{1}{2}\begin{pmatrix} W_\mu^3 & W_\mu^1 - iW_\mu^2 \\ W_\mu^1 + iW_\mu^2 & -W_\mu^3 \end{pmatrix}, \tag{4}$$

where $\underline{\tau} = (\tau^1, \tau^2, \tau^3)$ are the Pauli spin matrices. The transformation law for the gauge potentials is

$$\mathbf{W}_\mu(t,\vec{x}) \to M^{-1}(t,\vec{x})\mathbf{W}_\mu(t,\vec{x})M(t,\vec{x}) + \frac{1}{g}M^{-1}(t,\vec{x})\nabla_\mu M(t,\vec{x}) \tag{5}$$

where $M(t,\vec{x})$ is a $SU(2)$ matrix at each point in spacetime; note that if $M$ were independent of position, as in the case of spin, there would be no need for gauge potentials and the symmetry would be called a *global* $SU(2)$ symmetry.

The transformation is non-abelian and moreover the field strength, which has to be generalized from $F_{\mu\nu} = \partial_\mu A_\nu - \partial_\nu A_\mu$ to

$$\mathbf{G}_{\mu\nu} = \partial_\mu \mathbf{W}_\nu - \partial_\nu \mathbf{W}_\mu + g[\mathbf{W}_\mu \mathbf{W}_\nu - \mathbf{W}_\nu \mathbf{W}_\mu] \tag{6}$$

or

$$\underline{G}_{\mu\nu} = \partial_\mu \underline{W}_\nu - \partial_\nu \underline{W}_\mu + g\underline{W}_\mu \times \underline{W}_\nu \tag{7}$$

(with the cross product taken in internal space) is no longer invariant but changes like

$$\underline{G}_{\mu\nu} \cdot \underline{\tau} \to M^{-1}(t,\vec{x})\; \underline{G}_{\mu\nu} \cdot \underline{\tau}\; M(t,\vec{x}) \tag{8}$$

The transformation law of the scalars depends on the group representation to which they belong and we list here two that will be relevant later. The fundamental representation of $SU(2)$ is a doublet of complex fields $\Phi = (\phi_1, \phi_2)$ whose transformation law and covariant derivative are

$$\Phi \to M^{-1}(t,\vec{x})\Phi \qquad \text{and} \qquad D_\mu\Phi \equiv \nabla_\mu\Phi + g\mathbf{W}_\mu\Phi \qquad \text{respectively.} \tag{9}$$

The adjoint representation is a triplet of real scalars $\underline{\phi} = (\phi^1, \phi^2, \phi^3)$ which, like the gauge potentials, can be assembled into a matrix $\mathbf{\Phi} \equiv \underline{\phi} \cdot \underline{\tau}$ with transformation law

$$\mathbf{\Phi}(t,\vec{x}) \to M^{-1}(t,\vec{x})\mathbf{\Phi}(t,\vec{x})M(t,\vec{x}) \tag{10}$$

and covariant derivative

$$D_\mu\mathbf{\Phi} \equiv \nabla_\mu\mathbf{\Phi} + g[\mathbf{W}_\mu\mathbf{\Phi} - \mathbf{\Phi}\mathbf{W}_\mu] \tag{11}$$

or, equivalently,

$$D_\mu \underline{\Phi} \equiv \nabla_\mu \underline{\Phi} + g \underline{W}_\mu \times \underline{\Phi} \ . \tag{12}$$

In these expressions, $\mu = 0, 1, 2, 3$. In what follows, we will "work in temporal gauge", setting the time component of the gauge fields to zero. Thus, the vector potential (or gauge potential) will be a three vector and we will use the notation $\vec{\nabla} \times \underline{\vec{W}} + g \underline{\vec{W}} \times \underline{\vec{W}}$ for the field strength; the expressions above should serve to clarify whether the cross product is taken in internal space, in real space or in both.

## 3. Do-it-yourself magnetic monopoles

This section is "adapted" (*i.e.* taken) from Coleman's 1974 Erice lectures [8]. He refers to this as "the monopole hoax", a joke to be played (or at least attempted) by a cunning theorist on a gullible experimenter. In view of the number of cunning experimenters in this audience I will refrain from comments and just describe here how to build your own magnetic monopole.

1) Take a solenoid. It has to be very long and very thin so as to be invisible; as Coleman says, *it helps if the solenoid is many miles long and considerably thinner than a fermi (this is very much a gedanken hoax).*

2) Put one end at the experimenter's laboratory,

3) hide the other end, and

4) turn on the current.

For a gullible theorist this may pass as a magnetic monopole, but of course there is a way in which the solenoid could be detected: through Aharonov-Bohm scattering.

The interference pattern in a double-slit experiment is shifted when a solenoid is placed between the slits and the screen. Even if the particle trajectories remain well outside the solenoid, their wave functions $\psi_1$, $\psi_2$ acquire a phase $exp\,[ie \int \vec{A} \cdot \vec{dl}]$ (with $e$ the electric charge of the particle and the integral taken along the particle's path); if the paths are on either side of the solenoid, the interference pattern changes because the probability amplitude $|\psi_1 + \psi_2|^2$ becomes

$$|\, e^{ie \int_1 \vec{A} \cdot \vec{dl}} \psi_1 + e^{ie \int_2 \vec{A} \cdot \vec{dl}} \psi_2|^2 = |\psi_1 + e^{ie \oint \vec{A} \cdot \vec{dl}} \psi_2|^2 \ , \tag{13}$$

where $\oint \vec{A} \cdot \vec{dl}$, taken around the solenoid, measures its magnetic flux.

Notice that, if the flux is an integer multiple of $2\pi/e$, the solenoid becomes undetectable even quantum mechanically in our gedanken experiment – it is called a *Dirac string.*

The vector potential of a monopole whose Dirac string is along the negative $z$-axis is given (in spherical coordinates $(r, \theta, \varphi)$ centred on the monopole) by

$$\vec{A}_N \cdot d\vec{x} = \frac{q_m}{4\pi}(1 - \cos\theta)d\varphi, \qquad \theta \neq \pi \tag{14}$$

giving rise to a radial magnetic flux[1]

$$\vec{B} \cdot d\vec{S} = \frac{q_m}{4\pi} \sin\theta d\theta d\varphi \qquad \text{or} \qquad \vec{B} = \frac{q_m}{4\pi}\frac{1}{r^2}\hat{r} \tag{15}$$

where $q_m$ is the magnetic charge. The vector potential is singular on the Dirac string, which is located at $\theta = \pi$, but regular everywhere else.

Since the magnetic flux of the monopole is supplied by the Dirac string, the condition that the flux through the string should be a multiple of $2\pi/e$ gives rise to the famous Dirac quantization condition for the magnetic charge $q_m$ of the monopole:

$$q_m = \frac{2\pi N}{e} \qquad \text{or, reintroducing } \hbar \text{ and } c, \qquad \frac{eq_m}{4\pi} = \frac{N}{2}(\hbar c) \tag{16}$$

Thus, the existence of one monopole would be enough to force electric charge to be quantized!

Note that we can use gauge invariance to change the position of the Dirac string; an equivalent description of this monopole is given by the vector potential

$$\vec{A}_S \cdot d\vec{x} = -\frac{q_m}{4\pi}(1 + \cos\theta)d\varphi, \qquad \theta \neq 0 \tag{17}$$

which is singular only at $\theta = 0$ (the gauge transformation between $\vec{A}_N$ and $\vec{A}_S$ is singular in the position of both the old and new strings, of course). You should not worry about these singularities – in the next section we will eliminate the Dirac string altogether.

## 4. The Wu-Yang construction of Dirac monopoles

Electromagnetism is not a theory of gauge potentials *per se*, but rather of equivalence classes of gauge potentials. This was exploited by Wu and Yang [9] to give a non-singular description of magnetic monopoles by "patching up" vector potentials that are regular in different regions, provided they are equivalent on the overlaps.

Consider any sphere with non-zero radius surrounding the monopole. In the northern hemisphere $\theta \in [0, \pi/2 + \epsilon]$ take $\vec{A} = \vec{A}_N$ and in the southern hemisphere $\theta \in [\pi/2 - \epsilon, \pi]$ take $\vec{A} = \vec{A}_S$. In the overlap region,

1 Note that $\vec{B}$ does not include the singular contribution from the Dirac string.

$\pi/2 - \epsilon < \theta < \pi/2 + \epsilon$, the two descriptions are related by a *regular* gauge transformation,

$$(\vec{A}_N - \vec{A}_S) \cdot d\vec{x} = \vec{\nabla}\chi \cdot d\vec{x} \qquad \text{where } \chi(\varphi) = \frac{q_m}{2\pi}\,\varphi \ . \tag{18}$$

This object has magnetic charge $q_m$, as can be seen by computing the magnetic flux through the two hemispheres (using Stokes' theorem). Note that, even though $\chi$ is not singlevalued, $\vec{\nabla}\chi$ is, and therefore the gauge transformation is well defined on the gauge potentials. Moreover, singlevaluedness of the gauge transformation on the wave functions, $e^{ie\chi}$, in the overlap equatorial region implies the Dirac quantization condition! $q_m \equiv \chi(2\pi) - \chi(0) = 2\pi N/e$.

We have eliminated the need for a Dirac string – note that the only singularity in this description is at the origin, $r = 0$. But this must be a real singularity because the energy of the monopole diverges as $r \to 0$ due to the $1/r^2$ behaviour of the magnetic field,

$$E = \int d^3x \, \frac{1}{2} \, (|\vec{E}|^2 + |\vec{B}|^2) \sim \int_0^\infty \frac{dr}{r^2} \tag{19}$$

Electromagnetism is not the only force in nature. There are the weak and strong nuclear forces, and also gravity. We think that forces may become unified into one kind of interaction (a Unified Theory) at high energies [2]. When theorists started to investigate possible unified theories they found a surprise...

## 5. 't Hooft–Polyakov monopoles

One of the first attempts to unify the electromagnetic and weak interactions was the $O(3)$ Georgi-Glashow model [10] in which the fundamental fields are a triplet of (real) scalars and a triplet of gauge potentials.

$$\begin{aligned} \underline{\phi} &= (\phi^1, \phi^2, \phi^3) \\ \underline{\vec{W}} &= (\vec{W}^1, \vec{W}^2, \vec{W}^3) \end{aligned} \tag{20}$$

The most important aspect from our point of view is the energy, since we are seeking to remove the divergence at $r = 0$. I am making many simplifying assumptions here (no time dependence, no electric fields), and only writing the terms in the energy that are relevant for the argument:

$$E = \int d^3x \, [(\vec{\nabla}\underline{\phi} + g\underline{\vec{W}} \times \underline{\phi})^2 + (\vec{\nabla} \times \underline{\vec{W}} + g\underline{\vec{W}} \times \underline{\vec{W}})^2 + \lambda(\underline{\phi}\cdot\underline{\phi} - \eta^2)^2] \tag{21}$$

2 A special class are so–called Grand Unified Theories, or GUTs, where the electromagnetic, weak and strong interactions are described by a single simple group.

First of all, the Georgi–Glashow model includes electromagnetism as a subset: the configuration

$$\begin{aligned} \phi^1 = \phi^2 = 0\,, \qquad & \qquad \phi^3 = \eta\,, \\ \vec{W}^1 = \vec{W}^2 = 0\,, \qquad & \qquad \vec{W}^3 = \vec{A}\,, \end{aligned} \tag{22}$$

where $\vec{A}$ is any solution to Maxwell's equations, is also a solution of the full non–abelian field equations of the Georgi–Glashow model. In particular, the Wu-Yang (or Dirac) monopole is a solution. However, it is not a stable solution!

Indeed, the $1/r^2$ divergence in the monopole energy is now coming from $\vec{\nabla}\times\vec{W}^3$ in the $\int(\vec{\nabla}\times\vec{W}^3+g\vec{W}^1\times\vec{W}^2)^2$ term and could be controlled if $\vec{W}^1$ and $\vec{W}^2$ acquired non-zero values $\sim 1/r$. This is consistent with the fact that $\vec{W}^\pm \equiv (\vec{W}^1 \mp i\vec{W}^2)/\sqrt{2}$ are charged fields (the *W–bosons*) with charge $\pm g$ respectively and a magnetic moment $ig\vec{W}^-\times\vec{W}^+$ which couples to the $\vec{W}^3$ magnetic field, so their presence can reduce the magnetic energy. On the other hand, such "W-condensation" has two immediate effects: one is an increase in energy coming from the new, non–zero $(\vec{\nabla}\times\vec{W}^{1,2}\pm g\vec{W}^3\times\vec{W}^{2,1})^2$ terms; the other is that the scalar gradients $D_\mu\phi^{1,2} \sim g\underline{\vec{W}}^{2,1}\times\underline{\phi}^3$ now diverge as $1/r$. However this problem is eliminated if $\phi^3\sim r$ as $r\to 0$. The condition $\underline{\phi}(r=0)=0$ imposes a penalty in energy from the $\int\lambda(\underline{\phi}\cdot\underline{\phi}-\eta^2)^2$ term, but this is *finite* – thus, the result is always energetically favourable to the singular abelian monopole that we started with.

In the case when magnetic charge is two Dirac units, $q_m = 4\pi/g$ (see below for an explanation of this condition), this instability leads to the 't Hooft–Polyakov monopole [6], a spherically symmetric configuration describing a non-singular magnetic monopole of finite mass [11],

$$\begin{aligned} &\phi^1 = \phi^2 = 0\,, && \phi^3 = \eta\rho(r) \\ &\vec{W}^1 - i\vec{W}^2 = \frac{1}{g}\frac{f(r)}{r}e^{i\varphi}\left(\frac{\hat{\varphi}}{\sin\theta} - i\hat{\theta}\right), && \vec{W}^3 = \frac{1}{g}(1-\cos\theta)\frac{1}{r\sin\theta}\hat{\varphi}\,, \end{aligned} \tag{23}$$

with $f(r)\to 1$, $f'(r)\sim -r$ and $\rho(r)\sim r$ as $r\to 0$ and $f(r)\to 0$, $\rho(r)\to 1$ as $r\to\infty$.

Note that only the small $r$ behaviour of the fields has changed; in particular, the magnetic charge of the monopole remains the same. After a (singular) gauge transformation it reduces to the more familiar form [6]

$$\begin{aligned} \underline{\phi} &= \eta\rho(r)\hat{r} \\ \underline{\vec{W}} &= \frac{1}{g}(f(r)-1)\hat{r}\times\vec{\nabla}\hat{r}\ , \end{aligned} \tag{24}$$

which shows that the 't Hooft–Polyakov monopole is a topological defect (usually called a *hedgehog* because of the way the scalar field points radially outwards). The zero value of the scalar field at the origin $r = 0$ is forced by the non-trivial winding of the scalar field.

It remains to explain why the restriction to two units of magnetic charge. In the Wu-Yang construction we started with a sphere of non–zero radius, say $R$, divided into two hemispheres overlapping at the equator. If the monopole has $N$ units of magnetic charge the gauge transformation in the overlap region is a phase rotation by $2\pi N$. W–condensation replaces the singularity at the origin by an everywhere regular core. Since nothing changes outside the core, the patching condition for the 't Hooft– Polyakov monopole remains a $2\pi N$ rotation for all $R > r_{\text{core}}$. But this cannot be true inside the core: since there is no singularity, the gauge transformation must also change continuously so that it becomes the identity when we reach $r = 0$. If the gauge group is $U(1)$, this is simply not possible, and all monopoles are singular. But in $SU(2)$, a $2\pi$ rotation is not continuously connected to the identity whereas a $4\pi$ rotation is! Thus, only monopoles with even $N$ can be non-singular. Of all these, it turns out that only $N = 2$ remains spherically symmetric after W–condensation.

The existence of magnetic monopoles is a very generic prediction for a large class of theories containing electromagnetism. Moreover, they should be produced in large numbers in the early Universe [12]. The fact that we do not observe those monopoles is a serious challenge to cosmologists, and has become known as *the monopole problem.*

## 6. Magnetic monopoles in the Weinberg–Salam model; electroweak strings and dumbells

The standard model of electroweak interactions has a $SU(2) \times U(1)$ gauge symmetry, corresponding to weak isospin and hypercharge respectively. Its bosonic sector comprises a neutral scalar field $\phi^0$, a charged scalar field $\phi^+$, and the vector potentials corresponding to the massless photon $\vec{A}$ and three massive vector bosons: the charged W–bosons ($\vec{W}^{\pm}$) and the neutral $\vec{Z}$. The fermionic sector consists of the three families of quarks and leptons

$$\begin{pmatrix} \nu_e \\ e \\ u \\ d \end{pmatrix} \quad \begin{pmatrix} \nu_\mu \\ \mu \\ c \\ s \end{pmatrix} \quad \begin{pmatrix} \nu_\tau \\ \tau \\ t \\ b \end{pmatrix} \tag{25}$$

In the Weinberg-Salam model, the electromagnetic and $Z$-fields are combinations of the $SU(2)$ and $U(1)$ gauge potentials ($\underline{\vec{W}}$ and $\vec{Y}$ respectively):

$$\vec{Z} \equiv \cos\theta_w \vec{W}^3 - \sin\theta_w \vec{Y} \ , \qquad \vec{A} \equiv \sin\theta_w \vec{W}^3 + \cos\theta_w \vec{Y} \ , \tag{26}$$

where $\theta_w$ is called the weak mixing, or Weinberg, angle. Its measured value is $\sin^2\theta_w \approx 0.23$.

In this case, the field that satisfies Maxwells equations at low energy is $\vec{A}$ and, being massless, it is the only vector potential that can give rise to long-range electric and magnetic fields and thus magnetic monopoles. Its configuration far from the monopole will be exactly like what was discussed in sections 3 and 4. Very close to the monopole, though, we would expect other fields to condense due to the intense magnetic field, changing the core structure.

But there is a problem. Note that, since the electromagnetic field has a hypercharge component and hypercharge is an abelian field, isolated magnetic monopoles are always singular at the origin. We are back to square one!

In order to have monopoles with regular cores one has to embed the $SU(2)\times U(1)$ symmetry of the Weinberg–Salam model into larger symmetry groups. We already mentioned Grand Unified Theories (GUTs, for short), where one simple group not only contains the electroweak interaction, but also the strong interaction. These monopoles are very heavy, because the unification of these forces occurs at very high energies ($\sim 10^{16}$ GeV) and the fields that condense at the core are very massive. Far too heavy to be produced in a particle accelerator.

But there are also lighter magnetic monopoles in the Weinberg–Salam model: they occur as monopole–antimonopole pairs connected by a vortex (the vortex carries magnetic flux of the $Z$-boson, and it is usually called a *$Z$–string* or an electroweak string). Such configurations were called *dumbells* by Nambu, who first considered them in 1977 [13]. The dumbell is rotating to avoid longitudinal collapse, and its mass is estimated at a few TeV.

Their internal structure is rather interesting. The $SU(2)$ fields are those of a 't Hooft–Polyakov monopole–antimonopole pair, while the hypercharge $U(1)$ field configuration resembles that of a solenoid joining the monopole and antimonopole. As a result, the combination inside the solenoid is precisely the magnetic part of the $Z$ field, whereas the magnetic field that emanates from the solenoid ends is the massless electromagnetic field, *and there are no singularities anywhere.*

In some respects, the structure of the $Z$–string is similar to that of a magnetic vortex in an Abrikosov lattice that appears in a type II superconductor subjected to an external magnetic field. In the Weinberg–Salam model, the role of the vector potential is taken by the $Z$–field, and the order parameter is the neutral Higgs field $\phi^0$. But there is a very important difference: electroweak strings are non–topological, and therefore not necessarily stable.

A review of electroweak strings where the stability issue is discussed in some detail can be found in [14]. Stability depends on the value of the Weinberg angle and the masses of the various fields. In the physical range of these parameters, it is found that infinitely long, straight, bare strings are classically unstable (they are only stable for $\theta_w \sim \pi/2$ and $m_{\text{Higgs}} < m_Z$). On the other hand, stability improves for short segments and in the presence of magnetic fields. Finally, the strings are superconducting and fermion modes on the string might also stabilize them, although this is still under discussion. The stability of dumbells remains an open question.

Even in the region of parameter space where strings are (classically) stable, the fact that stability is not topological immediately raises the question of whether a network of strings would form in a phase transition via the Kibble mechanism. We now turn to this question, and we will focus on the (unphysical) limit of the Weinberg–Salam model in which the $SU(2)$ symmetry becomes global, while keeping the hypercharge $U(1)$ symmetry local. In this limit, known as the semilocal model, electroweak strings are classically stable.

## 7. Semilocal strings

The semilocal model is obtained when the complex scalar field $\phi$ in the Abelian Higgs (or Landau–Ginzburg) model is replaced by an $SU(2)$ doublet of complex fields $(\phi, \psi)$. [3]

It is also (the bosonic sector of) the Weinberg–Salam model in the limit in which the $SU(2)$ gauge coupling is set to zero; in this limit, the $W$-bosons and the photon decouple and the symmetry is $SU(2)_{\text{global}} \times U(1)_{\text{local}}$. The only gauge field is the neutral $Z$ boson, which coincides with the hypercharge field. The model has vortices (*semilocal strings*) whose properties are intermediate between electroweak strings and Abrikosov–Nielsen–Olesen vortices (see [14]).

The energy per unit length of cylindrically symmetric configurations is

$$E = \int d^2x [|(\vec{\nabla} - iq\vec{Y})\phi|^2 + |(\vec{\nabla} - iq\vec{Y})\psi|^2 + \frac{1}{2}(\vec{\nabla} \times \vec{Y})^2 + \lambda(\phi^2 + \psi^2 - \frac{\eta^2}{2})^2] \tag{27}$$

and it turns out that, even though the vacuum manifold is a three-sphere,

$$\phi^2 + \psi^2 = \frac{\eta^2}{2} \tag{28}$$

which is simply connected, $\pi_1(S^3) = 1$, there *are* stable strings if the scalar mass is smaller than the vector mass. These strings are not only stable

3 Similar systems have been considered in the condensed matter literature [15, 16].

classically, they are also stable to semiclassical tunnelling and to breaking by monopole pairs. If $m_{\text{scalar}} > m_{\text{vector}}$ the strings are classically unstable, and if the masses are equal there is a two-parameter family of configurations with the same energy where the quantized magnetic flux $2\pi/q$ spreads over an arbitrarily large core width.

Semilocal strings, like electroweak strings, can have open ends; but the monopoles at the ends are global monopoles and have a long–range interaction.

The stability of semilocal strings is well understood in terms of the competition between gradient energy and potential energy. The gauge field can compensate gradients in the (complex) phase of either of the two scalar fields, but it can only compensate both gradients *simultaneously* if there is a correlation between the phases of the two scalars; this correlation is present in the string solution, which optimizes gradient energy, but at the expense of a large concentration of potential energy at the core. If $\lambda/q^2$ is very large, it becomes energetically favourable to break the phase correlation in order to reduce potential energy, and the string is destroyed. If $\lambda$ is small, the cost in gradient energy to dissolve the string becomes too high and the string is stable.

Since the strings are non-topological, the question immediately arises as to whether they would form at all in a phase transition. In particular, stability depends on the mass ratio $m_{\text{scalar}}/m_{\text{vector}}$ and we would expect the formation rates to reflect this dependence.

We now turn to a first attempt to answer these questions through numerical simulations. Note that setting $\psi = 0$ in the semilocal model gives the Abelian Higgs (or Landau–Ginzburg) model, thus making it possible to compare the rates of formation of semilocal strings to those of topological strings in a similar environment. The conclusion seems to be that semilocal strings do form, and in some cases with number densities comparable to those of their topological counterparts.

## 8. Numerical simulations of semilocal string formation

For details of the simulations we refer the reader to [17, 18]. Here we will just point out the main features and results, summarized in figure 5.

Space is discretized into a lattice with periodic boundary conditions. The equations of motion are solved numerically using a standard staggered leapfrog method, and a dissipation term ($\eta\dot{\phi}$, $\eta\dot{\psi}$ or $\eta\dot{Y}_i$) is added to each equation to reduce the relaxation time. A range of strengths of dissipation was tested, and it did not significantly affect the number densities obtained; the simulations displayed in the figures all have have $\eta = 0.5$. Note that in an expanding Universe the expansion rate would act as a sort of viscosity,

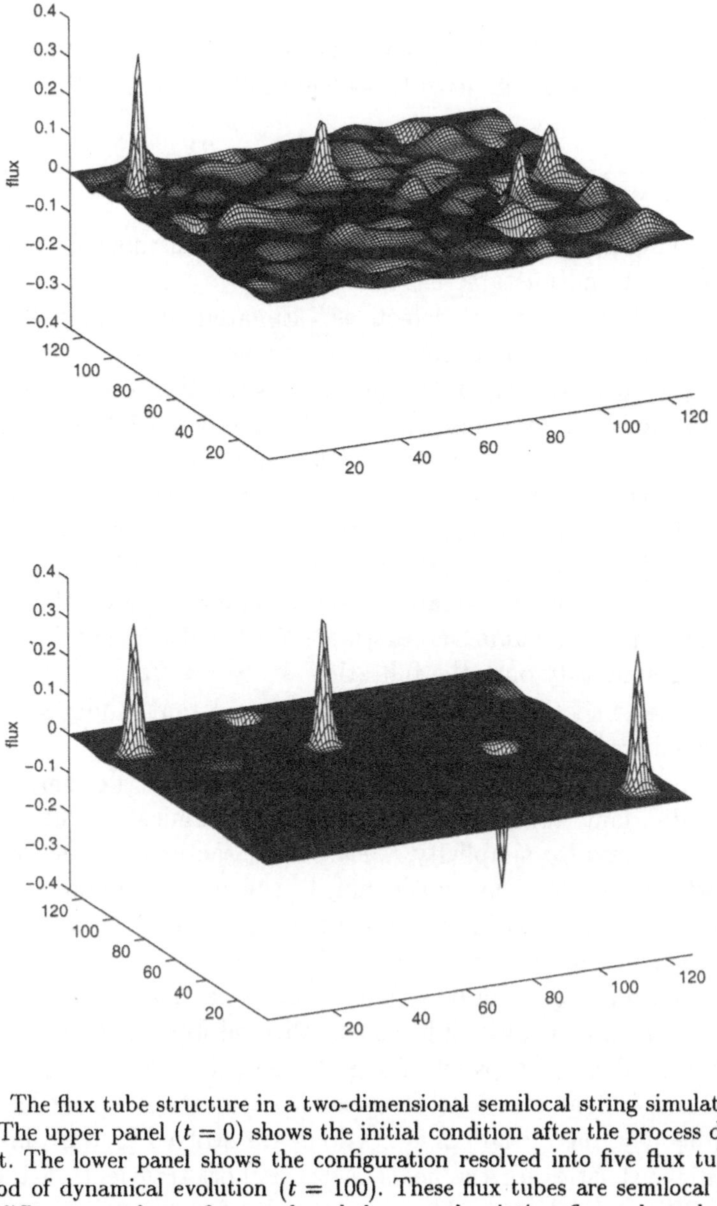


*Figure 1.* The flux tube structure in a two-dimensional semilocal string simulation with $\beta = 0.05$. The upper panel ($t = 0$) shows the initial condition after the process described in the text. The lower panel shows the configuration resolved into five flux tubes by a short period of dynamical evolution ($t = 100$). These flux tubes are semilocal vortices. Note the different numbers of upward and downward pointing flux tubes, despite the zero net flux boundary condition. The missing flux resides in the smaller 'nodules', made long-lived by the numerical viscosity; the expansion of the universe could have a similar effect and preserve these 'skyrmionic' configurations [20].

though $\eta$ would typically not be constant. However, this is not meant to be a cosmological simulation since spacetime is flat.

We work in temporal gauge. Then Gauss' law becomes a constraint derived from the gauge choice $Y_0 = 0$, and is used to test the stability of the code.

The inverse vector mass is taken as the unit of length and time, and $\eta^2$ as the unit of energy. In these units the dynamics is governed by a single parameter,

$$\beta = m^2_{\mathrm{scalar}}/m^2_{\mathrm{vector}} = \frac{2\lambda}{q^2} \tag{29}$$

which also determines the stability of the straight string solutions: they are stable if $\beta < 1$ and unstable if $\beta > 1$. [4]

The number density of defects is estimated by a multi-step process. The initial conditions are obtained with a generalization to non-topological strings of what is known to cosmologists as the Vachaspati–Vilenkin algorithm [19], followed by a short period of dynamical evolution.

• First, we generate a random initial configuration for the scalar fields drawn from the vacuum manifold, which is not discretised. If space is a grid of dimension $N^3$, the correlation length is chosen to be some number $p$ of grid points ($p = 16$ in [17, 18]; the size of the lattice is either $N = 64$ or $N = 256$). To obtain a reasonably smooth configuration for the scalar fields, we assign random vacuum values on a $(N/p)^3$ subgrid and interpolate the scalar field smoothly onto the full grid.

• We then find the gauge field configuration that minimizes the energy in this fixed scalar background.

Two dimensional test simulations have shown that the energy minimization is redundant, since the early stages of dynamical evolution carry out this role anyway; for simplicity, we used in practice a gauge field configuration which was close, but not equal, to the real minimum.

• An example of the initial conditions generated with this algorithm can be seen in Figure 1 in the case of a two-dimensional toy model with translational invariance in one dimension, say $z$. The plot shows magnetic field on the $x-y$ plane, which has a complicated flux structure with extrema of different values (top panel of Fig. 1), and it is far from clear which of these, if any, might evolve to form semilocal vortices; in order to resolve this ambiguity, the initial configurations are evolved forward in time with zero initial velocities for the fields. After a short transient, in the unstable regime $\beta > 1$ the flux quickly dissipates leaving no strings. By contrast, in the stable regime $\beta < 1$ stringlike features emerge when configurations in the "basin of attraction"of the semilocal string relax unambiguously into vortices (bottom panel of Fig. 1). We only count vortices after this relaxation process.

[4] $\beta$ is also the parameter that distinguishes type I ($\beta < 1$) from type II ($\beta > 1$) superconductors.

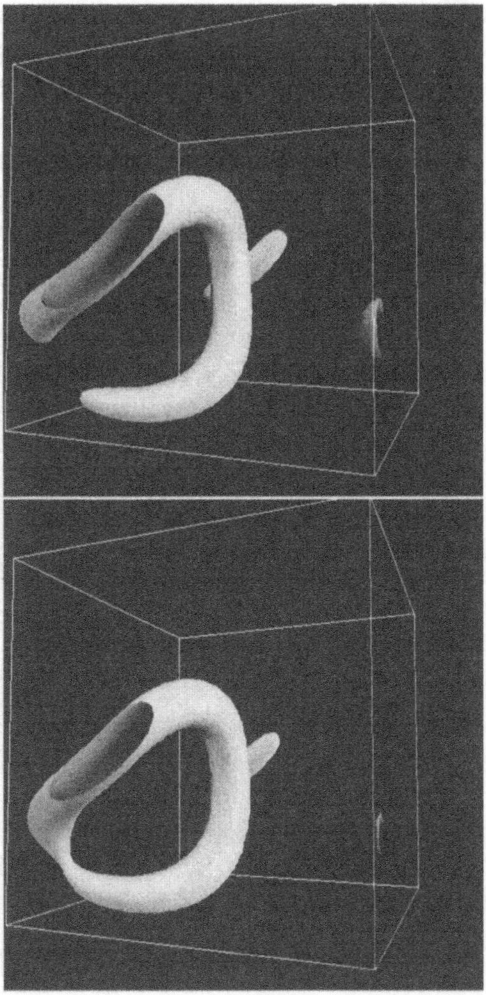

*Figure 2.* Loop formation from semilocal string segments. The figure shows two snapshots, at $t = 70$ and $t = 80$, of a $64^3$ numerical simulation of a network of semilocal strings with $\beta = 0.05$, where the ends of an open segment of string join up to form a closed loop. Subsequently the loops seem to behave like those of topological cosmic string, contracting and disappearing.

• One important point is that strings are always identified with the location of magnetic flux tubes, rather than by the zeroes of the scalar field. Figure 3 shows two snapshots of a simulation on a $256^3$ lattice. In order to make the figure, and also to compute the number density, we need to set a magnetic flux threshold – the strings that can be seen in that figure are made of those points in which the magnetic field exceeds half the maximum value in an Abrikosov-Nielsen-Olesen vortex. After the short transient, this threshold can be modified without significantly affecting the results, which are ratios of semilocal to topological string number densities (shown in figure 5). The error bars include, among other things, this threshold dependence as well as the dispersion between runs.

Since the initial conditions are somewhat artificial, the results were checked against various other choices of initial conditions, in particular different initial conditions for the gauge field and also initial conditions for the scalar field closer to a thermal environment (see Fig. 4). However, it cannot be sufficiently stressed that it is not a realistic thermal simulation of the phase transition. There is no thermal noise and all the initial conditions in [17, 18] had zero initial velocities for the fields – initial conditions with non-zero field momenta have not yet been investigated –. What the simulations show is that the rate of formation of these non-topological vortices is not only non-zero but in certain cases it could even be comparable to that of their topological counterparts. Obviously the details of the transition can and should now be investigated in full.

## 9. Discussion and outlook

The physical mechanism behind the observed string formation by accretion of magnetic field in Figure 1 or by growth of string segments in Figures 2 and 3 is not very different from the formation of Abrikosov lattices in a type II superconductor, even though in a system such as the early Universe it makes no sense to talk about external magnetic fields. Because of the $Y^2\Phi^2$ term in the energy, the magnetic field does not like to coexist with the superconducting ($\phi \neq 0$) phase of the scalars.

In the topological strings case one usually argues that the non-zero winding in the phase of the scalar field forces a zero of the Higgs, and magnetic flux gathers there in a vortex. Conversely, what we are observing in the semilocal model is that if there is a sufficiently large concentration of magnetic flux in a small region, for instance near the end of a string segment or maybe due to a fluctuation, a line of zeroes of the Higgs can develop there. Small segments of string, on the other hand, will tend to contract and disappear.

It seems clear that these results should extend to electroweak strings with $g \neq 0$ as long as we remain in the region of parameters where the strings are classically stable [5]. In the region of stability, we expect a non-zero density of electroweak vortices to form in a phase transition. Preliminary

5 Note that the monopoles at the ends of semilocal strings are global, whereas those at $g \neq 0$ will have finite size cores. However, the core size grows as $1/g$ and, for sufficiently small $g$, it may be larger than the average distance between string segments, causing the strings to grow

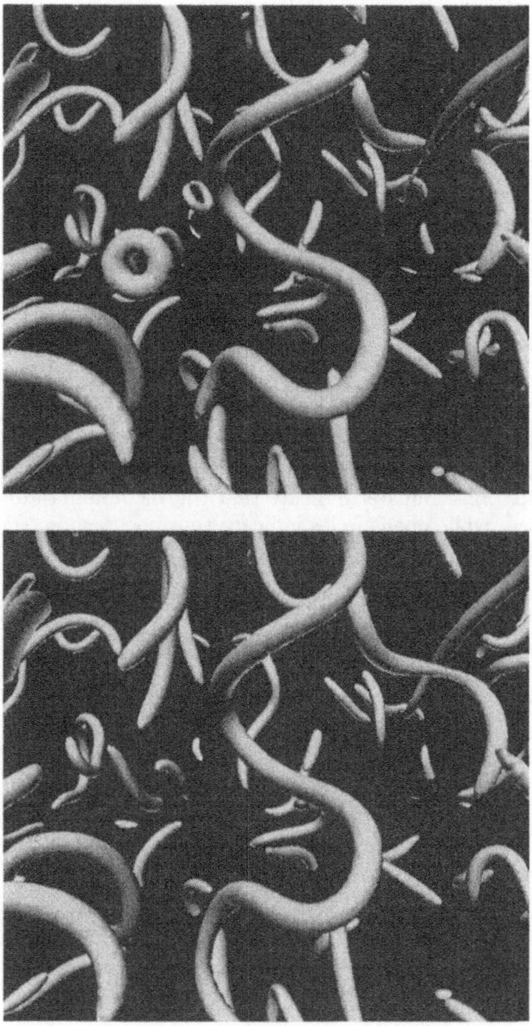

*Figure 3. The growth of string segments to form longer strings.* The figure shows two snapshots, at time $t = 60$ and $t = 70$ of a large $256^3$ numerical simulation of a network of semilocal strings with $\beta = 0.05$. Note several joinings of string segments, e.g. two separate joinings on the long central string, and the disappearance of some loops. The different apparent thickness of strings is entirely an effect of perspective. The simulation was performed on the Cray T3E at the National Energy Research Scientific Computing Center (NERSC) in Berkeley.

results seem to confirm this picture [21], so obviously this problem deserves attention.

We may be a long way from understanding the formation of magnetic monopoles in a phase transition, but it is possible that particle accelerators will show signatures of monopole-antimonopole pairs in the not too distant

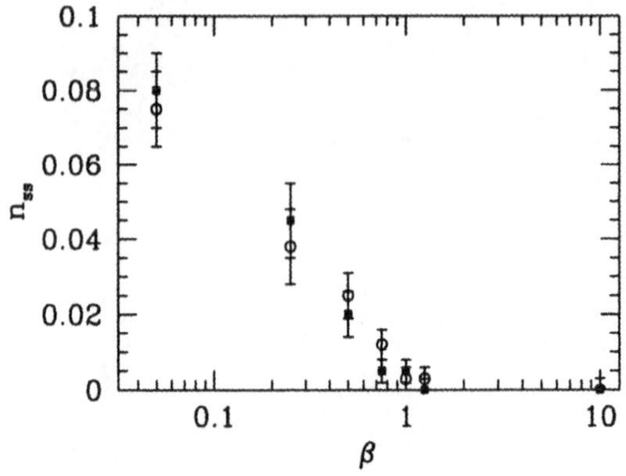


*Figure 4.* The number of semilocal strings formed per initial two-dimensional correlation volume in a toy model with translational invariance in one direction. Each point is an average over ten simulations. Squares indicate that the vacuum initial conditions described in the text were used, while open circles indicate that non-vacuum (more thermal) initial conditions were used.

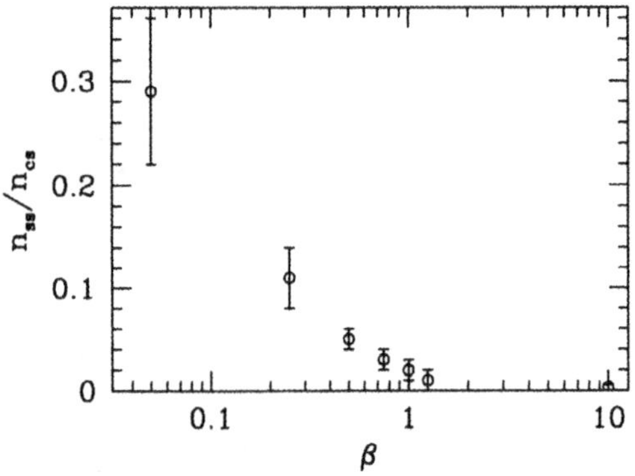


*Figure 5.* The ratio of lengths of semilocal and cosmic strings.

future. If Nambu's prediction is correct, dumbells could be the first soliton-like objects in the standard model of particle physics to be observed.

## 10. Acknowledgments

I am very grateful to Henri Godfrin and Yuriy Bunkov for organizing such a stimulating school, and to Brigitte Rousset and the Les Houches secretariat for their help at various points. Thanks also to my collaborators Julian Borrill and Andrew Liddle, and to Paul Saffin for a preview of his results on electroweak string formation.

## References


1. Dirac, P.A.M. (1931) Quantised singularities in the electromagnetic field, *Proc. Roy. Soc. London* **A133**, 60; *ibid* (1948) The theory of magnetic poles, *Phys. Rev.* **74**, 817.
2. Preskill, J. (1987) Vortices and monopoles, in *Architecture of Fundamental Interactions at Short Distances*, lectures delivered at Les Houches session XLIV, Elsevier. See also (1984) Magnetic monopoles, *Ann. Rev. Nucl. Part. Sci* **34**, 461.
3. Caso, C. *et al* (1998) Review of Particle Properties by the Particle Data Group, *The Eur. Phys. Jour.* **C3**, 1.
4. Cabrera, B. (1982) First results from a superconductive detector for moving magnetic monopoles, *Phys. Rev. Lett.* **48**, 1378. An interesting feature of this event is that the magnetic charge measured by the superconducting loop was within 5% of one Dirac unit.
5. Giacomelli, G. *et al* (1998) Magnetic monopole bibliography, preprint DFUB 9/98.
6. 't Hooft, G. (1974) Magnetic monopoles in unified gauge theories, *Nucl. Phys.* **B79**, 276; Polyakov, A. (1974) Particle spectrum in quantum field theory, *JETP Lett.* **20**, 194.
7. Harvey, J.A. (1997) *Magnetic monopoles, duality and supersymmetry*, lectures delivered at the 1995 ICTP Summer School on High Energy Physics and Cosmology, p. 66, World Scientific.
8. Coleman, S. (1981) *The magnetic monopole fifty years later*, lectures delivered at the Ettore Majorana school, Erice, 1981.
9. Wu, T.T. and Yang, C.N. (1975) Concept of non-integrable phase factors and global formulation of gauge fields, *Phys. Rev.* **D12**, 3845; *ibid* (1976) Dirac's monopole without strings: classical lagrangian, *Phys. Rev.* **D14**, 437; *ibid* (1976) Dirac monopole without strings: monopole harmonics, *Nucl. Phys.* **B107**, 365.
10. Georgi, H. and Glashow, S.L. (1972) Unified weak and electromagnetic interactions without neutral currents, *Phys. Rev. Lett.* **28**, 1494.
11. Cho, Y.M. and Maison, D. (1997) Monopole configuration in the Weinberg–Salam model, *Phys. Lett.* **B391**, 360.
12. Preskill, J. (1979) Cosmological production of superheavy magnetic monopoles *Phys. Rev. Lett.* **43**, 1365.
13. Nambu, Y. (1977) String–like configurations in the Weinberg–Salam theory, *Nucl. Phys.* **B130**, 505.
14. Achúcarro, A. and Vachaspati, T. (1999) Semilocal and electroweak strings, hep-th/9904229, to appear in *Physics Reports*.
15. Burlachkov, I. and Kopnin, N.B. (1987) *Sov. Phys. JETP* **65**, 630.
16. G.E. Volovik (1984) *Soviet Phys. Usp.***27**, 363.
17. Achúcarro, A., Borrill, J. and Liddle, A.R. (1998) Semilocal string formation in two dimensions, *Phys. Rev.* **D57**, 3742; *ibid.* (1998) On the formation of non-topological strings networks, *Physica* **B255**, 116.
18. Achúcarro, A., Borrill, J. and Liddle, A.R. (1999) The formation rate of semilocal strings, *Phys. Rev. Lett.* **82**, 3742.
19. Vachaspati, T. and Vilenkin, A. (1984) Formation and evolution of cosmic strings, *Phys. Rev.* **D30**, 2036.
20. Benson, K. and Bucher, M. (1993) Skyrmions and semilocal strings in cosmology, *Nucl. Phys.* **B 406**, 355.
21. P. Saffin, private communication.

# TIME-DEPENDENT GINZBURG-LANDAU THEORY AND DUALITY

ADRIAAN M. J. SCHAKEL
*Institut für Theoretische Physik*
*Freie Universität Berlin*
*Arnimallee 14, 14195 Berlin*

In the first part of this review paper, the time-dependent Ginzburg-Landau theory is derived starting from the microscopic BCS model with the help of a derivative expansion. Special attention is paid to two space dimensions, where the entire crossover from the weak-coupling BCS limit to the strong-coupling BEC limit of tightly bound fermion pairs is accessible analytically. The second part deals with the dual approach to the time-independent Ginzburg-Landau theory in three space dimensions. In this approach, the magnetic vortices of a superconductor play the central role, and the superconductor-to-normal phase transition is understood as a proliferation of these vortices.

## 1. Introduction

The role of topological defects in symmetry-breaking phase transitions cannot be underestimated. The prime example of a defect-driven (equilibrium) transition is the superfluid-to-normal phase transition in a $^4$He film, which can be understood as the unbinding of vortex-antivortex pairs [1, 2]. Since the very existence of an ordered state and the presence of topological defects are both manifestations of spontaneously broken symmetries, it doesn't come as a surprise that various phase transitions can be understood in terms of defects. To appreciate this point of view, note that defects are regions, usually macroscopic in size, where the symmetry is broken differently from that in the rest of the system. Often, the symmetry is completely restored in the defect cores, meaning that they are in the normal state. (In systems like superfluid $^3$He with a large symmetry group, the defect cores may be ordered states themselves, but of a different character.) Now, when defects with a normal-state core proliferate, the ordered state gets disordered and converted to the normal state.

In recent years the dynamics of phase transitions and the accompany-



*Y.M. Bunkov and H. Godfrin (eds.),*
*Topological Defects and the Non-Equilibrium Dynamics of Symmetry Breaking Phase Transitions*, 213–238.

ing formation of defects has also received much attention (see Ref. [3] for reviews). An important role is played here by the so-called time-dependent Ginzburg-Landau theory, which provides a phenomenological approach. Ideally, one would like to start with a microscopic model of the system under consideration and derive the time-dependent Ginzburg-Landau theory by integrating out irrelevant degrees of freedom, thus turning the phenomenological theory into an effective one. For a superconductor this program, at least in the time-independent case, has been carried out by Gorkov [4]. Starting from the microscopic theory of Bardeen, Cooper and Schrieffer (BCS) [5], he derived by means of a Green function method the Ginzburg-Landau theory of superconductivity which had been proposed as a phenomenological theory seven years before the BCS theory was formulated [6].

In the first part of this review paper, we reformulate Gorkov's derivation (extended to include time dependence) in terms of the functional-integral approach to quantum field theory. We will not only study the weak-coupling BCS limit of loosely bound Cooper pairs, but consider arbitrary values of the coupling constant $\lambda_0$, including the limit $\lambda_0 \to -\infty$, where the fermions form tightly bound pairs. When going from the weak-coupling to the large-$\lambda_0$ limit, the dynamics changes from dissipative to non-dissipative behavior.

In the second part, we discuss the dual approach to the time-independent Ginzburg-Landau theory. It is a formulation directly in terms of magnetic vortices, in which the superconductor-to-normal phase transition is described as a proliferation of vortices.

### 1.1. NOTATION

We adopt Feynman's notation and denote a spacetime point by $x = x_\mu = (t, \mathbf{x})$, $\mu = 0, 1, \cdots, d$, with $d$ the number of space dimensions, while the energy $k_0$ and momentum $\mathbf{k}$ will be denoted by $k = k_\mu = (k_0, \mathbf{k})$. The time derivative $\partial_0 = \partial/\partial t$ and the gradient $\nabla$ are sometimes combined in a single vector $\tilde{\partial}_\mu = (\partial_0, -\nabla)$. The tilde on $\partial_\mu$ is to alert the reader for the minus sign appearing in the spatial components of this vector. We define the scalar product $k \cdot x = k_\mu x_\mu = k_\mu g_{\mu\nu} k_\nu = k_0 t - \mathbf{k}\cdot\mathbf{x}$, with $g_{\mu\nu} = \mathrm{diag}(1, -1, \cdots, -1)$ and use Einstein's summation convention. Because of the minus sign in the definition of the vector $\tilde{\partial}_\mu$, it follows that $\tilde{\partial}_\mu a_\mu = \partial_0 a_0 + \nabla \cdot \mathbf{a}$, with $a_\mu$ an arbitrary vector.

Integrals over spacetime are denoted by

$$\int_x = \int_{t,\mathbf{x}} = \int \mathrm{d}t\, \mathrm{d}^d x,$$

while those over energy and momentum by

$$\int_k = \int_{k_0,\mathbf{k}} = \int \frac{\mathrm{d}k_0}{2\pi} \frac{\mathrm{d}^d k}{(2\pi)^d}.$$

When no integration limits are indicated, the integrals are assumed to be over all possible values of the integration variables.

Natural units $\hbar = k_\mathrm{B} = 1$ are adopted throughout.

## 2. Time-Dependent Ginzburg-Landau Theory

In this section, we derive the time-dependent Ginzburg-Landau theory for a superconductor, starting from the microscopic BCS model. Following Ref. [7], we integrate out the fermionic degrees of freedom in favor of a bosonic field describing the fermion pairs and use a derivative expansion method. We wish to obtain the Ginzburg-Landau theory not only in the weak-coupling BCS limit of loosely bound and overlapping Cooper pairs, but for arbitrary values of the coupling constant, in particular in the limit $\lambda_0 \to -\infty$ where the fermions form tightly bound pairs [8, 9]. This crossover between the two limits has recently been studied in detail [10, 11, 12, 13]. Since in two space dimensions this region becomes accessible by analytical methods, we shall, unless stated otherwise, restrict ourselves to $d = 2$ when carrying out momentum integrals.

### 2.1. BCS MODEL

The BCS Lagrangian reads [14]

$$\begin{aligned} \mathcal{L} &= \psi_\uparrow^*[i\partial_0 - \xi(-i\nabla)]\psi_\uparrow + \psi_\downarrow^*[i\partial_0 - \xi(-i\nabla)]\psi_\downarrow - \lambda_0 \psi_\uparrow^* \psi_\downarrow^* \psi_\downarrow \psi_\uparrow \\ &=: \mathcal{L}_0 + \mathcal{L}_\mathrm{i}, \end{aligned} \tag{1}$$

where $\mathcal{L}_\mathrm{i} = -\lambda_0 \psi_\uparrow^* \psi_\downarrow^* \psi_\downarrow \psi_\uparrow$ is a contact interaction term, representing the effective attraction between electrons with bare coupling constant $\lambda_0 < 0$, and $\mathcal{L}_0$ is the remainder. In Eq. (1), the field $\psi_{\uparrow(\downarrow)}$ is an anticommuting field describing the electrons with mass $m$ and spin up (down); $\xi(-i\nabla) = \epsilon(-i\nabla) - \mu_0$, with $\epsilon(-i\nabla) = -\nabla^2/2m$, is the kinetic energy operator with the bare chemical potential $\mu_0$ subtracted.

For computational convenience we introduce Nambu's notation and rewrite the Lagrangian (1) in terms of a two-component field

$$\psi = \begin{pmatrix} \psi_\uparrow \\ \psi_\downarrow^* \end{pmatrix} \qquad \psi^\dagger = (\psi_\uparrow^*, \psi_\downarrow). \tag{2}$$

In order to integrate out the fermionic degrees of freedom, the zero-temperature partition function represented as a functional integral

$$Z = \int \mathrm{D}\psi^\dagger \mathrm{D}\psi \exp\left(i \int_x \mathcal{L}\right), \tag{3}$$

must be written in a form bilinear in the electron fields. This is achieved by rewriting the quartic interaction term as a functional integral over auxiliary fields $\Delta$ and $\Delta^*$:

$$\exp\left(-i\lambda_0 \int_x \psi_\uparrow^* \psi_\downarrow^* \psi_\downarrow \psi_\uparrow\right) = \tag{4}$$
$$\int \mathrm{D}\Delta^* \mathrm{D}\Delta \exp\left[-i \int_x \left(\Delta^* \psi_\downarrow \psi_\uparrow + \psi_\uparrow^* \psi_\downarrow^* \Delta - \frac{1}{\lambda_0}|\Delta|^2\right)\right].$$

The field equation for $\Delta^*$,

$$\Delta = \lambda_0 \psi_\downarrow \psi_\uparrow, \tag{5}$$

shows that the auxiliary field describes electron pairs. We will therefore refer to it as pair field. The partition function thus becomes

$$Z = \int \mathrm{D}\psi^\dagger \mathrm{D}\psi \int \mathrm{D}\Delta^* \mathrm{D}\Delta \exp\left(\frac{i}{\lambda_0} \int_x |\Delta|^2\right) \tag{6}$$
$$\times \exp\left[i \int_x \psi^\dagger \begin{pmatrix} i\partial_0 - \xi(-i\nabla) & -\Delta \\ -\Delta^* & i\partial_0 + \xi(-i\nabla) \end{pmatrix} \psi\right].$$

Changing the order of integration and performing the Gaussian integral over the Grassmann fields, we obtain

$$Z = \int \mathrm{D}\Delta^* \mathrm{D}\Delta \exp\left(iS_{\text{eff}}[\Delta^*, \Delta] + \frac{i}{\lambda_0} \int_x |\Delta|^2\right), \tag{7}$$

with $S_{\text{eff}}$ the one-loop effective action which, using the identity $\mathrm{Det}(A) = \exp[\mathrm{Tr}\ln(A)]$, can be rewritten as

$$S_{\text{eff}}[\Delta^*, \Delta] = -i \,\mathrm{Tr} \ln \begin{pmatrix} p_0 - \xi(\mathbf{p}) & -\Delta \\ -\Delta^* & p_0 + \xi(\mathbf{p}) \end{pmatrix}, \tag{8}$$

where $p_0 = i\partial_0$ and $\xi(\mathbf{p}) = \epsilon(\mathbf{p}) - \mu_0$, with $\epsilon(\mathbf{p}) = \mathbf{p}^2/2m$.

The trace Tr appearing here needs some explanation. Explicitly, it is defined as

$$S_{\text{eff}} = -i\mathrm{Tr}\ \ln\left[K(p,x)\right] = -i\mathrm{tr}\ln\left[K(p,x)\delta(x-y)\big|_{y=x}\right], \tag{9}$$

where the trace tr is the usual one over discrete indices. We abbreviated the matrix appearing in (8) by $K(p,x)$ so as to cover the entire class of actions of the form

$$S = \int_x \psi^\dagger(x) K(p,x) \psi(x). \tag{10}$$

The delta function in (9) arises because $K(p,x)$ is obtained as a second functional derivative of the action

$$\frac{\delta^2 S}{\delta\psi^\dagger(x)\,\delta\psi(x)} = K(p,x)\,\delta(x-y)|_{y=x}, \tag{11}$$

each of which gives a delta function. Since the action has only one integral $\int_x$ over spacetime, one delta function remains. Because it is diagonal, it may be taken out of the logarithm and the effective action (9) can be written as

$$\begin{aligned} S_{\rm eff} &= -i\mathrm{tr}\int_x \ln\left[K(p,x)\right]\delta(x-y)|_{y=x} \\ &= -i\mathrm{tr}\int_x\int_k \mathrm{e}^{ik\cdot x}\ln\left[K(p,x)\right]\mathrm{e}^{-ik\cdot x}. \end{aligned} \tag{12}$$

In the last step, we used the integral representation of the delta function:

$$\delta(x) = \int_k \mathrm{e}^{-ik\cdot x}, \tag{13}$$

shifted the exponential function $\exp(ik\cdot y)$ to the left, which is justified because the derivative $p_\mu$ does not operate on it, and, finally, set $y_\mu$ equal to $x_\mu$. We thus see that the trace Tr in (9) stands for the trace over discrete indices as well as the integration over spacetime and over energy and momentum. The integral $\int_k$ arises because the effective action (9) is a one-loop result with $k_\mu$ the loop energy and momentum.

In the mean-field approximation, the functional integral (7) is approximated by the saddle point:

$$Z = \exp\left(iS_{\rm eff}[\Delta^*_{\rm mf},\Delta_{\rm mf}] + \frac{i}{\lambda_0}\int_x |\Delta_{\rm mf}|^2\right), \tag{14}$$

where $\Delta_{\rm mf}$ is the solution of the mean-field equation

$$\frac{\delta S_{\rm eff}}{\delta\Delta^*(x)} = -\frac{1}{\lambda_0}\Delta(x). \tag{15}$$

If we assume the system to be spacetime independent, so that $\Delta_{\rm mf}(x)=\bar\Delta$, Eq. (15) yields the celebrated BCS gap equation [5]:

$$\begin{aligned} \frac{1}{\lambda_0} &= -i\int_k \frac{1}{k_0^2 - E^2(k) + i\eta} \\ &= -\frac{1}{2}\int_{\mathbf{k}} \frac{1}{E(\mathbf{k})}, \end{aligned} \tag{16}$$

where $\eta$ is an infinitesimal positive constant that is to be set to zero at the end of the calculation, and

$$E(\mathbf{k}) = \sqrt{\xi^2(\mathbf{k}) + |\bar{\Delta}|^2} \tag{17}$$

is the spectrum of the elementary fermionic excitations.

## 2.2. BCS TO BEC

For a constant pair field, the effective action can be calculated in closed form. Writing

$$\begin{pmatrix} k_0 - \xi(\mathbf{k}) & -\bar{\Delta} \\ -\bar{\Delta}^* & k_0 + \xi(\mathbf{k}) \end{pmatrix} = \begin{pmatrix} k_0 - \xi(\mathbf{k}) & 0 \\ 0 & k_0 + \xi(\mathbf{k}) \end{pmatrix} - \begin{pmatrix} 0 & \bar{\Delta} \\ \bar{\Delta}^* & 0 \end{pmatrix}, \tag{18}$$

and expanding the second logarithm in a Taylor series, we recognize the form

$$S_{\text{eff}}[\bar{\Delta}^*, \bar{\Delta}] = \\ -i \operatorname{Tr} \ln \begin{pmatrix} k_0 - \xi(\mathbf{k}) & 0 \\ 0 & k_0 + \xi(\mathbf{k}) \end{pmatrix} - i \operatorname{Tr} \ln \left( 1 - \frac{|\bar{\Delta}|^2}{k_0^2 - \xi^2(\mathbf{k})} \right), \tag{19}$$

where we ignored an irrelevant constant. The integral over the loop energy $k_0$ can be carried out to yield for the effective Lagrangian

$$\mathcal{L}_{\text{eff}} = \int_{\mathbf{k}} \left[ E(\mathbf{k}) - \xi(\mathbf{k}) \right]. \tag{20}$$

To this one-loop result we have to add the tree term $|\bar{\Delta}|^2/\lambda_0$. Expanding $E(\mathbf{k})$ in a Taylor series, we see that the effective Lagrangian also contains a term quadratic in $\bar{\Delta}$. This term diverges in the ultraviolet. To render the theory finite to this order, we have to introduce a renormalized coupling constant $\lambda$ defined by:

$$\frac{1}{\lambda} = \frac{1}{\lambda_0} + \frac{1}{2} \int_{\mathbf{k}} \frac{1}{|\xi(\mathbf{k})|}. \tag{21}$$

To this order in the loop expansion there is no renormalization of the chemical potential, so that we can write $\mu = \mu_0$. We regularize the diverging integral in Eq. (21) by introducing a momentum cutoff $\Lambda$. In, for example, $d = 3$, we then obtain

$$\frac{1}{\lambda} = \frac{1}{\lambda_0} + \frac{m}{2\pi^2} \Lambda, \tag{22}$$

where we omitted the (irrelevant) finite part of the integral. It should be remembered that the bare coupling constant $\lambda_0$ is negative, so that the interaction between the fermions is attractive. We can distinguish two limits. One, the famous weak-coupling BCS limit, which is obtained by taking the bare coupling constant to zero, $\lambda_0 \to 0^-$. Second, the limit which is obtained by letting $\lambda_0 \to -\infty$. In this limit, the two-particle interaction is such that the fermions form tightly bound pairs of mass $2m$ [8, 9].

To explicate this so-called Bose-Einstein condensation (BEC) limit in $d = 2$, we swap the bare coupling constant for a more convenient parameter, namely the binding energy $\epsilon_a$ of a fermion pair in vacuum [15]. Both parameters characterize the strength of the contact interaction. To see the connection between the two, let us consider the Schrödinger equation for the problem at hand. In reduced coordinates, it reads

$$\left[-\frac{\nabla^2}{m} + \lambda_0\, \delta(\mathbf{x})\right] \psi(\mathbf{x}) = -\epsilon_a \psi(\mathbf{x}), \tag{23}$$

where the reduced mass is $m/2$ and the delta-function potential, with $\lambda_0 < 0$, represents the attractive contact interaction $\mathcal{L}_\mathrm{i}$ in the BCS Lagrangian (1). We stress that this is a two-particle problem in vacuum; it is not the famous Cooper problem of two interacting fermions on top of a filled Fermi sea. The equation is most easily solved by Fourier transforming it. This yields the bound-state equation

$$\psi(\mathbf{k}) = -\frac{\lambda_0}{\mathbf{k}^2/m + \epsilon_a}\psi(0), \tag{24}$$

or

$$-\frac{1}{\lambda_0} = \int_\mathbf{k} \frac{1}{\mathbf{k}^2/m + \epsilon_a} = \frac{1}{2}\nu(0)\ln\left(\frac{2\epsilon_\Lambda}{\epsilon_a}\right), \tag{25}$$

where $\nu(0) = m/2\pi$ is the two-dimensional density of states (per spin degree of freedom), and $\epsilon_\Lambda = \Lambda^2/2m$. This equation allows us to replace the bare coupling constant $\lambda_0$ with the binding energy $\epsilon_a$. When substituted in the gap equation (16), the latter becomes

$$\int_\mathbf{k} \frac{1}{\mathbf{k}^2/m + \epsilon_a} = \frac{1}{2}\int_\mathbf{k} \frac{1}{E(\mathbf{k})}. \tag{26}$$

By inspection, it is easily seen that this equation has a solution [9]

$$\bar{\Delta} \to 0, \qquad \mu_0 \to -\tfrac{1}{2}\epsilon_a, \tag{27}$$

with a negative chemical potential. This is the strong-coupling BEC limit. To appreciate the physical significance of the specific value found for the chemical potential in this limit, we note that the spectrum $E_\mathrm{b}(\mathbf{q})$ of the two-fermion bound state measured relative to the pair chemical potential $2\mu_0$ reads

$$E_b(\mathbf{q}) = -\epsilon_a + \frac{\mathbf{q}^2}{4m} - 2\mu_0. \tag{28}$$

The negative value for $\mu_0$ found in (27) is precisely the condition for a Bose-Einstein condensation of the composite bosons in the $\mathbf{q} = 0$ state—whence the name BEC limit.

Since there are two unknowns contained in the theory, viz. $\bar{\Delta}$ and $\mu$, a second equation is needed to determine these variables in the mean-field approximation [9]. It is provided by the requirement that the average fermion number $N$, which is obtained by differentiating the effective action (8) with respect to $\mu$

$$N = \frac{\partial S_{\text{eff}}}{\partial \mu}, \tag{29}$$

be fixed. If the system is spacetime independent, this reduces to

$$\bar{n} = -i \,\text{tr} \int_k G(k)\tau_3, \tag{30}$$

where $\bar{n} = N/V$, with $V$ the volume of the system, is the constant fermion number density, $\tau_3$ is the diagonal Pauli matrix in Nambu space,

$$\tau_3 = \begin{pmatrix} 1 & 0 \\ 0 & -1 \end{pmatrix}, \tag{31}$$

and $G(k)$ is the Green function,

$$\begin{aligned} G(k) &= \begin{pmatrix} k_0 - \xi(\mathbf{k}) & -\bar{\Delta} \\ -\bar{\Delta}^* & k_0 + \xi(\mathbf{k}) \end{pmatrix}^{-1} \\ &= \frac{1}{k_0^2 - E^2(\mathbf{k}) + i\eta} \begin{pmatrix} k_0 \, e^{ik_0\eta} + \xi(\mathbf{k}) & \bar{\Delta} \\ \bar{\Delta}^* & k_0 \, e^{-ik_0\eta} - \xi(\mathbf{k}) \end{pmatrix}. \end{aligned} \tag{32}$$

Here, $\eta$ is an infinitesimal positive constant that is to be set to zero at the end of the calculation. The exponential functions in the diagonal elements of the Green function are an additional convergence factor needed in nonrelativistic theories [16]. If the integral over the loop energy $k_0$ in the particle number equation (30) is carried out, it takes the familiar form

$$\bar{n} = \int_{\mathbf{k}} \left(1 - \frac{\xi(\mathbf{k})}{E(\mathbf{k})}\right). \tag{33}$$

The two equations (16) and (30) determine $\bar{\Delta}$ and $\mu$.

## 2.3. DERIVATIVE EXPANSION

When the pair field $\Delta$ is spacetime-dependent, the integrals in (8) cannot be evaluated in closed form because the logarithm contains energy-momentum operators and spacetime-dependent functions in a mixed order. To disentangle the integrals we have to resort to a derivative expansion [17] in which the logarithm is expanded in a Taylor series. Each term contains powers of the energy-momentum operator $p_\mu$ which acts on every spacetime-dependent function to its right. All these operators are shifted to the left by repeatedly applying the identity

$$f(x)p_\mu g(x) = (p_\mu - i\tilde{\partial}_\mu)f(x)g(x), \tag{34}$$

where $f(x)$ and $g(x)$ are arbitrary functions of spacetime and the derivative $\tilde{\partial}_\mu = (\partial_0, -\nabla)$ acts *only* on the next object to the right. One then integrates by parts, so that all the $p_\mu$'s act to the left where only a factor $\exp(ik \cdot x)$ stands. Ignoring total derivatives and taking into account the minus signs that arise when integrating by parts, one sees that all occurrences of $p_\mu$ (an operator) are replaced with $k_\mu$ (an integration variable). The exponential function $\exp(-ik \cdot x)$ can at this stage be moved to the left where it is annihilated by the function $\exp(ik \cdot x)$. The energy-momentum integration can now in principle be carried out and the effective action be cast in the form of an integral over a local density $\mathcal{L}_{\text{eff}}$:

$$S_{\text{eff}} = \int_x \mathcal{L}_{\text{eff}}. \tag{35}$$

This is in a nutshell how the derivative expansion works [17].

To apply it to the BCS model and derive the Ginzburg-Landau theory we use the following decomposition in (8):

$$\begin{pmatrix} p_0 - \xi(\mathbf{p}) & -\Delta \\ -\Delta^* & p_0 + \xi(\mathbf{p}) \end{pmatrix} = G_0^{-1}\left[1 - G_0 \begin{pmatrix} 0 & \Delta \\ \Delta^* & 0 \end{pmatrix}\right], \tag{36}$$

where $G_0$ is the Green function (32) with $\bar{\Delta} = 0$. Apart from an irrelevant constant, this leads to the expression for the effective action

$$\begin{aligned} S_{\text{eff}} &= -i \operatorname{Tr} \ln \left[1 - G_0 \begin{pmatrix} 0 & \Delta \\ \Delta^* & 0 \end{pmatrix}\right] \\ &= i \operatorname{Tr} \sum_{\ell=1}^{\infty} \frac{1}{\ell} \left[G_0 \begin{pmatrix} 0 & \Delta \\ \Delta^* & 0 \end{pmatrix}\right]^\ell =: \sum_{\ell=1}^{\infty} S_{\text{eff}}^{(\ell)}. \end{aligned} \tag{37}$$

For $\ell = 1$, the trace over the $2 \times 2$ matrix immediately yields zero. In a similar fashion all terms $S_{\text{eff}}^{(\ell)}$, with $\ell$ odd, give zero. For the quadratic term, we obtain

$$S_{\rm eff}^{(2)} = i \, {\rm Tr} \, \frac{1}{p_0 + \xi(\mathbf{p})} \, \Delta^* \, \frac{1}{p_0 - \xi(\mathbf{p})} \, \Delta, \tag{38}$$

where we recall the definition of the derivative $p_\mu$ as operating on everything that appears to the right. Applying the derivative expansion rules outlined above, we can cast the quadratic term in the effective action in the form

$$S_{\rm eff}^{(2)} = i \, {\rm Tr} \, \frac{1}{k_0 + \xi(\mathbf{k})} \, \frac{1}{k_0 - i\,\partial_0 - \xi(\mathbf{k} + i\nabla)} \, \Delta^* \Delta. \tag{39}$$

Usually, the quartic terms are included in the Ginzburg-Landau theory without derivatives. We then can treat $p_0 \pm \xi(\mathbf{p})$ as a c-number and

$$S_{\rm eff}^{(4)} = \frac{i}{2} \, {\rm Tr} \, \frac{1}{[\, k_0^2 - \xi^2(\mathbf{k})\,]^2} \, |\Delta|^4. \tag{40}$$

We may truncate the series in (37) here, provided the pair field $\Delta$ is small as is the case in the vicinity of the phase transition.

To include the temperature $T$ in the theory, we adopt the imaginary-time approach to thermal field theory [18, 19]. Very briefly, it can be derived from the corresponding quantum field theory at zero temperature simply by going over to imaginary times, $t \rightarrow -i\tau$, and substituting

$$\int \frac{{\rm d}k_0}{2\pi} \, g(k_0) \rightarrow i\,\beta^{-1} \sum_n g(i\,\omega_n), \tag{41}$$

where $g$ is an arbitrary function, while $\omega_n$ denotes the Matsubara frequencies,

$$\omega_n = \begin{cases} \pi\beta^{-1} 2n, & \text{(bosonic)} \\ \pi\beta^{-1}(2n+1), & \text{(fermionic)} \end{cases} \tag{42}$$

with $n$ an integer and $\beta = 1/T$. With these rules, the Minkowski action $S$ goes over into

$$S = \int_x \mathcal{L}(t, \mathbf{x}) \rightarrow -i \int_0^\beta {\rm d}\tau \int_{\mathbf{x}} \mathcal{L}(-i\tau, \mathbf{x}) =: i S^{\rm E}, \tag{43}$$

where the superscript E on the action at the right-hand side is to indicate that it pertains to Euclidean rather than to Minkowski spacetime.

For the case at hand we obtain in this way the finite-temperature action:

$$S_{\rm eff}^{\rm E} = {\rm Tr} \left[ \frac{1}{i\omega_n + \xi(\mathbf{k})} \, \frac{1}{i\omega_n + \partial_\tau - \xi(\mathbf{k} + i\nabla)} \Delta^* \Delta + \frac{1}{2} \frac{1}{[\omega_n^2 + \xi^2(\mathbf{k})]^2} |\Delta|^4 \right], \tag{44}$$

where the trace at finite temperature reads explicitly

$$\mathrm{Tr} = \int_0^\beta \mathrm{d}\tau \int \mathrm{d}^d x \frac{\mathrm{d}^d k}{(2\pi)^d} \beta^{-1} \sum_n . \tag{45}$$

Let us for the moment consider only the time-independent part of the effective action (44) and expand in gradients. The sums over the Matsubara frequencies are carried out with the help of the formulas

$$\begin{aligned}
\beta^{-1} \sum_n \frac{1}{\omega_n^2 + \xi^2} &= \frac{1}{2}\frac{X}{\xi} \\
\beta^{-1} \sum_n \frac{1}{(\omega_n^2 + \xi^2)^2} &= -\frac{1}{4\xi}\frac{\partial}{\partial \xi}\left(\frac{X}{\xi}\right) \\
\beta^{-1} \sum_n \frac{1}{i\omega_n + \xi}\frac{1}{(i\omega_n - \xi)^2} &= -\frac{1}{4}\frac{\partial}{\partial \xi}\left(\frac{X}{\xi}\right) \\
\beta^{-1} \sum_n \frac{1}{i\omega_n + \xi}\frac{1}{(i\omega_n - \xi)^3} &= -\frac{X}{8\xi^3} - \frac{\beta}{16\xi}\frac{\partial}{\partial \xi}\left(\frac{Y}{\xi}\right),
\end{aligned} \tag{46}$$

where $X$ and $Y$ abbreviate the functions

$$X = \tanh(\beta\xi/2), \qquad Y = 1/\cosh^2(\beta\xi/2). \tag{47}$$

As was first shown by Drechsler and Zwerger [10], most of the momentum integrals can be performed analytically for arbitrary values of the coupling constant in two space dimensions. In this way, they arrived at the time-independent Ginzburg-Landau theory describing the crossover from the weak-coupling BCS limit to the strong-coupling BEC limit:

$$\mathcal{L}_{\text{eff}} = c\Delta^* \nabla^2 \Delta + a|\Delta|^2 - \tfrac{1}{2} b |\Delta|^4, \tag{48}$$

with the coefficients

$$\begin{aligned}
a = \frac{\nu(0)}{2} \Bigg[ & 2\ln\left(\frac{4\mathrm{e}^\gamma}{\pi}\right)\theta(\mu) + \ln(\beta\epsilon_a/4) + \ln(\beta|\mu|/2)\tanh(\beta\mu/2) \\
& + \mathrm{sgn}(\mu) \int_{\beta|\mu|/2}^\infty \mathrm{d}x \, \ln(x) Y(x) \Bigg],
\end{aligned} \tag{49}$$

$$b = \frac{\nu(0)}{4} \left[ \frac{7\zeta(3)}{2\pi^2}\beta^2 \theta(\mu) - \frac{1}{\mu^2}\tanh(\beta\mu/2) + \mathrm{sgn}(\mu)\frac{\beta^2}{4}\int_{\beta|\mu|/2}^\infty \mathrm{d}x \frac{X(x)}{x^3}\right], \tag{50}$$

and

$$c = \frac{\nu(0)}{8m} \left[ \frac{7\zeta(3)}{2\pi^2}\beta^2 \mu\theta(\mu) + \frac{\beta^2}{4}|\mu| \int_{\beta|\mu|/2}^\infty \mathrm{d}x \frac{X(x)}{x^3}\right]. \tag{51}$$

Here, we introduced the integration variable $x = \beta\xi/2$, $\gamma = 0.577216\cdots$ is Euler's constant, $\theta(x)$ is the Heaviside unit step function, $\mathrm{sgn}(x)$ the sign function, and $\zeta(x)$ Riemann's zeta function, with $\zeta(3) = 1.20206\cdots$. (The second term at the right-hand side of Eq. (50) differs slightly from the corresponding term in Ref. [10], but is consistent with a later work by Zwerger and collaborator [20].)

When studying the time-dependence of the effective action (44), special care has to be taken with analytic continuation. We follow Sá de Melo, Randeria and Engelbrecht [12] and first analytic continue, using the formula

$$\beta^{-1}\sum_n \frac{1}{i\omega_n + \xi}\frac{1}{i\omega_n - i\omega_l - \xi} = -\left[\mathrm{P}\frac{1}{2\xi + q_0} - i\pi\delta(2\xi + q_0)\right] X, \quad (52)$$

before expanding in time derivatives. (In Ref. [10], the expansion was done first, leading to results which are not consistent with those known in the BCS limit [21, 22].) In Eq. (52), P stands for the principal part, while $\omega_l$ is a bosonic Matsubara frequency, which at the right-hand side is analyticly continued to the real axes by replacing $i\omega_l$ with $q_0 + i\eta$. In this way, we find for the dynamic part (in Minkowski spacetime)

$$\mathcal{L}_{\mathrm{dyn}} = [Q'(i\partial_0) - i\pi Q''(i\partial_0)]\Delta^*\Delta, \quad (53)$$

with

$$\begin{aligned} Q'(q_0) &= -\mathrm{P}\int_{\mathbf{k}} \frac{q_0}{2\xi(2\xi + q_0)} X \\ Q''(q_0) &= \tfrac{1}{2}\nu(0)\tanh(\beta q_0/4)\theta(\mu - q_0/2). \end{aligned} \quad (54)$$

The most important result to be noted here is that the low-energy dynamics is dissipative when the chemical potential $\mu$ is positive [12]. This is because the pairs can break up and decay into a continuum of fermionic excitations. On the other hand, for negative values of $\mu$, where the pairs are more tightly bound, the time-dependent Ginzburg-Landau theory describes a purely propagating pair mode. To investigate this point further, let us consider the two limits in detail.

## 2.4. BCS LIMIT

In the weak-coupling BCS limit, where the chemical potential is well approximated by the Fermi energy $\mu = k_{\mathrm{F}}^2/2m$, with $k_{\mathrm{F}}$ the Fermi momentum, and $\epsilon_a/\epsilon_{\mathrm{F}} \to 0$, we recover the standard result (adjusted for the reduced space dimensionality) [22]

$$\mathcal{L}_{\mathrm{eff}} = \nu(0)\left\{\Delta^*\left[\ln\left(\frac{T_0}{T}\right) - \frac{\pi}{8T_0}\partial_0 + \frac{3\xi_0^2}{v_{\mathrm{F}}^2}\left(\partial_0^2 + \frac{v_{\mathrm{F}}^2}{2}\nabla^2\right)\right]\Delta - \frac{3\xi_0^2}{v_{\mathrm{F}}^2}|\Delta|^4\right\} \quad (55)$$

with $v_{\mathrm{F}} = k_{\mathrm{F}}/m$ the Fermi velocity, $\xi_0$ the BCS correlation length

$$\xi_0^2 = \frac{7\zeta(3)}{48\pi^2}\frac{v_{\mathrm{F}}^2}{T_0^2}, \tag{56}$$

and $T_0$ the BCS transition temperature [10]

$$T_0 = \frac{e^\gamma}{\pi}(2\mu\epsilon_a)^{1/2} \tag{57}$$

expressed in terms of the binding energy $\epsilon_a$. Comparing the two terms in Eq. (55) involving time derivatives, we recognize a series expansion in powers of $q_0/T$; the terms without derivatives constitute an expansion in powers of $|\Delta|/T$ [23]. It therefore follows that Eq. (55) represents the first terms in a high-temperature expansion. Since we neglected time derivatives in the quartic and higher-order terms, we are implicitly assuming that $q_0 >> |\Delta|$ [22]. The time-dependent Ginzburg-Landau theory (55) obtained in the weak-coupling BCS limit is of the dissipative type often used in the context of defect formation in a symmetry-breaking phase transition (for reviews see Ref. [3]). For applications in the context of superconductivity see, for example, Refs. [24, 25, 26].

## 2.5. BEC LIMIT

In the strong-coupling BEC limit, the general form of the effective action (48) and (53) reduces to [10]

$$\mathcal{L}_{\mathrm{eff}} = \hat{\Delta}^* \left( i\partial_0 + \mu_{\mathrm{b}} + \frac{\nabla^2}{4m} \right) \hat{\Delta} - \lambda_{\mathrm{b}} |\hat{\Delta}|^4, \tag{58}$$

where we introduced a rescaled pair field

$$\hat{\Delta} = \left( \frac{\nu(0)}{4|\mu|} \right)^{1/2} \Delta. \tag{59}$$

The effective theory (58) is precisely of the form of a Gross-Pitaevski theory [27], describing a weakly interacting composite Bose gas with a mass $2m$ (as expected), a small chemical potential which vanishes on approaching the critical temperature from below [see Eq. (27)]

$$\mu_{\mathrm{b}} = 2|\mu| \ln \left( \frac{\epsilon_a}{2|\mu|} \right), \tag{60}$$

or using Eq. (26),

$$\mu_{\mathrm{b}} = \frac{|\bar{\Delta}|^2}{\epsilon_a}, \tag{61}$$

and a repulsive contact interaction

$$\lambda_{\mathrm{b}} = \frac{1}{\nu(0)} \tag{62}$$

independent of the binding energy $\epsilon_a$ characterizing the interaction between the electrons. This is special to two dimensions; in $d$ space dimensions we have instead [28]

$$\lambda_{\mathrm{b}} = (4\pi)^{d/2} \frac{1 - d/4}{\Gamma(2 - d/2)} \frac{\epsilon_a^{1-d/2}}{m^{d/2}}, \tag{63}$$

with $\Gamma(x)$ the Gamma function. Note the absence of any temperature-dependence in the effective theory (58). At zero temperature, exactly the same effective theory was obtained by Hausmann, using a self-consistent Green function method [11]. (See Ref. [28] for a derivation along the lines presented here.) This is, as that author argued, because in the BEC limit, the critical temperature $T_0$ is much smaller than the dissociation temperature $T_{\mathrm{diss}} \approx \epsilon_a$ at which the tightly bound fermion pairs are broken up by thermal fluctuations. Hence, for all temperatures in the range $T \leq T_0 << T_{\mathrm{diss}}$ we are effectively in the zero-temperature regime [11].

The Gross-Pitaevski theory (58) describes a gapless, purely propagating mode, viz. the Goldstone mode associated with the spontaneously broken global U(1) symmetry—the so-called Anderson-Bogoliubov mode [23].

The spacetime-dependent effective theory of a superconductor can only be derived in a few special cases [22]: one being close to the transition temperature where the expansion parameter is $1/T$ and another being close to the absolute zero of temperature where the expansion parameter is $1/|\bar{\Delta}|$. Outside these regimes, the effective theory depends on the ratio $|\nabla|/\partial_0$, and an expansion in both time derivatives and gradients is not possible. It is amusing to note that in going from the weak-coupling BCS limit to the strong-coupling BEC limit, we move from one valid regime to the other.

## 3. Dual Theory

In this section we investigate the dual formulation of the time-independent Ginzburg-Landau theory. This approach, in which the magnetic vortices of a superconductor play the central role, originates from lattice studies that started more than two decades ago [29, 30, 31, 32, 33, 34, 35, 36]. These were in turn instigated by the success of the Kosterlitz-Thouless theory describing the phase transition in a superfluid film as the unbinding of vortex-antivortex pairs [1, 2]. In the dual formulation of the Ginzburg-Landau theory, the superconductor-to-normal phase transition is understood as a proliferation of magnetic vortices. A detailed presentation of these matters as well as an extensive list of references to the literature can be found in Ref. [37].

3.1. ELECTRIC CURRENT LOOPS

To account for the magnetic interaction we couple the Ginzburg-Landau theory in the usual, minimal way to a vector potential **A**. We also rescale the pair field such that $\beta$ times the Hamiltonian becomes

$$\mathcal{H}_{\rm GL} = |(\nabla - 2ie\mathbf{A})\phi|^2 + m_\phi^2|\phi|^2 + \lambda|\phi|^4 + \frac{1}{2}(\nabla \times \mathbf{A})^2 + \frac{1}{2\alpha}(\nabla \cdot \mathbf{A})^2, \quad (64)$$

where we added a gauge-fixing term with parameter $\alpha$. To acquire a physical understanding of what this Hamiltonian describes [38], we recall that a $|\phi|^4$ - theory gives a field theoretic description of strings with contact repulsion [39]. This equivalence rests on Feynman's observation [40] that the Green function

$$G(\mathbf{x}) = \int_\mathbf{k} \frac{e^{i\mathbf{k}\cdot\mathbf{x}}}{\mathbf{k}^2 + m_\phi^2} \quad (65)$$

can be expressed as a path integral. An easy way to see this is to invoke Schwinger's proper-time method, which is based on Euler's form

$$\frac{1}{a^z} = \frac{1}{\Gamma(z)} \int_0^\infty \frac{d\tau}{\tau} \tau^z \, e^{-\tau a}, \quad (66)$$

to write the right-hand side of (65) as [41]:

$$\begin{aligned} G(\mathbf{x}) &= \int_0^\infty d\tau \, e^{-\tau m_\phi^2} \int_\mathbf{k} e^{i\mathbf{k}\cdot\mathbf{x}} e^{-\tau \mathbf{k}^2} \\ &= \int_0^\infty d\tau \, e^{-\tau m_\phi^2} \left(\frac{1}{4\pi\tau}\right)^{3/2} e^{-\frac{1}{4}\mathbf{x}^2/\tau}. \end{aligned} \quad (67)$$

The factor $(1/4\pi\tau)^{3/2}\exp(-\frac{1}{4}\mathbf{x}^2/\tau)$ appearing here can be interpreted as describing a Brownian string trajectory, showing that one endpoint of the string (located at $\mathbf{x}$) has a Gaussian distribution with respect to its other endpoint (located at the origin) [42]. If we imagine the string to be composed of $N$ links, each of length $a$, then in the limit where $N \to \infty$, $a \to 0$, the variable $\tau$ parameterizing the string stands for $\tau = Na^2/6$. The integration over $\tau$ in (67) indicates that the string can be arbitrary long, but the weighing factor $\exp(-m_\phi^2\tau)$ exponentially suppresses long ones in the normal phase where $m_\phi^2 > 0$. To understand the physical meaning of this factor, let us return to the discrete string model and write

$$m_\phi^2 = 6[\sigma(a) - \sigma_{\rm cr}]/a, \quad (68)$$

so that it becomes

$$e^{-m_\phi^2\tau} = e^{-\sigma_{\rm eff}(a)L}, \tag{69}$$

with $\sigma_{\rm eff}(a) = \sigma(a) - \sigma_{\rm cr}$ the effective string tension, and $L = Na$ the length of the string. The continuum limit is obtained by simultaneously letting $a \to 0$ and $\sigma(a) \to \sigma_{\rm cr}$, in such a way that the right-hand side of (68) tends to the finite value $m_\phi^2$. We thus see that the factor $\exp(-m_\phi^2\tau)$ weighs strings according to their lengths.

Following Feynman [40], we write the right-hand side of (67) as a path-integral, i.e., as a sum over all possible string trajectories having one end-point at $\mathbf{x}(0) = 0$ and the other at $\mathbf{x}(\tau) = \mathbf{x}$ [43]:

$$G(\mathbf{x}) = \int_0^\infty \mathrm{d}\tau \int_{\mathbf{x}(0)=0}^{\mathbf{x}(\tau)=\mathbf{x}} \mathrm{D}\mathbf{x}(\tau')\, e^{-S_0}, \tag{70}$$

with the (Euclidean) action

$$S_0 = \int_0^\tau \mathrm{d}\tau' \left[\tfrac{1}{4}\dot{\mathbf{x}}^2(\tau') + m_\phi^2\right], \tag{71}$$

where $\dot{\mathbf{x}}(\tau) = \mathrm{d}\mathbf{x}(\tau)/\mathrm{d}\tau$.

In a similar way, the partition function of the free theory, which written as a functional integral reads

$$Z_0 = \int \mathrm{D}\phi^*\mathrm{D}\phi \exp\left(-\int_{\mathbf{x}} \mathcal{H}_0\right), \tag{72}$$

can also be represented as a *path* integral, this time involving only closed strings:

$$\begin{aligned} \ln(Z_0) &= -\ln[\mathrm{Det}(\mathbf{p}^2 + m_\phi^2)] = -\mathrm{Tr}\ln(\mathbf{p}^2 + m_\phi^2) \qquad (73) \\ &= \int_0^\infty \frac{d\tau}{\tau} e^{-\tau m_\phi^2} \int_{\mathbf{k}} e^{-\tau \mathbf{k}^2} = \int_0^\infty \frac{d\tau}{\tau} \oint \mathrm{D}\mathbf{x}(\tau') e^{-S_0}. \end{aligned}$$

Here, we used Euler's form (66) in the limit of small $z$. An additional factor $1/\tau$ arises because one can start traversing a closed string anywhere along the loop. The $|\phi|^4$-interaction in the Ginzburg-Landau model can be shown to result in the additional term [44]

$$S_\lambda = -\lambda \int_0^\tau \mathrm{d}\tau_l' \mathrm{d}\tau_k'\, \delta\left[\mathbf{x}(\tau_l') - \mathbf{x}(\tau_k')\right] \tag{74}$$

in the action, which gives an extra weight each time two strings, one parameterized by $\tau_l'$ and one by $\tau_k'$, intersect. Physically, it represents a repulsive contact interaction between strings. Finally, the coupling of the field $\phi$ to the magnetic vector potential $\mathbf{A}$ via the electric current

$$\mathbf{j}_e = -2ei(\phi^* \nabla\phi - \phi\nabla\phi^*) - 2(2e)^2 \mathbf{A}|\phi|^2 \tag{75}$$

with a charge $2e$, results in the extra term [37]

$$S_e = 2ie \int_0^T \mathrm{d}\tau' \, \dot{\mathbf{x}}(\tau') \cdot \mathbf{A}[\mathbf{x}(\tau')], \tag{76}$$

showing that the strings described by the Ginzburg-Landau theory carry an electric current. Pasting the pieces together, we conclude that the partition function of the Ginzburg-Landau theory can be equivalently represented as a grand canonical ensemble of fluctuating electric current loops, of arbitrary length and shape [45]:

$$Z = \int \mathrm{D}\mathbf{A} \, \mathrm{e}^{-\frac{1}{2}\int_\mathbf{x}[(\nabla\times\mathbf{A})^2 + \frac{1}{\alpha}(\nabla\cdot\mathbf{A})^2]} \sum_{N=0}^{\infty} \frac{1}{N!} \prod_{l=1}^{N} \left[ \int_0^\infty \frac{\mathrm{d}\tau_l}{\tau_l} \oint \mathrm{D}\mathbf{x}(\tau_l') \right] \mathrm{e}^{-S_{\mathrm{GL}}}, \tag{77}$$

with the action

$$\begin{aligned} S_{\mathrm{GL}} &= \sum_{l=1}^{N} \int_0^{\tau_l} \mathrm{d}\tau_l' \left\{ \tfrac{1}{4}\dot{\mathbf{x}}^2(\tau_l') + m_\phi^2 + 2ie\dot{\mathbf{x}}(\tau_l') \cdot \mathbf{A}[\mathbf{x}(\tau_l')] \right\} \\ &\quad + \lambda \sum_{l,k=1}^{N} \int_0^{\tau_l} \mathrm{d}\tau_l' \int_0^{\tau_k} \mathrm{d}\tau_k' \, \delta\left[\mathbf{x}(\tau_l') - \mathbf{x}(\tau_k')\right]. \end{aligned} \tag{78}$$

On entering the superconducting phase, characterized by a sign change in the mass term of the Ginzburg-Landau theory, the effective string tension approaches zero as

$$\sigma_{\mathrm{eff}}(a) \propto m_\phi^2 \propto (T - T_c)^{2\nu}, \tag{79}$$

with $\nu$ the correlation length exponent, and the electric current loops proliferate. This proliferation is also signaled by the absolute value $|\phi|$ of the pair field which then develops a vacuum expectation value, identifying it as the order parameter.

### 3.2. DISORDER PARAMETER

The dual formulation of the Ginzburg-Landau theory gives a similar representation of the partition function as the path integral (77); however this time not in terms of electric current, but of magnetic vortex loops. To derive it, we start with a trick [46] and introduce a (hypothetical) magnetic monopole at some point $\mathbf{z}$ inside the superconductor. A monopole is a source of magnetic flux. Due to the Meissner effect, the flux lines emanating from the monopole are squeezed into a flux tube. In this way, we managed to create a magnetic vortex at zero external field. Electrodynamics in the presence of a monopole was first described by Dirac [47] who argued that the physical local magnetic induction **h** is given by the combination $\nabla \times \mathbf{A}(\mathbf{x}) - \mathbf{B}^{\mathrm{P}}(\mathbf{x})$. The subtracted plastic field

$$B_i^{\mathrm{P}}(\mathbf{x}) = \Phi_0 \int_{L_\mathbf{z}} \mathrm{d}y_i \, \delta(\mathbf{x} - \mathbf{y}), \tag{80}$$

with $\Phi_0 = \pi/e$ the magnetic flux quantum, removes the field of the so-called Dirac string running along some path $L_\mathbf{z}$ from the location $\mathbf{z}$ of the monopole to infinity. On account of Stokes' theorem, the plastic field satisfies the equation

$$\nabla \cdot \mathbf{B}^{\mathrm{P}}(\mathbf{x}) = -\rho_{\mathrm{m}}(\mathbf{x}), \tag{81}$$

with $\rho_{\mathrm{m}}(\mathbf{x}) = \Phi_0 \, \delta(\mathbf{x} - \mathbf{z})$ the monopole density.

We continue by writing $\phi(\mathbf{x})$ as

$$\phi(\mathbf{x}) = \bar{\phi} \, \mathrm{e}^{2i\varphi(\mathbf{x})}, \tag{82}$$

where $\bar{\phi}$ is a constant solution of the mean-field equation. This approximation, where the phase of the order field is allowed to vary in space while the modulus is kept fixed, is called the London limit. The Ginzburg-Landau Hamiltonian then becomes after integrating out the phase field $\varphi$:

$$\mathcal{H}_{\mathrm{GL}}^{\mathrm{P}} = \frac{1}{2}(\nabla \times \mathbf{A} - \mathbf{B}^{\mathrm{P}})^2 + \frac{1}{2} m_A^2 A_i \left( \delta_{ij} - \frac{\partial_i \partial_j}{\nabla^2} \right) A_j + \frac{1}{2\alpha} (\nabla \cdot \mathbf{A})^2, \tag{83}$$

where

$$m_A = 2e|\bar{\phi}| \tag{84}$$

is the inverse magnetic penetration depth $m_A = \lambda_{\mathrm{L}}^{-1}$. We gave $\mathcal{H}$ the superscript P to indicate the presence of the monopole. While the Dirac string is immaterial in the normal phase, Nambu [48] argued that it acquires physical relevance in the superconducting phase where it serves as the core of the magnetic vortex or Abrikosov flux tube originating at the monopole.

The energy $E_V$ of this configuration is easily obtained by substituting the solution of the field equation for the gauge field

$$A_i(\mathbf{x}) = \int_{\mathbf{y}} G_{ij}(\mathbf{x} - \mathbf{y}) \left[ \nabla \times \mathbf{B}^{\mathrm{P}}(\mathbf{y}) \right]_j, \tag{85}$$

with $G_{ij}$ the gauge-field Green function

$$Gij(\mathbf{x}) = \int_{\mathbf{k}} \left( \frac{\delta_{ij} - k_i k_j / \mathbf{k}^2}{\mathbf{k}^2 + m_A^2} + \alpha \frac{k_i k_j}{\mathbf{k}^4} \right) \mathrm{e}^{i\mathbf{k} \cdot \mathbf{x}}, \tag{86}$$

back into the Hamiltonian (83). The energy is divergent in the ultraviolet because in the London limit, where the mass $|m_\phi|$ of the $\phi$-field is taken to be infinite, the vortices are considered to be ideal lines. For a finite mass, a vortex core has a typical width of the order of the coherence length $\xi = 1/|m_\phi|$. This mass therefore provides a natural ultraviolet cutoff to the theory. Omitting the irrelevant (diverging) monopole self-interaction, one finds [48]

$$E_V = \frac{1}{2} g^2 \int_{L_z} \mathrm{d}x_i \int_{L_z} \mathrm{d}y_i \, G(\mathbf{x} - \mathbf{y}) = \sigma_V L_z. \tag{87}$$

Here,

$$G(\mathbf{x}) = \int_{\mathbf{k}} \frac{\mathrm{e}^{i\mathbf{k}\cdot\mathbf{x}}}{\mathbf{k}^2 + m_A^2} = \frac{1}{4\pi} \frac{\mathrm{e}^{-m_A|\mathbf{x}|}}{|\mathbf{x}|} \tag{88}$$

is the scalar Green function, $g$ abbreviates the combination

$$g = \Phi_0 m_A, \tag{89}$$

while $L_z$ denotes the (infinite) length of the Dirac string and

$$\sigma_V = \frac{1}{8\pi} g^2 \ln\left(\frac{|m_\phi|^2}{m_A^2}\right) = \frac{1}{4\pi} g^2 \ln(\kappa_{\mathrm{GL}}) \tag{90}$$

its line tension, first calculated by Abrikosov [49]. The value $\kappa_{\mathrm{GL}} = 1/\sqrt{2}$ of the Ginzburg-Landau parameter

$$\kappa_{\mathrm{GL}} = \frac{|m_\phi|}{m_A} = \frac{\lambda_{\mathrm{L}}}{\xi}, \quad \text{or} \quad \kappa_{\mathrm{GL}}^2 = \frac{\lambda}{2e^2}, \tag{91}$$

separates the type-II regime ($\kappa_{\mathrm{GL}} > 1/\sqrt{2}$), where isolated vortices can exist, from the type-I regime ($\kappa_{\mathrm{GL}} < 1/\sqrt{2}$), where a partial penetration of an external field is impossible. Remembering that $L_z$ was the Dirac string, we see from (87) that in the superconducting phase it does indeed become the core of the magnetic vortex originating at the monopole, as was first observed by Nambu [48].

The operator $V(L_z)$ describing the monopole with its emerging flux tube [50, 51] is easily constructed by noting that in the functional-integral approach, a given field configuration is weighted with a Boltzmann factor $\exp\left(-\int_{\mathbf{x}} \mathcal{H}_{\mathrm{GL}}^{\mathrm{P}}\right)$, with the Hamiltonian given by Eq. (83). From this we infer that the explicit form of the vortex operator is

$$V(L_z) = \exp\left\{\int_{\mathbf{x}} \left[(\nabla \times \mathbf{A}) \cdot \mathbf{B}^{\mathrm{P}} - \tfrac{1}{2}\left(\mathbf{B}^{\mathrm{P}}\right)^2\right]\right\}. \tag{92}$$

We next demonstrate that this operator can be used to distinguish the superconducting from the normal phase [50, 51]. To this end, let us consider the correlation function $\langle V(L_{\mathbf{z}})V^*(L_{\bar{\mathbf{z}}})\rangle$, where $V^*(L_{\bar{\mathbf{z}}})$ describes an additional antimonopole brought into the system at $\bar{\mathbf{z}}$, with $L_{\bar{\mathbf{z}}}$ being the accompanying Dirac string running from infinity to $\bar{\mathbf{z}}$. Since all the integrals involved are Gaussian, this expectation value can be evaluated directly. We proceed, however, in an indirect way for reasons that will become clear when we proceed, and first linearize the functional integral over the gauge field by introducing an auxiliary field $\tilde{\mathbf{h}}$. In the gauge $\nabla \cdot \mathbf{A} = 0$, which corresponds to setting $\alpha = 0$ in the Hamiltonian (83), we find [52]

$$\langle V(L_{\mathbf{z}})V^*(L_{\bar{\mathbf{z}}})\rangle = \int \mathrm{D}\mathbf{A}\mathrm{D}\tilde{\mathbf{h}} \exp\left\{\int_{\mathbf{x}}\left[-\frac{1}{2}\tilde{\mathbf{h}}^2 + i\tilde{\mathbf{h}}\cdot(\nabla\times\mathbf{A} - \mathbf{B}^{\mathrm{P}}) - \frac{m_A^2}{2}\mathbf{A}^2\right]\right\}, \tag{93}$$

where the plastic field satisfies (81) with the monopole density given by

$$\rho_{\mathrm{m}}(\mathbf{x}) = \Phi_0[\delta(\mathbf{x}-\mathbf{z}) - \delta(\mathbf{x}-\bar{\mathbf{z}})]. \tag{94}$$

To appreciate the physical relevance of the auxiliary field, let us consider its field equation

$$\tilde{\mathbf{h}} = i(\nabla\times\mathbf{A} - \mathbf{B}^{\mathrm{P}}) = i\mathbf{h}. \tag{95}$$

It tells us that apart from a factor $i$, the auxiliary field $\tilde{\mathbf{h}}$ can be thought of as representing the local magnetic induction $\mathbf{h}$.

The integral over the vector potential is easily carried out by substituting the field equation for $\mathbf{A}$,

$$\mathbf{A} = \frac{i}{m_A^2}\nabla\times\tilde{\mathbf{h}}, \tag{96}$$

back into (93), with the result

$$\langle V(L_{\mathbf{z}})V^*(L_{\bar{\mathbf{z}}})\rangle = \int \mathrm{D}\tilde{\mathbf{h}} \exp\left\{-\frac{1}{2}\int_{\mathbf{x}}\left[\frac{1}{m_A^2}(\nabla\times\tilde{\mathbf{h}})^2 + \tilde{\mathbf{h}}^2\right] - i\int_{\mathbf{x}}\tilde{\mathbf{h}}\cdot\mathbf{B}^{\mathrm{P}}\right\}. \tag{97}$$

This shows that the magnetic vortex described by the plastic field $\mathbf{B}^{\mathrm{P}}$ couples to the fluctuating massive vector field $\tilde{\mathbf{h}}$, with a coupling constant $g$ introduced in Eq. (89). As the temperature approaches the critical temperature from below, $\bar{\phi}$ tends to zero, so that the vortex decouples from the auxiliary field $\tilde{\mathbf{h}}$. The finite magnetic penetration depth in the superconducting phase is reflected by the mass term in (97).

*Figure 1.* Biot-Savart interaction (wiggly line) between two line elements $dx_i$ and $dy_i$ of a magnetic vortex (straight line).

After carrying out the integral over $\tilde{\mathbf{h}}$ in (97), we obtain for the correlation function

$$\langle V(L_{\mathbf{z}})V^*(L_{\bar{\mathbf{z}}})\rangle = \exp\left\{-\frac{1}{2}\int_{\mathbf{x},\mathbf{y}} [\rho_{\rm m}(\mathbf{x})\, G(\mathbf{x}-\mathbf{y})\, \rho_{\rm m}(\mathbf{y}) + m_A^2 B_i^{\rm P}(\mathbf{x})\, G(\mathbf{x}-\mathbf{y})\, B_i^{\rm P}(\mathbf{y})]\right\}. \tag{98}$$

The first term in the argument of the exponential function contains a diverging monopole self-interaction for $\mathbf{x} = \mathbf{y}$. This divergence is irrelevant and can be eliminated by defining a renormalized operator

$$V_{\rm r}(L_{\mathbf{z}}) = V(L_{\mathbf{z}}) \exp\left[\tfrac{1}{2}\Phi_0^2 G(0)\right]. \tag{99}$$

The second term in the argument is the most important one for our purposes. It represents a Biot-Savart interaction between two line elements $dx_i$ and $dy_i$ of the magnetic vortex (see Fig. 1). For the renormalized operators we find

$$\langle V_{\rm r}(L_{\mathbf{z}})V_{\rm r}^*(L_{\bar{\mathbf{z}}})\rangle = \exp(-\sigma_V L_{\mathbf{z}\bar{\mathbf{z}}}) \exp\left(\frac{\Phi_0^2}{4\pi} \frac{{\rm e}^{-m_A L_{\mathbf{z}\bar{\mathbf{z}}}}}{L_{\mathbf{z}\bar{\mathbf{z}}}}\right), \tag{100}$$

where $L_{\mathbf{z}\bar{\mathbf{z}}}$ denotes the length of the flux tube connecting the monopole at $\mathbf{z}$ with the antimonopole at $\bar{\mathbf{z}}$. Initially, the two Dirac strings may run to infinity along two arbitrary paths. Due to the string tension, however, they join on the shortest path $L_{\mathbf{z}\bar{\mathbf{z}}}$ connecting the monopoles.

The correlation function (100) behaves differently in the two phases. In the superconducting phase, where $m_A \neq 0$, the first factor dominates. It represents the confining linear potential between the monopole and the antimonopole. As a result, the correlation function decays exponentially for distances larger than $1/\sigma_V$:

$$\langle V_{\rm r}(L_{z\bar{z}})V_{\rm r}^*(L_{z\bar{z}})\rangle \to 0. \tag{101}$$

On the other hand, in the normal phase, where the gauge field is massless, the confinement factor in the correlation function (100) disappears, while the argument of the second exponential turns into a pure Coulomb potential. The correlation function remains, consequently, finite for large distances:

$$\langle V_{\mathrm{r}}(L_{z\bar{z}})V_{\mathrm{r}}^{*}(L_{z\bar{z}})\rangle \to 1, \tag{102}$$

thus indicating a proliferation of magnetic vortices in the normal phase. This demonstrates that the vortex operator can be used as an order parameter to distinguish the two phases. Since it develops a vacuum expectation value not in the superconducting, but in the normal phase it is referred to as a *disorder* parameter.

It is interesting to consider in detail the limit $T \to T_{\mathrm{c}}$, where the coupling constant $g = \Phi_0 m_A$ tends to zero and the magnetic vortex decouples from the massive vector field [52]. In Eq. (97), this limit yields the constraint $\nabla \times \tilde{\mathbf{h}} = 0$ which can be solved by setting $\tilde{\mathbf{h}} = \nabla\gamma$. The correlation function then takes the simple form

$$\langle V_{\mathrm{r}}(\mathbf{z})V_{\mathrm{r}}^{*}(\bar{\mathbf{z}})\rangle = \int \mathrm{D}\gamma \exp\left[-\frac{1}{2}\int_{\mathbf{x}}(\nabla\gamma)^2 + i\int_{\mathbf{x}}\gamma\rho_{\mathrm{m}}\right]. \tag{103}$$

In the absence of monopoles, the theory reduces to that of a free gapless mode $\gamma$ that may be thought of as representing the magnetic scalar potential:

$$\nabla\gamma = i\nabla \times \mathbf{A}. \tag{104}$$

This follows from combining the physical interpretation of the vector field $\mathbf{h}$ (95) with the equation $\tilde{\mathbf{h}} = \nabla\gamma$.

The correlation function (103) can be put in the form

$$\langle V_{\mathrm{r}}(\mathbf{z})V_{\mathrm{r}}^{*}(\bar{\mathbf{z}}) = \left\langle \mathrm{e}^{i\Phi_0[\gamma(\mathbf{z})-\gamma(\bar{\mathbf{z}})]}\right\rangle, \tag{105}$$

where the average at the right-hand side is taken with respect to the free scalar theory. This equation shows that in the normal phase ($T > T_{\mathrm{c}}$), the Dirac string looses its physical relevance, the right-hand side depending only on the end points $\mathbf{z}$ and $\bar{\mathbf{z}}$, not on the path $L_{\mathbf{z}\bar{\mathbf{z}}}$ connecting these points. The notion of a magnetic vortex is of no relevance in this phase because the vortices proliferate and carry no energy. This is the reason for omitting any reference to vortex lines in the argument of $V$ in Eqs. (103) and (105).

## 3.3. MAGNETIC VORTEX LOOPS

From the above results for a single vortex, we can now easily infer the dual formulation of the Ginzburg-Landau theory. In this formulation, the partition function is written as a grand canonical ensemble of fluctuating magnetic vortex loops, of arbitrary length and shape:

$$Z = \int \mathrm{D}\tilde{\mathbf{h}}\, \mathrm{e}^{-\frac{1}{2}\int_{\mathbf{x}}[(\nabla\times\tilde{\mathbf{h}})^2/m_A^2+\frac{1}{2}\tilde{\mathbf{h}}^2]} \sum_{N=0}^{\infty} \frac{1}{N!} \prod_{l=1}^{N} \left[\int_0^{\infty} \frac{\mathrm{d}\tau_l}{\tau_l} \oint \mathrm{D}\mathbf{x}(\tau_l')\right] \mathrm{e}^{-S_{\mathrm{dual}}} \tag{106}$$

with the dual action, [cf. Eq. (78)]

$$\begin{aligned} S_{\mathrm{dual}} &= \sum_{l=1}^{N} \int_0^{\tau_l} \mathrm{d}\tau_l' \left\{\tfrac{1}{4}\dot{\mathbf{x}}^2(\tau_l') + m_\psi^2 + i\Phi_0 \dot{\mathbf{x}}(\tau_l') \cdot \tilde{\mathbf{h}}[\mathbf{x}(\tau_l')]\right\} \\ &\quad + u \sum_{l,k=1}^{N} \int_0^{\tau_l} \mathrm{d}\tau_l' \int_0^{\tau_k} \mathrm{d}\tau_k'\, \delta\left[\mathbf{x}(\tau_l') - \mathbf{x}(\tau_k')\right]. \end{aligned} \tag{107}$$

We have included here a mass term $m_\psi^2$ representing the intrinsic vortex line tension [cf. Eq. (68)] and also a contact repulsion between the vortices parameterized by $u$.

The equivalent field representation of this dual theory reads [53, 54, 37, 55, 52, 46],

$$Z = \int \mathrm{D}\tilde{\mathbf{h}} \mathrm{D}\psi^* \mathrm{D}\psi \exp\left(-\int_{\mathbf{x}} \mathcal{H}_{\mathrm{dual}}\right) \tag{108}$$

with the Hamiltonian

$$\mathcal{H}_{\mathrm{dual}} = \frac{1}{2m_A^2}(\nabla\times\tilde{\mathbf{h}})^2 + \frac{1}{2}\tilde{\mathbf{h}}^2 + |(\nabla - i\Phi_0\tilde{\mathbf{h}})\psi|^2 + m_\psi^2|\psi|^2 + u|\psi|^4, \tag{109}$$

where the disorder field $\psi$ is minimally coupled to the massive vector field $\tilde{\mathbf{h}}$, representing the local magnetic induction. The dual formulation contains the same information as the original Ginzburg-Landau formulation. For example, the vortex line tension (90) appears in the dual theory as a one-loop on-shell mass correction stemming from the graph depicted in Fig. 1 which we now interpret as a Feynman graph of the dual theory (109), with the straight and wiggly line denoting respectively the $\psi$- and $\tilde{\mathbf{h}}$-field Green function. Also the fixed points of the Ginzburg-Landau theory map onto those of the dual field theory [55]. And both formulations can be used to study the critical behavior of the superconductor-to-normal phase transition (see Refs. [30, 34, 37, 56, 46, 57, 58] for the dual approach).

## Acknowledgments

It is a pleasure to thank N. Antunes, L. Bettencourt, A. Leggett, D. Steer, and G. Volovik for useful discussions during the NATO Winter School and European Science Foundation (ESF) Workshop *Topological Defects and the*

*Non-Equilibrium Dynamics of Symmetry Breaking Phase Transitions*, Les Houches, February 16-26, 1999, and M. Crisan for helpful correspondence.

This work is performed as part of a scientific network supported by the ESF (see the network's URL, http://www.physik.fu-berlin.de/~defect).

## References

1. Berezinskii, V.L. (1972) Destruction of Long-Range Order *Sov. Phys. JETP* **34**, pp. 610–616
2. Kosterlitz, J.M. and Thouless, D. J. (1973) Metastability and Phase Transitions in Two-Dimensional Systems *J. Phys. C* **6**, pp. 1181–1203
3. Davis, A.-C. and Brandenberger, R. (Eds.) (1995) *Formation and Interactions of Topological Defects* Plenum Press, New York
4. Gorkov, L.P. (1959) Microscopic Derivation of the Ginzburg-Landau Equations in the Theory of Superconductivity *Sov. Phys. JETP* **9** pp. 1364–1367
5. Bardeen, J., Cooper, L. N. and Schrieffer, J. R. (1957) Theory of Superconductivity *Phys. Rev.* **108** pp. 1175–1204
6. Ginzburg, V.L. and Landau, L.D. (1950) *Zh. Eksp. Teor. Fiz.* **20** pp. 1064
7. Schakel, A.M.J. (1989) *On Broken Symmetries in Fermi Systems*, Ph. D. Thesis, University of Amsterdam
8. Eagles, D.M. (1969) Possible Pairing without Superconductivity at Low Carrier Concentrations in Bulk and Thin-Film Superconducting Semiconductors *Phys. Rev.* **186** pp. 456–463
9. Leggett, A.J. (1980) *in:* Pekalski, A. and Przystawa, J. (Eds.) *Modern Trends in the Theory of Condensed Matter* Springer-Verlag, Berlin pp. 13–27
10. Drechsler, M. and Zwerger, W. (1992) Crossover from BCS-Superconductivity to Bose-Condensation *Ann. Phys. (Germany)* **1** pp. 15–23
11. Haussmann, R. (1993) Crossover from BCS Superconductivity to Bose-Einstein Condensation—A Self-Consistent Theory *Z. Phys. B* **91** pp. 291–308
12. Sá de Melo, C.A.R., Randeria, M. and Engelbrecht, J.R. (1993) Crossover from BCS to Bose Superconductivity—Transition-Temperature and Time-Dependent Ginzburg-Landau Theory *Phys. Rev. Lett.* **71** pp. 3202–3205
13. Marini, M., Pistolesi, F. and Strinati, G.C. (1998) Evolution from BCS Superconductivity to Bose Condensation: Analytic Results for the Crossover in Three Dimensions *Eur. Phys. J. B* **1** pp. 151–159
14. Popov, V.N. (1987) *Functional Integrals and Collective Excitations* Cambridge University Press, Cambridge
15. Randeria, M. Duan, J.-M. and Shieh L.-Y. (1990) Superconductivity in a Two-Dimensional Fermi Gas: Evolution from Cooper Pairing to Bose Condensation *Phys. Rev. B* **41** pp. 327–343
16. Mattuck, R.D. (1976) *A Guide to Feynman Diagrams in the Many-Body Problem* McGraw-Hill, New York
17. Fraser, C.M. (1985) Calculation of Higher Derivative Terms in the One Loop Effective Lagrangian *Z. Phys. C* **28** pp. 101–106; Aitchison, I.J.R. and Fraser, C.M. (1985) Derivative Expansions of Fermion Determinants: Anomaly-Induced Vertices, Goldstone-Wilczek Currents, and Skyrme Terms *Phys. Rev. D* **31** pp. 2605–2615
18. Rivers, R.J (1987) *Path Integrals in Quantum Field Theory* Cambridge University Press, Cambridge
19. Kapusta, J.I. (1989) *Finite-Temperature Field Theory* Cambridge University Press, Cambridge
20. Stintzing, S. and Zwerger, W. (1997) Ginzburg-Landau Theory of Superconductors with Short Coherence Length *Phys. Rev. B* **56** pp. 9004-9014
21. Schmid, A. (1966) A Time-Dependent Ginzburg-Landau Equation and its Appli-

cations to the Problem of Resistivity in the Mixed State *Phys. Kond. Materie* **5** pp. 302–317
22. Abrahams, E. and Tsuneto, T. (1966) Time Variation of the Ginzburg-Landau Order Parameter *Phys. Rev.* **152** pp. 416–432
23. Schakel, A.M.J. (1994) Effective Theory of Bosonic Superfluids *Int. J. Mod. Phys. B* **8** pp. 2021–2039
24. de Gennes, P.G. (1966) *Superconductivity in Metals and Alloys* Benjamin, New York
25. Tinkham, M. (1975) *Introduction to Superconductivity* McGraw-Hill, New York
26. Crisan, M. (1989) *Theory of Superconductivity* World Scientific, Singapore
27. Gross, E.P. (1961) Structure of a Quantized Vortex in Boson Systems *Nuovo Cimento* **20** pp. 454–477 ; Pitaevskii, L.P. (1961) Vortex Lines in an Imperfect Bose Gas *Sov. Phys. JETP* **13** pp. 451–454
28. Schakel, A.M.J. (1999) *in*: Shopova, D.V. and Uzunov, D.I. (Eds.) *Correlations, Coherence, and Order* Plenum Press, New York pp. 295–382
29. Banks, T. Meyerson, B. and Kogut, J. (1977) Phase Transitions in Abelian Lattice Gauge Theories *Nucl. Phys. B* **129** pp. 493–510
30. Peskin, M. (1978) Mandelstam-'t Hooft Duality in Abelian Lattice Models *Ann. Phys.* **113** pp. 122–152
31. Thomas, P.R. and Stone, M. Nature of the Phase Transition in a Nonlinear $O(2)_3$ Model (1978) *Nucl. Phys. B* **144** pp. 513–524
32. Helfrich, W. and Müller, W. (1980) Concentrated Thermally Equilibrated Polymer Solutions *in*: *Continuum Models of Discrete Systems* Waterloo University Press, Waterloo pp. 753–760
33. Dasgupta, C. and Halperin, B. I. (1981) Phase Transition in a Lattice Model of Superconductivity *Phys. Rev. Lett.* **47** pp. 1556–1560
34. Kleinert, H. (1982) Disorder Version of the Abelian Higgs Model and the Order of the Superconductive Phase Transition *Lett. Nuovo Cimento* **35** pp. 405–412
35. Bartholomew, J. (1983) Phase Structure of a Lattice Superconductor *Phys. Rev. B* **28** pp. 5378–5381
36. Savit, R. (1989) Duality in Field Theory and Statistical Systems *Rev. Mod. Phys.* **52** pp. 453–487
37. Kleinert, H. (1989) *Gauge Fields in Condensed Matter*, **1**, World Scientific, Singapore
38. Schakel, A.M.J. (1998) *Boulevard of Broken Symmetries*, e-print cond-mat/9805152
39. Symanzik, K. (1969) *in*: Jost, R. (Ed.) *Euclidean Quantum Field Theory* Academic Press, New York
40. Feynman, R.P. (1950) Mathematical Formulation of the Quantum Theory of Eletromagnetic Interactions *Phys. Rev.* **80** pp. 440–457
41. Schwinger, J. (1951) On Gauge Invariance and Vacuum Polarization *Phys. Rev.* **82** pp. 664–679
42. Kleinert, H. (1995) *Path Integrals in Quantum Mechanics, Statistics and Polymer Physics*, 2nd Edition, World Scientific, Singapore
43. Feynman, R.P. (1948) Space-Time Approach to Non-Relativistic Quantum Mechanics *Rev. Mod. Phys.* **20** pp. 367–387
44. Parisi, G. (1988) *Statistical Field Theory* Addison-Wesley, New-York
45. Copeland, E., Haws, D., Holbraad, S. and Rivers, R. (1990) The Statistical Properties of Strings. 1. Free Strings *in*: Gibbons, G.W., Hawking, S.W. and Vachaspati, T. (Eds.) *The Formation and Evolution of Cosmic Strings* Cambridge University Press, Cambridge pp. 35–47; The Statistical Properties of Strings. 2. Interacting Strings *ibid* pp. 49–69
46. Kiometzis, M., Kleinert, H. and Schakel, A.M.J. (1995) Dual Description of the Superconducting Phase Transition *Fortschr. Phys.* **43** pp. 697–732
47. Dirac, P.A.M. (1948) The Theory of Magnetic Poles *Phys. Rev.* **74** pp. 817–830
48. Nambu, Y. (1974) Strings, Monopoles and Gauge Fields *Phys. Rev. D* **10** pp. 4262–4268

49. Abrikosov, A.A. (1957) On the Magnetic Properties of Superconductors of the Second Group *Sov. Phys. JETP* **5** pp. 1174–1182
50. Marino, E.C. (1988) Quantum Theory of Nonlocal Vortex Fields *Phys. Rev. D* **38** pp. 3194–3198 (1988); Marino, E.C., Marques, G.C., Ramos, R.O. and Ruiz, J.S. (1992) Mass Spectrum and Correlation Function of Quantum Vortices in the Abelian Higgs Model *Phys. Rev. D* **45** pp. 3690–3700; Marino, E.C. (1993) Duality, Quantum Vortices and Anyons in Maxwell-Chern-Simons-Higgs Theories *Ann. Phys. (NY)* **224** pp. 225–274
51. Kovner, A., Rosenstein, B. and Eliezer, D. (1990) Photon as Goldstone Boson in (2+1)-Dimensional Higgs Model *Mod. Phys. Lett. A* **5** pp. 2733–2740; (1991) Photon as a Goldstone Boson in (2+1)-Dimensional Abelian Gauge Theories *Nucl Phys. B* **350** pp. 325–354; Kovner, A. and Rosenstein, B. (1991) Topological Interpretation of Electric Charge and the Aharonov-Bohm Effect in 2+1 Dimensions *Phys. Rev. Lett.* **67** pp. 1490–1493
52. Kiometzis, M. and Schakel, A.M.J. (1993) Landau Description of the Superconducting Phase-Transition *Int. J. Mod. Phys. B* **7** pp. 4271–4288
53. Bardakci, K. and Samuel, S. (1978) Local Field Theory for Solitons *Phys. Rev. D* **18** pp. 2849–2860
54. Kawai, H. (1981) A Dual Transformation of the Nielsen-Olesen Model *Prog. Theor. Phys.* **65** pp. 351–364
55. Kovner, A., Kurzepa, P. and Rosenstein, B. (1993) A Candidate for Exact Continuum Dual Theory for Scalar $QED_3$ *Mod. Phys. Lett. A.* **8** pp. 1343–1355
56. Kiometzis, M., Kleinert, H. and Schakel, A.M.J. (1994) Critical Exponents of the Superconducting Phase-Transition *Phys. Rev. Lett.* **73** pp. 1975–1977
57. Herbut, I.F. (1996) Continuum Dual Theory of the Transition in 3D Lattice Superconductor *J. Phys. A: Math. Gen.* **29** pp. 423–429
58. de Calan, C. and Nogueira, F.S. (1999) e-print `cond-mat/9903247`

# BROKEN SYMMETRY & NONEQUILIBRIUM SUPERFLUID $^3$HE

*Circular Birefringence of Propagating Transverse Currents*

J. A. SAULS

*Northwestern University, Evanston, Illinois 60208, USA*

The superfluid phases of $^3$He provide a unique physical system to study the dynamical effects of spontaneous symmetry breaking in condensed matter. The theory of superfluid $^3$He is grounded in two of the most successful theories of strongly interacting matter, Landau's Fermi-liquid theory and the BCS pairing theory of superconductivity. These two theories were placed into a common theoretical framework in the late 60's. I discuss applications of this theory to nonequilibrium dynamics of superfluid $^3$He. In 1957 Landau predicted that liquid $^3$He would support propagating shear waves at low temperatures, i.e. a *transverse sound* mode. Such waves have recently been observed at low temperatures in the superfluid B-phase of liquid $^3$He. These observations provide a beautiful example of the effect of spontaneous symmetry breaking in condensed matter. I discuss the theory of transverse wave propagation in $^3$He and the recent detection of these waves by magneto-acoustic rotation of the polarization in a magnetic field.

## 1. Introduction

The physics of superfluid $^3$He is sufficiently rich that it can provide interesting analogues of theoretical models in astrophysics, high-energy physics and cosmology [36], and more recently in the field of ultra-low-temperature atomic gases [14]. But the significance of liquid $^3$He to theoretical physics may well be that it provides the model system for developing and extending one of the most successful theories of strongly interacting matter - *the Fermi-liquid theory of superconductivity.*[1] These lecture notes provide an introduction to non-equilibrium superfluid $^3$He, with applications to high-frequency excitations, including aspects of symmetry breaking in $^3$He and its effects on collective mode dynamics.

1 For a detailed description see [33].



*Y.M. Bunkov and H. Godfrin (eds.),*
*Topological Defects and the Non-Equilibrium Dynamics of Symmetry Breaking Phase Transitions*, 239–265.

The density and current modes of liquid $^3$He are governed by conservation laws for mass (or particle number) and momentum,

$$\frac{\partial n}{\partial t} + \nabla \cdot \mathbf{J} = 0 \quad , \quad \frac{\partial J_i}{\partial t} + \frac{1}{m}\frac{\partial \Pi_{ij}}{\partial x_j} = 0 \, , \tag{1}$$

where $n(\mathbf{R}, t)$ is the particle density, $\mathbf{J}(\mathbf{R}, t)$ is the particle current density, $m$ is the atomic mass of $^3$He and $\Pi_{ij}(\mathbf{R}, t)$ is the momentum stress tensor of the liquid. The conservation laws are supplemented by equations relating the density, current and stress tensor. In the hydrodynamic limit these constitutive equations are obtained from local thermodynamics. The assumption of hydrodynamics is that the characteristic timescale for a disturbance from equilibrium is long compared with the timescale for the restoration of equilibrium locally on the scale of a typical wavelength of the disturbance. The equations of hydrodynamics provide an accurate description of the long wavelength, low frequency ($\omega \ll 1/\tau$) dynamics of $^3$He. However, this description breaks down at sufficiently high frequencies or low temperatures. The regime of applicability of hydrodynamics to liquid $^3$He at low temperatures is severely restricted by Fermi statistics. The time-scale for the restoration of local equilibrium increases rapidly below 100 mK, $\tau(T) \simeq 1\,\mu$sec-mK$^2/T^2$. Thus, for excitation frequencies of order $\omega/2\pi \simeq 20$ MHz the hydrodynamic regime is restricted to temperatures above $T \simeq 15 - 20$ mK. The physics of $^3$He at lower temperatures, or higher frequencies, departs radically from the predictions of hydrodynamics. Liquid $^3$He at these excitation energies is governed by coupled dynamical equations for quasiparticle (distribution function) and Cooper pair (pair correlation amplitude) excitations.

*Transport Theory*

In his seminal papers on Fermi liquid theory [17, 18] Landau explained how a system of strongly interacting Fermions could exhibit both a spectrum of low-lying Fermionic excitations with a well-defined Fermi surface, as well as low lying Bosonic modes associated with deformations of the Fermi surface. The stability of the Fermi surface requires that interactions between the Fermionic excitations ("quasiparticles") act to *restore* the Fermi surface to its equilibrium configuration. Under these conditions the "Fermi sea" behaves as an elastic medium in which restoring forces lead to natural oscillations, or free vibrations of the Fermi surface, called "zero sound modes". The dynamical variables describing these modes are related to the *deformation* of the Fermi surface, which is defined in terms of the deviation of the quasiparticle distribution function from its equilibrium form,

$$\phi(\hat{\mathbf{p}}, \mathbf{R}; t) = \int d\varepsilon\, \Phi(\hat{\mathbf{p}}, \mathbf{R}; \varepsilon, t)\,, \tag{2}$$

where $\Phi(\hat{\mathbf{p}}, \mathbf{R}; \varepsilon, t)$ is the distribution function for quasiparticles with momentum near the Fermi surface, $\mathbf{p}_f = p_f \hat{\mathbf{p}}$, and excitation energies, $\varepsilon$, that are small compared with the Fermi energy, $|\varepsilon| \ll E_f$.[2] BCS pairing in $^3$He is an instability of the Fermi surface; thus, one should expect that a description of collisionless modes in the superfluid phases requires new dynamical variables describing the condensed phase, which are coupled to the dynamics of the distribution function.

The transport equation for the distribution function is the central equation of Landau's theory of normal $^3$He,

$$\left(\frac{\partial}{\partial t} + \mathbf{v}_f \cdot \boldsymbol{\nabla}_{\mathbf{R}}\right) \Phi(\hat{\mathbf{p}}, \mathbf{R}; \varepsilon, t) - \left(\frac{\partial \Phi_0}{\partial \varepsilon}\right) \frac{\partial}{\partial t} \mathcal{U}_{\text{tot}}(\hat{\mathbf{p}}, \mathbf{R}; t) = I[\Phi]\,. \tag{3}$$

The first two terms describe the ballistic propagation of quasiparticles with a group velocity given by the Fermi velocity, $\mathbf{v}_f = v_f \hat{\mathbf{p}}$. The third term represents the action of external and internal forces acting on the quasiparticles; $\mathcal{U}_{\text{tot}}(\hat{\mathbf{p}}, \mathbf{R}; t) = u_{\text{ext}}(\mathbf{R}; t) + \mathcal{E}(\hat{\mathbf{p}}, \mathbf{R}; t)$ is the sum of the external potential energy ($u_{\text{ext}}$) and the interaction energy ($\mathcal{E}$) of a quasiparticle with momentum $\mathbf{p}$ on the Fermi surface with the distribution of other nonequilibrium quasiparticles. The derivative of the equilibrium distribution, $\partial\Phi_0(\varepsilon)/\partial\varepsilon$, restricts the dynamics to low-energy excitations near the Fermi level. Both external and internal forces accelerate the quasiparticles leading to smooth changes in the distribution function in space and time. The right side of the transport equation determines the rate of change of the distribution function from quasiparticle collisions. This term leads to irreversibility and relaxation of the distribution function on the timescale of the mean time between collisions, i.e. $I[\Phi] \sim -1/\tau\, \delta\Phi$. The collision terms are generally small for the typical frequencies of interest, i.e. $\omega \gg 1/\tau$. Thus, we can often neglect collisional effects, except for collisional broadening of otherwise sharp collective modes.

The distribution function determines the density and momentum fluctuations associated with nonequilibrium states of $^3$He,

$$n(\mathbf{R}, t) = \frac{p_f^3}{3\pi^2\hbar^3} + \frac{N_f}{1 + F_0^s} \int \frac{d\Omega_{\mathbf{p}}}{4\pi} \int d\varepsilon\, \Phi(\hat{\mathbf{p}}, \mathbf{R}; \varepsilon, t)\,, \tag{4}$$

$$\mathbf{J} = N_f \int \frac{d\Omega_{\mathbf{p}}}{4\pi} \mathbf{v}_f \int d\varepsilon\, \Phi(\hat{\mathbf{p}}, \mathbf{R}; \varepsilon, t)\,, \tag{5}$$

2 In the microscopic formulation of Fermi liquid theory the distribution function is defined in terms of a correlation function in the limit, $\hbar q \ll p_f$, $\hbar\omega \ll E_f$ [33].

where $N_f$ is the density of states at the Fermi level and $F_0^s$ is the $\ell = 0$ Landau interaction parameter (defined below). The conservation laws for mass and momentum follow from the transport equation and conservation of fermion number and momentum in collision processes: $\int d\Omega_{\mathbf{p}} \int d\varepsilon \, I[\Phi] = 0$ and $\int d\Omega_{\mathbf{p}} \hat{\mathbf{p}} \int d\varepsilon \, I[\Phi] = 0$. The continuity equation for $\mathbf{J}$ determines the stress tensor,

$$\Pi_{ij} = N_f \int \frac{d\Omega_{\mathbf{p}}}{4\pi} [\mathbf{v}_f]_i \, [\mathbf{p}_f]_j \int d\varepsilon \, \Phi(\hat{\mathbf{p}}, \mathbf{R}; \varepsilon, t) \, . \tag{6}$$

The result for $\Pi_{ij}$ follows from the microscopic theory of a Fermi liquid and is valid above and below the superfliud transition, as well as in the nonlinear response regime, so long as $u_{\text{ext}} \ll E_f$ [24].

Normal $^3$He is isotropic so the eigenmodes of the Fermi surface are given by the amplitudes, $\phi_{\ell,m}$, of the spherical harmonic expansion of the distribution function. For example, $\phi_{0,0}$ is the amplitude for an isotropic expansion or contraction of the Fermi surface, and determines the density fluctuation,

$$\delta n(\mathbf{R}, t) = \frac{N_f}{1 + F_0^s} \, \phi_{0,0} \, , \tag{7}$$

while the $\ell = 1$ modes determine the current density,

$$\mathbf{J}(\mathbf{R}, t) = \frac{1}{3} N_f v_f \sum_{m=0,\pm 1} \phi_{1,m} \, \mathbf{e}^{(m)} \, , \tag{8}$$

where $\mathbf{e}^{(m)}$ are unit vectors defining the three linearly independent current modes: the longitudinal current, $\mathbf{J}_l \sim \mathbf{e}^{(0)} = \hat{\mathbf{q}}$, and the two circularly polarized transverse modes, $\mathbf{J}_\pm \sim \mathbf{e}^{(\pm)}$. The circularly polarized basis vectors are defined by, $\mathbf{e}^{(0)} = \mathbf{e}_3$ and $\mathbf{e}^{(\pm)} = (\mathbf{e}_1 \pm i\mathbf{e}_2)/\sqrt{2}$, where $\{\mathbf{e}_1, \mathbf{e}_2, \mathbf{e}_3\}$ is a Cartesian triad with $\mathbf{e}_3 = \hat{\mathbf{q}}$. The $\ell = 2$ amplitudes are related to the five traceless and symmetric components of the stress tensor,

$$\pi_{ij} = \Pi_{ij} - \frac{1}{3} \mathrm{Tr}[\mathbf{\Pi}] \delta_{ij} = \frac{2}{15} N_f p_f v_f \sum_{m=0,\pm 1,\pm 2} \phi_{2,m} \, \mathbf{t}_{ij}^{(2,m)} \, , \tag{9}$$

where $\mathbf{t}_{ij}^{(2,m)}$ are the $\ell = 2$ spherical tensors with the quantization axis chosen as $\mathbf{e}^{(0)} = \hat{\mathbf{q}}$ (see Eqs. (18) below). The stress tensor plays a central role in the nonequilibrium response of liquid $^3$He. The solution of the transport equation leads to constitutive equations relating the stress tensor, the density and current response, and in the superfluid phase to additional variables representing the dynamics of the condensate.

In the collisionless limit restoring forces for spin-independent deformations of the Fermi surface originate from

$$\mathcal{E}(\hat{\mathbf{p}}, \mathbf{R}; t) = \int \frac{d\Omega_{\mathbf{p}'}}{4\pi} A^s(\hat{\mathbf{p}}, \hat{\mathbf{p}}') \, \phi(\hat{\mathbf{p}}', \mathbf{R}; t) \,. \tag{10}$$

This term represents the interaction of a quasiparticle of momentum $p_f \hat{\mathbf{p}}$ with the distribution of quasiparticles that represent the deformation of the Fermi surface. The function $A^s(\hat{\mathbf{p}} \cdot \hat{\mathbf{p}}')$ is the spin-independent amplitude for forward scattering of two quasiparticles with momenta $\mathbf{p}$ and $\mathbf{p}'$ on the Fermi surface, and is related to the spin-independent interaction energy between two quasiparticles, $F^s(\hat{\mathbf{p}}, \hat{\mathbf{p}}')$, by [18]

$$A^s(\hat{\mathbf{p}} \cdot \hat{\mathbf{p}}') = F^s(\hat{\mathbf{p}} \cdot \hat{\mathbf{p}}') + \int \frac{d\Omega_{\mathbf{p}''}}{4\pi} F^s(\hat{\mathbf{p}} \cdot \hat{\mathbf{p}}'') A^s(\hat{\mathbf{p}}'' \cdot \hat{\mathbf{p}}') \,. \tag{11}$$

The underlying force between $^3$He atoms is rotationally invariant. Thus, $A^s$ is a function only of the relative orientation of $\mathbf{p}$ and $\mathbf{p}'$, which allows us to parametrize $F^s$ (or $A^s$) by a set of interactions for quasiparticles in relative angular momentum channels, i.e. $F^s = \sum_\ell F^s_\ell \, P_\ell(\hat{\mathbf{p}} \cdot \hat{\mathbf{p}}')$, where the $F^s_\ell$ are the dimensionless Landau parameters. These interaction parameters determine many of the physical properties of the Fermi liquid, e.g. the change in the interaction energy induced by an *isotropic deformation* of the Fermi surface, a dilatation, is determined by the $\ell = 0$ interaction, $F^s_0$. Similarly, the interaction parameter in the $\ell = 1$ channel represents the current-current interaction between quasiparticles. Galilean invariance of the interactions in liquid $^3$He provides a relation connecting the effective mass of a quasiparticle to the $\ell = 1$ interaction, $m^*/m = (1 + F^s_1/3)$. Since the effective mass can be obtained from the low-temperature heat capacity we obtain from fundamental measurements the quasiparticle interaction in the current-current channel. Both interactions determine the compressibility and hydrodynamic sound velocity of $^3$He. Strong repulsive interactions in liquid $^3$He lead to a sound velocity that is significantly larger than the velocity of quasiparticle excitations, $c_1/v_f = \sqrt{\frac{1}{3}(1 + F^s_0)(1 + \frac{1}{3}F^s_1)}$. Measurements of the heat capacity and sound velocity yield values of $F^s_0 \simeq 10$ ($F^s_1 \simeq 6$) at $p = 1$ bar to $F^s_0 \simeq 100$ ($F^s_1 \simeq 15$) at $p = 34$ atm, corresponding to interaction energies $\sim 10 - 100$ times the Fermi energy. [3] Thus, although quasiparticles are long-lived single-particle excitations at low energies and low temperatures, liquid $^3$He can never be described as a "gas" of weakly interacting quasiparticles.

3 The spin-dependent interaction energy originates from exchange interactions which are described by the spin-dependent Landau parameters, $\{F^a_\ell\}$. These interactions are important for $^3$He in a magnetic field; they produce exchange enhancement of an applied magnetic field.

An important consequence of strong interactions is that zero sound modes exist at low temperatures in the collisionless regime, $\omega\tau(T) \gg 1$. The restoring forces are provided by strong repulsive quasiparticle interactions. The first observation of longitudinal zero sound was reported in Ref. [1], and has been studied extensively [13]. The signatures of the cross-over from hydrodynamic to collisionless sound are a small change in sound velocity in the collisionless regime, $(c_0 - c_1)/c_1 \simeq \frac{2}{5}(1+\frac{1}{5}F_2^s)/(1+F_0^s) \sim 10^{-2}$, and a dramatic change in the temperature dependence of the attenuation, $\alpha_1 \sim \omega^2/T^2$ for $\omega\tau < 1$ compared with $\alpha_0 \sim T^2$ for $\omega\tau > 1$ [1].

In addition to longitudinal zero sound, Landau predicted a transverse mode of the Fermi surface in the collisionless regime [18]. This mode should be observable as a propagating shear wave, or *transverse zero sound* (TZS). The velocity of TZS in normal $^3$He is expected to be close to the Fermi velocity, $c_t \simeq v_f$ [10], and the collisional damping is predicted to be serveral orders of magnitude larger than the damping of the longitudinal zero sound mode [7]. In contrast to the detection of longitudinal zero sound, the search for TZS has been more difficult. Early experimental efforts to observe this mode were inconclusive [28, 9]. The recent observation of a propagating transverse current mode in superfluid $^3$He-B [19] provides new insight into the dynamics of normal and superfluid $^3$He, as well as new information on many-body correlation effects. Before discussing the theory of TZS and order parameter dynamics, I review some basic facts about superfluid $^3$He.

## 2. Symmetry Breaking and the Order Parameter

The Hamiltonian that determines the properties of $^3$He has a high degree of symmetry. The ground state electronic configuration of the $^3$He atom is a filled $1s$ shell, so the atom is isotropic and the interaction energy between two $^3$He atoms is dominated by a central potential, $v(r)$, which is also isotropic and spin-independent. This interaction energy is of order $v(\bar{r}) \simeq -10\,\mathrm{K}$ at the minimum of the $^3$He-$^3$He potential, and has a hard repulsive core at $\bar{r} \approx 2.5\,\text{Å}$. By comparison, the nuclear dipolar energy at these densities is of order $v_{\mathrm{d}} = (\gamma\hbar)^2/\bar{r}^3 \simeq 10^{-7}\,\mathrm{K}$. The dipolar energy is tiny even in comparison to the energy scale of the superfluid transition, $T_c \simeq 10^{-3}\,\mathrm{K}$. Thus, to an excellent approximation the equilibrium phase of $^3$He above the superfluid transition is *separately* invariant under rotations in orbital space and in spin space.

Above $T_c$ $^3$He also posseses discrete symmetries under space-inversion ($\mathcal{P}$), time-inversion ($\mathcal{T}$), and an approximate symmetry under particle to hole conversion ($\mathcal{C}$), *i.e.* the transformation of a quasiparticle with energy $\varepsilon = \xi_{\mathbf{p}}\ (>0)$ and spin projection $\uparrow$ into a quasihole with the same excitation energy ($|\xi_{\underline{\mathbf{p}}}| = \xi_{\mathbf{p}}$) and spin projection $\downarrow$. The transformation is represented by a unitary operator, $C^\dagger a^\dagger_{\mathbf{p}\alpha} C = [i\sigma_y]_{\alpha\beta}\, a_{\underline{\mathbf{p}}\beta}$, where $a^\dagger_{\mathbf{p}\alpha}$ $(a_{\underline{\mathbf{p}}\beta})$ creates a quasiparticle (quasihole) with momentum $\mathbf{p}$ and spin projection $\alpha$.[4] Under the particle-hole transformation the low-energy effective Hamiltonian for quasiparticles is approximately invariant. Particle-hole symmetry in a Fermi liquid is analogous to charge conjugation. The order of magnitude of the particle-hole *asymmetry* is given by $\zeta = N'(E_f)\varepsilon/N(E_f) \sim (\varepsilon/E_f)$. Particle-hole symmetry becomes exact for $\varepsilon \to 0$; at $T \simeq T_c$, particle-hole asymmetry is $\zeta \lesssim 10^{-2}$.

The normal state of liquid $^3$He is also invariant under global gauge transformations, $\psi(\mathbf{r}) \to \psi'(\mathbf{r}) = \psi(\mathbf{r})e^{i\chi}$. When combined with the discrete symmetries under $\mathcal{P}$, $\mathcal{T}$ and $\mathcal{C}$, the symmetry group of liquid $^3$He below $T \sim 100\,\text{mK}$ and above $T_c \sim 1\,\text{mK}$ is

$$\mathcal{G} = SO(3)_{\mathrm{S}} \times SO(3)_{\mathrm{L}} \times U(1)_\chi \times \mathcal{P} \times \mathcal{T} \times \mathcal{C}\,. \tag{12}$$

This high degree of symmetry of liquid $^3$He is spontaneously broken at the superfluid transition. The *residual symmetry* of the superfluid phase(s) is reflected in the invariant subgroup of $\mathcal{G}$ defined by the order parameter; the latter is a particular realization of an orbital p-wave, spin-triplet amplitude for Cooper pairs. The order parameter is defined in terms of a superposition of quasiparticle pair states with zero total momentum in the low-energy shell ($|\xi_{\mathbf{p}}| \le \epsilon_c \ll E_f$) about the Fermi surface,

$$\Delta_{\alpha\beta}(\hat{\mathbf{p}}, \mathbf{R}; t) = \int \frac{d\Omega_{\hat{\mathbf{p}}'}}{4\pi} V_t(\hat{\mathbf{p}} \cdot \hat{\mathbf{p}}') \int_{-\epsilon_c}^{\epsilon_c} \frac{d\varepsilon}{2\pi i} f_{\alpha\beta}(\hat{\mathbf{p}}', \mathbf{R}; \varepsilon, t)\,, \tag{13}$$

where $f_{\alpha\beta} \sim \left\langle a_{\mathbf{p}\alpha} a^\dagger_{-\mathbf{p}\beta} \right\rangle$ is the amplitude for Cooper pairs in the spin state $|\,\alpha\beta\,\rangle$, and $V_t = 3V_1 \hat{\mathbf{p}} \cdot \hat{\mathbf{p}}'$ is the pairing interaction in the spin-triplet, p-wave channel. The matrix structure in spin-space represents the three spin-triplet amplitudes, each of which is a function of the orbital momentum of the pairs. The order parameter can be represented as [4]

$$\Delta_{\alpha\beta}(\hat{\mathbf{p}}, \mathbf{R}; t) = \sum_\mu \sum_i \,(i\sigma_\mu \sigma_y)_{\alpha\beta}\, d_{\mu i}(\mathbf{R}, t)\, \hat{\mathbf{p}}_i\,, \tag{14}$$

where the $3 \times 3$ complex matrix, $d_{\mu i}$ , transforms as a vector (with respect to $\mu$) under rotations in spin space and, independently, as a vector (with respect to $i$) under rotations in orbital space.

The equilibrium B-phase of $^3$He is identified [20] with an order parameter belonging to the Balian-Werthamer (BW) class of p-wave states [4].

4 The quasiparticle operators are defined on a restricted Hilbert space of states in the low-energy band $|\xi_{\mathbf{p}}| < \epsilon_c$ about the Fermi surface, and are constructed to generate the low-energy quasiparticle Green's function [27, 32].

The simplest BW state, $d_{\mu i} \sim \delta_{\mu i}$, corresponds to pairing with quantum numbers $S = 1$, $L = 1$ and $J = 0$, where $\mathbf{J} = \mathbf{L} + \mathbf{S}$ is the total angular momentum. The $J = 0$ state belongs to a continuous manifold of degenerate ground states; *i.e.* any order parameter obtained from the $J = 0$ state by a relative rotation of the spin and orbital coordinates and a uniform gauge transformation minimizes the free energy. Thus, a general BW state is,

$$d_{\mu i} = \frac{\Delta}{\sqrt{3}} e^{i\chi} \mathcal{R}_{\mu i}[\hat{\mathbf{n}}, \vartheta] , \tag{15}$$

where $\mathcal{R}[\hat{\mathbf{n}}, \vartheta]$ is an orthogonal matrix representing a relative rotation of the spin and orbital coordinates about the axis $\hat{\mathbf{n}}$ by the angle $\vartheta$. Note that the *relative spin-orbit symmetry* is spontaneously broken below $T_c$.[5] Since parity and gauge symmetry are also broken below $T_c$ the residual symmetry group for $^3$He-B includes the group of combined spin and orbital rotations and time-inversion,

$$\mathcal{G}_{\text{resid}} = SO(3)_{\text{L+S}} \times \mathcal{T} \times \mathcal{C} , \tag{16}$$

with the generator for the rotation group given by $\mathbf{J} = \mathbf{L} + \mathcal{R}^{-1} \cdot \mathbf{S}$.

The dynamics of the order parameter is governed by the time-dependent 'gap equation', Eq. (13), where the pairing amplitude and quasiparticle distribution function obey coupled quasiclassical transport equations. The coupling of the quasiparticle and condensate dynamics is rooted in particle-hole coherence, and is responsible for most of the novel physics associated with nonequilibrium superfluidity [16], including the coupling between the mass current and the collective modes of the order parameter. The rotational symmetry of the ground state of $^3$He-B allows us to classify the order parameter excitations, which are Bosonic modes, in terms of the eigenvalues of $J^2$ and $J_3$ [21]. Furthermore, there is a doubling of the spectrum for each $(J, M)$ labeled by $\mathcal{C}$ parity, $\zeta = \pm$. Within the spin-triplet, p-wave subspace we can expand the order parameter fluctuations in the eigenfunctions of $(J^2, J_3, \mathcal{C})$,

$$\delta d_{\mu i} = \sum_{J,M,\zeta=\pm} \mathcal{D}^{(\zeta)}_{J,M} \, \mathbf{t}^{(J,M)}_{\mu i} , \tag{17}$$

where $\mathbf{t}^{(J,M)}_{\mu i}$ are spherical tensors that transform according to the $J^{\text{th}}$ irreducible representation of the rotation group,

$$\begin{array}{ll} \mathbf{t}^{(0,0)}_{ij} = \frac{1}{\sqrt{3}}\delta_{ij} & \mathbf{t}^{(2,0)}_{ij} = \sqrt{\frac{3}{2}}(\mathbf{e}^{(0)}_i\mathbf{e}^{(0)}_j - \frac{1}{3}\delta_{ij}) \\ \mathbf{t}^{(1,M)}_{ij} = \frac{1}{\sqrt{2}}\varepsilon_{ijk}\,\mathbf{e}^{(M)}_k & \mathbf{t}^{(2,\pm1)}_{ij} = \frac{1}{\sqrt{2}}(\mathbf{e}^{(0)}_i\mathbf{e}^{(\pm)}_j + \mathbf{e}^{(\pm)}_i\mathbf{e}^{(0)}_j) \\ & \mathbf{t}^{(2,\pm2)}_{ij} = \mathbf{e}^{(\pm)}_i\mathbf{e}^{(\pm)}_j \end{array} . \tag{18}$$

5 The degeneracy of the homogeneous BW state is partially resolved by weak perturbations, e.g. the nuclear dipolar energy or the magnetic Zeeman energy.

| $\mathcal{D}^{(\zeta)}_{J,M}$ | Mode | Frequency | Degeneracy | Coupling to | $\Pi$ |
|---|---|---|---|---|---|
| $\mathcal{D}^{(-)}_{0,0}$ | phase | $\omega = \frac{1}{\sqrt{3}} v_f q$ | 1 | 1 | (L) |
| $\mathcal{D}^{(+)}_{0,0}$ | amplitude | $\omega = 2\Delta$ | 1 | $\zeta\left(\frac{qv_f}{\Delta}\right)$ | (L) |
| $\mathcal{D}^{(-)}_{1,M}$ | amplitude | $\omega = 2\Delta$ | 3 | - | - |
| $\mathcal{D}^{(+)}_{1,M}$ | spin-waves | $\omega = \frac{1}{\sqrt{3}} v_f q$ | 3 | - | - |
| $\mathcal{D}^{(-)}_{2,M}$ | Im squashing | $\omega = \sqrt{\frac{12}{5}}\Delta$ | 5 | $\left(\frac{\omega}{\Delta}\right)$ | (L,T) |
| $\mathcal{D}^{(+)}_{2,M}$ | Re squashing | $\omega = \sqrt{\frac{8}{5}}\Delta$ | 5 | $\zeta\left(\frac{qv_f}{\Delta}\right)$ | (L,T) |

TABLE 1. The order parameter collective mode spectrum for p-wave, spin-triplet pairing fluctuations in $^3$He-B. There are four Goldstone modes associated with the spontaneous breaking of gauge and relative spin-orbit rotational symmetry. All other modes are 'massive' and correspond to fluctuations of the order parameter that are not related by a long wavelength rotation or gauge transformation of the ground state order parameter.

The tensors are defined in terms of the circularly polarized basis vectors, $\mathbf{e}^{(0)} = \mathbf{e}_3$ and $\mathbf{e}^{(\pm)} = (\mathbf{e}_1 \pm i\mathbf{e}_2)/\sqrt{2}$, where $\{\mathbf{e}_1, \mathbf{e}_2, \mathbf{e}_3\}$ is a Cartesian triad of unit vectors defining the coordinate system for the Cooper pairs.

There are 2 (gauge) $\times$ 3 (spin) $\times$ 3 (orbital) $=$ 18 p-wave, spin-triplet Bosonic modes of $^3$He-B, which are listed in Table 2 and labeled by the quantum numbers, $(J^\zeta, M)$. The eigenfrequencies are given for the limit $q \to 0$ *neglecting interactions.* There are four Goldstone modes associated with the spontaneous breaking of gauge and relative spin-orbit rotational symmetry. These modes are observable because they couple to the mass and spin currents. The $J = 0^-$ mode is the phase mode [2, 5], which is essential for understanding the propagation of *longitudinal sound* in superfluid $^3$He. The Goldstone modes with $J = 1^+$ correspond to long wavelength fluctuations in the axis of rotation, $\mathbf{n}$, and the angle of rotation, $\vartheta$, and are related to the NMR and spin dynamics in $^3$He-B [20]. All other modes have an excitation energy of order $\Delta$ and correspond to deformations of the order parameter that are unrelated by a rotation or gauge transformation to the ground state order parameter.

Figure 1 shows the spectrum of uncoupled acoustic and $J = 2$ order parameter collective modes of superfluid $^3$He-B. The pairbreaking continuum onsets at $\omega = 2\Delta(T)$. Below the pairbreaking edge the zero sound mode crosses the dispersion curves for the $J = 2^\pm$ order parameter collective modes. Resonant absorption and anomalous dispersion of longitudinal sound result from the coupling of these modes to density, current and stress fluctuations [37, 21]. Quasiparticle interactions lead to renomalization of the mode frequencies [29], as well as phenomena that cannot be anticipated from a theory based on uncoupled bosonic modes and weakly interacting quasiparticles [24].

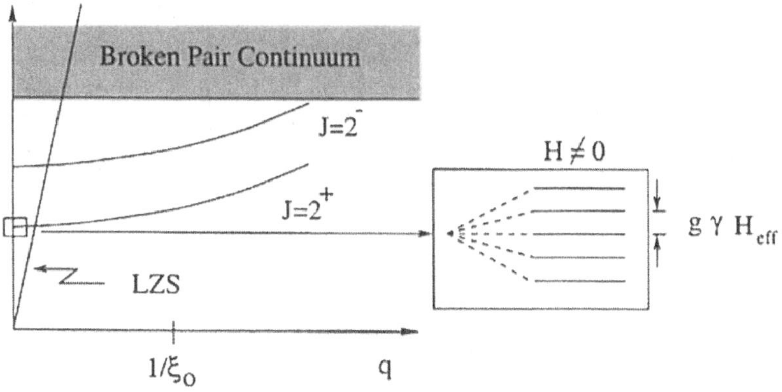


*Figure 1.* Dispersion relations for the uncoupled acoustic and $J = 2^{\pm}$ order parameter collective modes of $^3$He-B. Longitudinal zero sound (LSZ) has a steep dispersion relation that crosses the $J = 2^{\pm}$ modes that lie below the pairbreaking continuum at $2\Delta$. The nuclear Zeeman splitting of the $J = 2^+$ modes is also shown. The level splitting is determined by the effective Larmor frequency, $\omega_{\text{eff}} = \gamma H_{\text{eff}}$, and is small compared with the mode frequency for fields below $H \simeq 1\,\text{kG}$.

Sound propagates in normal liquid $^3$He because quasiparticle interactions and collisions provide restoring forces against density fluctuations. The restoring forces enter through the stress tensor. In the collisionless regime the main contributions to the induced stress fluctuations in normal $^3$He come from the interaction of a quasiparticle with the density and current fluctuations. Both channels contribute to the longitudinal zero sound velocity; however, only the current fluctuations contribute to the restoring force for transverse zero sound. In the superfluid phases new physics enters because dynamical order parameter fluctuations also contribute to the stress tensor, and therefore couple to density and current fluctuations.

The general form of the stress tensor follows from symmetry considerations. In $^3$He-B the stress fluctuations are described by a symmetric second-rank tensor under the group of joint rotations, $SO(3)_{\text{L+S}}$,

$$\begin{aligned} \delta\Pi_{ij} &\simeq \Pi_n\, \delta n\, \delta_{ij} + \Pi_J\,(J_i q_j + J_j q_i) \\ &\quad + \Pi_{d+}\left(\delta d^{(+)}_{ij} + \delta d^{(+)}_{ji}\right) + \Pi_{d-}\left(\delta d^{(-)}_{ij} + \delta d^{(-)}_{ji}\right), \quad (19) \end{aligned}$$

where $\delta d^{(\pm)}_{ij}$ represents order parameter fluctuations which are even (+) and odd (−) under particle-hole conversion, i.e. $\delta d^{(\pm)}_{ij} \overset{\mathcal{C}}{\to} \pm\, \delta d^{(\pm)}_{ij}$. In the limit of exact particle-hole symmetry the density, current and stress fluctuations are odd under $\mathcal{C}$: $\delta n \to -\delta n$, $\mathbf{J} \to -\mathbf{J}$ and $\delta\Pi_{ij} \to -\delta\Pi_{ij}$. This implies that coupling of the stress to the order parameter fluctuations, $\delta d_{ij}^{(+)}$, is non-vanishing only because of particle-hole *asymmetry* [15, 32]. Thus, the relative magnitude of the $J = 2^{\pm}$ coupling is $\Pi_{d+} \sim (\Delta/E_f)\Pi_{d-}$. The contributions of these dynamical fluctuations to the stress tensor are obtained by solving the transport equations for the quasiparticle distribution function and paring amplitude.

For the acoustic modes, the analysis of the transport equations is simplified by expressing the conservation laws for particle number and momentum in terms of the amplitudes of the distribution function, $\phi_{\ell m}$, using Eqs. (7)-(8). The generalization of the amplitudes, $\phi_{\ell,m}$, to the superfluid phases is discussed for example in [24].

The continuity equation for the number density is equivalent to,

$$\omega\,\phi_{0,0} - \frac{1}{3}\,(1+F_0^s)\,qv_f\,\phi_{1,0} = 0\,, \tag{20}$$

while the equations for momentum conservation can be expressed as,

$$\omega\,\phi_{1,0} - \left(1+\frac{1}{3}F_1^s\right)\,qv_f\,\left(\phi_{0,0} + \frac{2}{5}\sqrt{\frac{2}{3}}\,\phi_{2,0}\right) = 0\,, \tag{21}$$

for the longitudinal current, and

$$\omega\,\phi_{1,\pm1} - \left(1+\frac{1}{3}F_1^s\right)\,qv_f\,\frac{2}{5}\sqrt{\frac{1}{2}}\,\phi_{2,\pm1} = 0\,, \tag{22}$$

for the transverse current.

## 3. Longitudinal Modes

The dispersion relation for longitudinal sound is obtained by combining Eqs. (20) and (21),

$$\omega^2 - c_1^2 q^2 \left\{1 + \frac{2}{5}\sqrt{\frac{2}{3}}\,\xi(\omega,q)\right\} = 0\,, \tag{23}$$

where $c_1$ is the hydrodynamic sound velocity. The effects of mode coupling are contained in the response function,

$$\xi(\omega,q) = \left[\frac{\delta\phi_{2,0}}{\delta\phi_{0,0}}\right]_{\text{tot}} = \left(\frac{\delta\phi_{2,0}}{\delta\phi_{1,0}}\right)\left(\frac{\delta\phi_{1,0}}{\delta\phi_{0,0}}\right) + \ldots\,, \tag{24}$$

which represents the total stress induced by a density fluctuation. To calculate the response function we solve the coupled dynamical equations for the quasiparticle distribution function, time-dependent order parameter, and the Landau interactions self-consistently, *c.f.* [24]. The solution for the longitudinal component of the induced stress is

$$\begin{aligned}\phi_{2,0} &= A_0^s\,\alpha_0(q,\omega)\,\phi_{0,0} + \frac{1}{3}A_1^s\,\alpha_1(q,\omega)\,\phi_{1,0} \\ &+ \left(\frac{\omega}{2\Delta}\right)\left[\beta_0(q,\omega)\,\mathcal{D}_{0,0}^{(-)} + \beta_2(q,\omega)\,\mathcal{D}_{2,0}^{(-)}\right] \\ &+ \left(\frac{qv_f}{2\Delta}\right)\left[\zeta_0(q,\omega)\,\mathcal{D}_{0,0}^{(+)} + \zeta_2(q,\omega)\,\mathcal{D}_{2,0}^{(+)}\right], \end{aligned} \tag{25}$$

where $\alpha_0 = \xi_0$, $\alpha_1 = (qv_f/\omega)[\sqrt{\frac{2}{3}}\xi_2 + \frac{1}{3}\xi_1]$, $\beta_0 = \frac{15}{2\sqrt{2}}[\lambda_1 - \frac{1}{3}\lambda_0]$, $\beta_2 = \frac{45}{4}[\lambda_2 - \frac{2}{3}\lambda_1 + \frac{1}{9}lambda_0]$, and the functions, $\xi_0$, $\xi_1$, $\lambda_0$, $\lambda_1$ and $\lambda_2$ are given in Ref. [25]. In addition to the density, $\phi_{0,0}$, and longitudinal current fluctuation, $\phi_{1,0}$, the order parameter modes with $J = 0, 2$ and $M = 0$ contribute to the longitudinal stress. The stress induced by the $J = 0^+$ and $J = 2^+$ modes is small compared with the $J = 0^-$ and $J = 2^-$ modes by the particle-hole asymmetry parameter, $\zeta$. For the particle-hole asymmetric couplings, $\zeta_{0,2}$ see [15]. Unless stated otherwise, I will omit the contributions to the stress tensor coming from the Fermi-liquid interactions with $\ell \geq 2$ as well as higher angular momentum pairing channels.

The equations for the order parameter modes, $\mathcal{D}_{0,0}^{(-)}$ and $\mathcal{D}_{2,0}^{(-)}$, obtained from solutions of the time-dependent gap equation, reduce to,

$$\left[\omega^2 - \frac{1}{3}(qv_f)^2\right]\mathcal{D}_{0,0}^{(-)} = \left[2\sqrt{3}\,(\omega\Delta)\,\phi_{0,0} + \frac{2\sqrt{2}}{15}\,(qv_f)^2\,\mathcal{D}_{2,0}^{(-)}\right], \tag{26}$$

$$\begin{aligned}\left[(\omega + i\Gamma)^2 - \Omega_{2^-}^2 - \frac{7}{5}q^2v_f^2\right]\mathcal{D}_{2,0}^{(-)} &= \frac{2}{15}(qv_f)^2\mathcal{D}_{0,0}^{(-)} \\ &+ \frac{A_1^s}{3}\left(\frac{qv_f}{2\Delta}\right)\left(\frac{8}{5}\Delta^2\right)\sqrt{\frac{2}{3}}\,\phi_{1,0}\,. \end{aligned} \tag{27}$$

Equation (26) is the wave equation for the Anderson-Bogoliubov phase mode, $\mathcal{D}_{0,0}^{(-)}$ [2, 5]. For $q \to 0$ the driving term on the right side of Eq.(26) is proportional to the density fluctuation, and in this limit the equation for $\mathcal{D}_{0,0}^{(-)}$ is equivalent to the Josephson equation, $i\hbar\partial_t\chi = \delta\mu$ [24], for the order parameter phase, $\chi$, where $\delta\mu$ is the change in the chemical potential. For shorter wavelengths and higher frequencies, the phase fluctuations are also coupled to the high-frequency modes. Equation (27) describes driven oscillations of the $J = 2^-$, $M = 0$ order parameter mode by the change in the Landau interaction energy induced by a current fluctuation; note the Landau parameter $A_1^s$. The driving term includes the longitudinal current, $\phi_{1,0}$, and the phase fluctuation mode, $\mathcal{D}_{0,0}^{(-)}$. The stress induced by fluctuations of the $J = 2^+$ mode is much weaker than that of the $J = 2^-$ mode because of approximate particle-hole symmetry . The $J = 2^+$ mode is excited by density and current fluctuations, but the coupling is again small by a factor of $\zeta$,

$$\left[(\omega + i\Gamma)^2 - \Omega_{2+}^2 - \frac{7}{5}q^2 v_f^2\right] \mathcal{D}_{2,0}^{(+)} = 2\zeta\,\omega\Delta\,\mathcal{D}_{2,0}^{(-)}\,. \tag{28}$$

Combining Eqs. (24)-(28) we obtain the dispersion relation for longitudinal sound [37, 15] with

$$\begin{aligned}\xi(q,\omega) &= \frac{2}{5}\left(\frac{c_1 q}{\omega}\right)^2 \left(\frac{1}{1+F_0^s}\right) \\ \times\Bigg\{\rho_n(\omega) &+ \frac{3}{5}\rho_s(\omega)\left[\frac{\omega^2}{(\omega+i\Gamma)^2 - \Omega_{2-}^2 - \frac{7}{5}(q^2 v_f^2)}\right] \\ &+ \zeta^2\,\rho_s(\omega)\left[\frac{\omega^2}{(\omega+i\Gamma)^2 - \Omega_{2+}^2 - \frac{7}{5}(q^2 v_f^2)}\right]\Bigg\}.\end{aligned} \tag{29}$$

The response function exhibits the contribution from the quasiparticle restoring force $\sim \rho_n(\omega)$, as well as the $J = 2^\pm$, $M = 0$, order parameter modes, $\sim \rho_s(\omega)$. Note that $\rho_s$ and $\rho_n$ represent condensate and non-condensate response functions with $\rho_s + \rho_n = 1$; these functions reduce to the superfluid and normal fluid densites in the limit $\omega = 0$.[6] The contribution from the $J = 2^+$ mode is also proportional to the particle-hole asymmetry factor, $\zeta^2$. The mode couplings are enhanced near the resonance frequency of the modes. The resonance linewidth is determined by quasiparticle scattering rate which becomes exceedingly small at low temperatures, $\Gamma(T) \simeq \Gamma_n(T_c/T)^{\frac{3}{2}}\,e^{-\Delta/T} \ll \Omega_{2\pm} \sim \Delta$, with $\Gamma_n \simeq 1/\tau(T_c) \sim 0.1\,\text{MHz}$ [12]. Thus, even weakly coupled modes are observable near resonance as strong absorption features in the attenuation spectrum of longitudinal sound. In addition to the resonant contributions to the absorption of longitudinal sound, there is an absorption band for frequencies above the pair-breaking threshold, $\omega \geq 2\Delta(T)$. These features are shown in the calculated attenuation spectrum of Fig. 2 as a function of temperature for longitudinal sound at $P \simeq 0\,\text{bar}$ ($T_c = 0.93\,\text{mK}$) and frequency of $\omega/2\pi = 35.8\,\text{MHz}$. The sharp

6 The condensate response function $\rho_s(\omega)$ is equivalent to the Tsuneto function $\lambda(\omega)$ in [25].

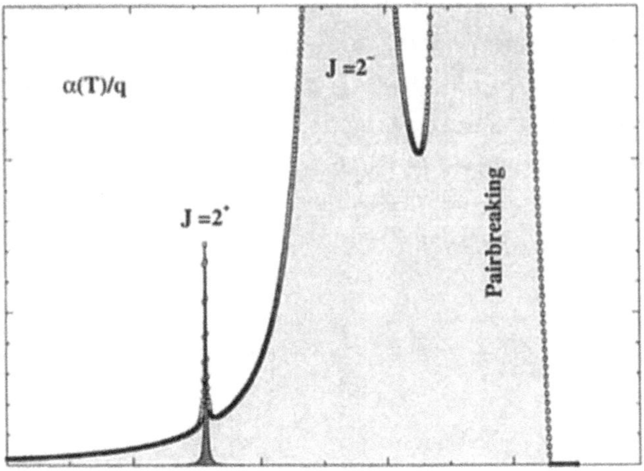


*Figure 2.* The calculated absorption spectrum (relative to the normal-state at $T_c$) for longitudinal sound in $^3$He-B at $P \simeq 0\,\mathrm{bar}$ ($T_c = 0.93\,\mathrm{mK}$) and a frequency of 35.8 MHz. The calculation assumes $\zeta = 10^{-2}$. The peaks correspond to resonant absorption by the $J = 2^{\pm}$ modes. The relative weight of the resonance peaks reflects the weak coupling of the $J = 2^{+}$ mode due to approximate particle-hole symmetry.

resonance peak at the lowest temperature is the $J = 2^{+}$ resonance, while the huge absorption peak at a higher temperature is $J = 2^{-}$ resonance. Pair-breaking gives rise to the broad absorption band onsetting at $T_c$. The $J = 2^{+}$ mode resonance was observed in sound absorption by [11] and [23]. The identification of this resonance with the $J = 2^{+}$ mode was made by [3] who observed the Zeeman splitting of the absorption line in a magnetic field.

### *Zeeman Splitting of the $J = 2^{\pm}$ Modes*

The $J = 2$ modes are five-fold degenerate in zero field; however, the nuclear magnetic moment of $^3$He leads to a Zeeman coupling in a magnetic field that lifts the degeneracy [35], producing five Zeeman levels in a field. For weak magnetic fields the level splitting is linear in the field and given by

$$\Omega_{2^{\pm},M} = \Omega_{2^{\pm}} + M\, g_{2^{\pm}}\, \omega_{\text{eff}} \,, \tag{30}$$

where $g_{2^{\pm}}$ is the g-factor for the $J = 2^{\pm}$ modes [35, 30] and $\omega_{\text{eff}}$ is the effective Larmor frequency of the excited pairs,

$$\omega_{\text{eff}} = \gamma H \frac{(1+\frac{1}{5}F_2^a)}{1+F_0^a(\frac{2}{3}+\frac{1}{3}Y(T))+\frac{1}{5}F_2^a[\frac{1}{3}+(\frac{2}{3}+F_0^a)Y(T)]}, \tag{31}$$

where $Y(T)$ is the Yosida function, and $\gamma$ is the gyromagnetic ratio for the $^3$He nucleus. Equation (31) includes the exchange field enhancement of the applied field, $H$, in the superfluid state. The g-factors for the excited pair states, including Fermi liquid effects and higher angular momentum pairing channels, were calculated by [30].

Measurements of the Zeeman splitting provide an experimental determination of the Lande g-factor. The g-factor is sensitive to exchange and pairing interactions, and is an excellent parameter to study many-body correlations and strong-coupling effects in superfluid $^3$He [30]. Measurements of the g-factor from the Zeeman splitting of the absorption spectrum of LZS for the $J = 2^+$ modes are in reasonable agreement with theoretical calculations. However, precision measurements for the $J = 2^-$ modes have not previously been possible because the coupling of the $J = 2^-$ modes is so strong that high magnetic fields are required to resolve different Zeeman levels. At high magnetic fields the $J = 2$ level shifts evolve nonlinearly with $H$ as a result of gap distortion by the magnetic field [34]. For a discussion of these effects see [31, 8, 13]. Nevertheless, [26] obtained a value of $g_{2^-} \simeq 0.04$ at $P = 19$ bar from their measurements at fields above 1 kG, in reasonable agreement with theoretical expectations. However, as I show below the g-factor for the $J = 2^-$ modes can be determined with high acccuracy from an analysis of *transverse* sound propagation in a magnetic field. But, first I describe the mechanisms that lead to a propagating transverse current mode.

## 4. Transverse Modes

The dispersion relation for a transverse current excitation is given by the momentum conservation equation, Eq. (22), and the response function for the stress induced by a transverse current fluctuation,

$$\left(\frac{\omega}{v_f q}\right)_{\pm} = \frac{2}{5}(1+F_1^s/3)\,\frac{1}{\sqrt{2}}\left[\frac{\delta\phi_{2,\pm 1}}{\delta\phi_{1,\pm 1}}\right]_{\text{tot}}, \tag{32}$$

where the subscripts $\pm$ refer to the two independent circular polarizations of the transverse current. In addition to the direct contribution to $\phi_{2,\pm 1}$ coming from the current fluctuations, order parameter modes with $J = 2$ and $M = \pm 1$ also induce transverse stress fluctuations and therefore couple to the transverse current. The result for the transverse stress is [25],

$$\phi_{2,\pm 1} = \frac{1}{3} A_1^s \left(\frac{q v_f}{\omega}\right) \xi_1(q,\omega) \frac{1}{\sqrt{2}} \phi_{1,\pm 1} + \left(\frac{\omega}{2\Delta}\right) \Lambda_1(q,\omega) \mathcal{D}^{(-)}_{2,\pm}, \qquad (33)$$

where the term $\sim A_1^s \phi_{1,\pm 1}$ represents the quasiparticle contribution to the transverse restoring force, while the condensate contribution, which also enters through the Landau interaction, is proportional to the amplitude of the $J = 2^-$, $M = \pm 1$ order parameter modes, $\mathcal{D}^{(-)}_{2,\pm}$. I omit the $J = 2^+$, $M = \pm 1$ modes because the coupling of these modes is smaller by a factor of the particle-hole asymmetry ratio, $\zeta$. Unlike the case of LZS, the $J = 2^+$ modes cannot be easily detected in the TZS spectrum because the TZS mode is suppressed for frequencies below the $J = 2^-$ mode, which lies above the $J = 2^+$ mode. The equations of motion for the $J = 2^-$, $M = \pm 1$ order parameter modes are given by

$$\left[(\omega + i\Gamma)^2 - \Omega^2_{2^-,\pm 1} - \frac{2}{5} q^2 v_f^2\right] \mathcal{D}_{2,\pm} = \left(\frac{8}{5}\Delta^2\right) \left(\frac{q v_f}{2\Delta}\right) \frac{A_1^s}{3} \frac{1}{\sqrt{2}} \phi_{1,\pm 1} . \qquad (34)$$

In zero field the dispersion relation for transverse current excitations is independent of the polarization,

$$\left(\frac{\omega}{q v_f}\right)^2 = \frac{F_1^s}{15} \rho_n(\omega) + \frac{2F_1^s}{75} \rho_s(\omega) \left\{ \frac{\omega^2}{(\omega + i\Gamma)^2 - \Omega^2_{2^-} - \frac{2}{5}(q^2 v_f^2)} \right\}, \qquad (35)$$

which displays contributions from the quasiparticle restoring force ($\sim \rho_n$), and from the condensate ($\sim \rho_s$) [22, 25]. Equation (35) is based on a long-wavelength expansion, i.e. $q v_f \ll \omega$, which is valid in the B-phase for frequencies in the range, $\Gamma \ll |\omega - \Omega_{2^-}| \ll \omega$.[7]

In normal $^3$He transverse sound propagates with a phase velocity $c_t \geq v_f$ provided the $\ell = 1$ and $\ell = 2$ Landau parameters satisfy, $F_1^s/3 + F_2^s/(1 + F_1^s/3) > 2$ [10]; in the limit $F_1^s \gg 15$, $c_t \simeq \sqrt{F_1^s/15}\, v_f$, which is the normal-state result in the limit $q v_f/\omega \ll 1$. The condition for a propagating TZS mode is obeyed for pressures above approximately 1 bar; however, the predicted velocity is close to the Fermi velocity at all pressures. As a result it is difficult to differentiate transverse oscillations of the Fermi surface from an incoherent current of quasiparticles; and early attempts to observe transverse sound in the normal phase of $^3$He were inconlusive, [28, 9].

The onset of superfluidity in $^3$He has dramatic effects on the collisionless transverse modes in $^3$He. The opening of a gap in the excitation spectrum

7 Results for shorter wavelengths and frequencies far off resonance can be found in [25].

suppresses the quasiparticle contribution to the restoring force. This suppression comes from the factor of $\rho_n$ in Eq. (35), which decreases rapidly below $T_c$. For this reason early theoretical investigations emphasized that the transverse zero sound mode, even if observable in the normal state, would disappear below $T_c$ [21, 6, 22]. The condensate also contributes to the dispersion relation through the coupling to the $J = 2^-$, $M = \pm 1$ order parameter modes. The condensate response suppresses TZS at low frequencies, as can be seen from Eq. (35). However, at intermediate frequencies, $\Omega_{2^-} < \omega < 2\Delta$, the condensate provides a strong restoring force through the Landau interaction energy, leading to a propagating transverse current mode at low temperatures. The calculated temperature dependences of the

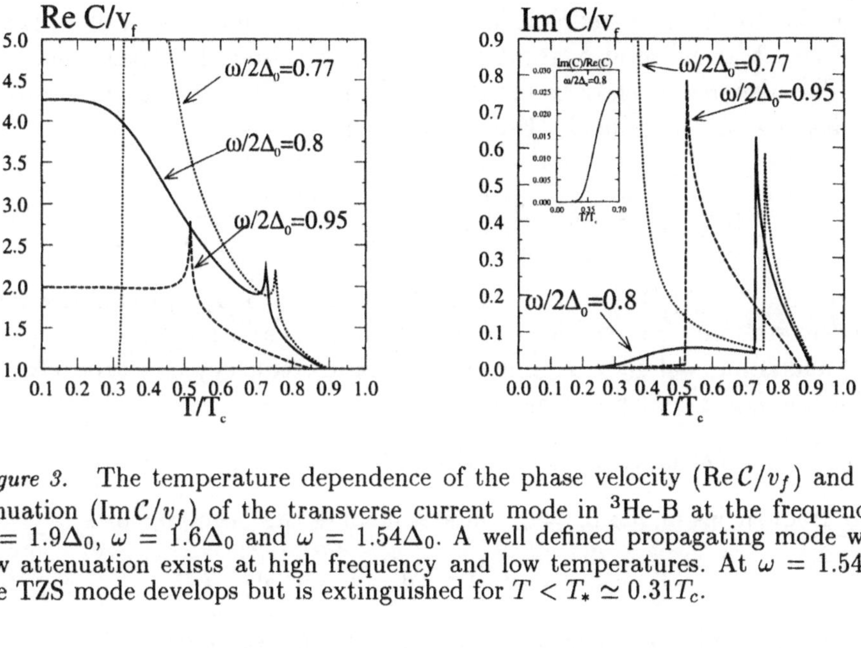


*Figure 3.* The temperature dependence of the phase velocity ($\mathrm{Re}\,C/v_f$) and attenuation ($\mathrm{Im}\,C/v_f$) of the transverse current mode in $^3$He-B at the frequencies $\omega = 1.9\Delta_0$, $\omega = 1.6\Delta_0$ and $\omega = 1.54\Delta_0$. A well defined propagating mode with low attenuation exists at high frequency and low temperatures. At $\omega = 1.54\Delta_0$ the TZS mode develops but is extinguished for $T < T_* \simeq 0.31 T_c$.

phase velocity and attenuation of a transverse current excitation in $^3$He-B are shown in Fig. 3 taken from Ref. [25]. At low frequency, $\omega \ll \Omega_{2^-}(0)$, the transverse current mode is highly attenuated over the full temperature range below $T_c$. At high frequencies, $\omega = 1.9\Delta_0 > \Omega_{2^-}(0)$, a propagating transverse current mode develops with a large phase velocity and low absorption that decreases exponentially at low temperature. For intermediate frequency, $\omega = 1.54\Delta_0$, TZS develops for $T < T_{\mathrm{pb}}$; the attenuation drops until $T = T_*$, at which point TZS is extinguished by resonant absorption from the $J = 2^-$ modes.

*Circular Birefringence of Transverse Waves*

The B-phase of $^3$He is symmetric under time-inversion, and consequently is non-birefringent for transverse wave propagation, i.e. right- and left-circularly polarized waves propagate at the same velocity. However, *circular birefringence*, i.e. $C_+ \neq C_-$, can be induced by a magnetic field. If the field is applied parallel to the direction of propagation of the waves, $\mathbf{H}||\mathbf{q}$, then axial symmetry is preserved and the eigenmodes for the transverse current are the right- and left-circular polarization states. However, the eigenfrequencies of these modes, or equivalently the complex phase velocities, are no longer degenerate. This leads to the acoustic analog of (circular) optical birefringence ($\mathrm{Re}\,C_+ \neq \mathrm{Re}\,C_-$) and dichroism ($\mathrm{Im}\,C_+ \neq \mathrm{Im}\,C_-$) [25]. At low temperatures and high frequencies, $\omega > \Omega_{2^-,\pm1}$, circular birefringence gives rise to the *acoustic Faraday effect* in which the direction of a linearly polarized transverse wave rotates along the direction of propagation.

The dispersion relations for RCP (+) and LCP (−) transverse current modes in a magnetic field, $\mathbf{H}||\mathbf{q}$, are given by

$$\left(\frac{\omega}{qv_f}\right)^2_\pm = \Lambda_n + \Lambda_s\left(\frac{\omega^2}{(\omega+i\Gamma)^2 - \Omega^2_{2^-,\pm1}(T,\omega,H)}\right) \tag{36}$$

where $\Lambda_n = (F_1^s/15)\,\rho_n(\omega,T)$ and $\Lambda_s = (2F_1^s/75)\,\rho_s(\omega,T)$ in the long-wavelength limit. The condensate term dominates at low temperature ($\Lambda_s \gg \Lambda_n$) and is anomalously large when the sound frequency is nearly resonant with the $J = 2^-$, $M = \pm1$ modes,

$$D^2_\pm(\omega,H,T) = (\omega+i\Gamma)^2 - \Omega^2_{2^-,\pm1}(\omega,H,T) \approx 0\,. \tag{37}$$

The frequencies of the $M = \pm1$ modes include the Zeeman shifts,

$$\Omega^2_{2^-,\pm1} = \Omega^2_{2^-}(T) \pm 2\,g_{2^-}(T)\,\omega\,\omega_{\text{eff}}(H,T)\,, \tag{38}$$

where $\Omega_{2^-}(T)$ is the frequency of the $J = 2^-$ modes in zero field.

Consider a transverse current excitation of frequency $\omega$ at $z = 0$, linearly polarized along $\mathbf{x}$ and propagating in the $z$-direction, $\mathbf{J}(\omega, z=0) = \mathcal{J}(\omega)\,\hat{\mathbf{x}}$. The RCP and LCP modes propagate with different phase velocities; thus the current evolves as,

$$\mathbf{J}(\omega,z) = \frac{\mathcal{J}}{\sqrt{2}}\,e^{iq_+(\omega)z}\,\hat{\mathbf{e}}_+ + \frac{\mathcal{J}}{\sqrt{2}}\,e^{iq_-(\omega)z}\,\hat{\mathbf{e}}_-\,, \tag{39}$$

where $\hat{\mathbf{e}}_\pm = (\hat{\mathbf{x}} \pm i\hat{\mathbf{y}})/\sqrt{2}$ are the polarization vectors for RCP and LCP current modes. This response corresponds to a pure Faraday rotation of the polarization if the phase velocities are real. This is the case for low temperatures, $T \ll T_c$, and frequencies above the collective mode resonances, $\omega > \Omega_{2^-,\pm 1}(T)$. The spatial period of the rotation of the polarization is identified by writing

$$\mathbf{J}(\omega, z) = \mathcal{J}\, e^{iq(\omega)z}\, (\cos(\delta q z)\hat{\mathbf{x}} - \sin(\delta q z)\hat{\mathbf{y}})\,, \tag{40}$$

where the sum and difference wavevectors are $q = (q_+ + q_-)/2$ and $\delta q = (q_+ - q_-)/2$. The linearly polarized transverse wave propagates with a mean phase velocity, $\bar{c}_t = \omega/q$, and the polarization *rotates* with the spatial period,

$$\Lambda(\omega, H, T) = \frac{2\pi}{|\delta q|} = \frac{4\pi}{q_- - q_+}\,. \tag{41}$$

Near the $M = +1$ mode crossing the wavenumber for the RCP wave vanishes; $q_+(T_+) = 0$, since $D_+(\omega, T_+) = 0$ and $\Lambda_n \ll \Lambda_s$. The temperature dependence of the Faraday rotation period is then determined by the response of this mode *off resonance*,

$$q_+(T) \simeq q_f \frac{D_+(T)}{\sqrt{\Lambda_s \omega}} = q_f \sqrt{s(T/T_+ - 1)}\,, \tag{42}$$

where $q_f = \omega/v_f$,

$$s \equiv \frac{2|\Omega'_+|T_+}{\Lambda_s \Omega_+}\,, \tag{43}$$

$\Omega_+ = \Omega_{2^-}(T_+) = \omega$, and $\Omega'_+ = d\Omega_{2^-}(T)/dT|_{T_+}$ are evaluated at the mode resonance, $T = T_+$. The field dependence originates from the linear Zeeman splitting of the $M = \pm 1$ modes, and enters via the wavenumber of the $M = -1$ mode,

$$q_- \simeq q_f \sqrt{\frac{2|\Omega'_+|(T - T_+) + 4 g_{2^-}\omega_{\text{eff}}}{\Lambda_s \Omega_+}}\,, \tag{44}$$

for $T_+ \lesssim T \ll T_c$. Thus, the Faraday rotation period varies with field on the scale,

$$B_+ \equiv \frac{\Lambda_s \Omega_+}{4\,\gamma_{\text{eff}}\, g_{2^-}}\,, \tag{45}$$

where $\gamma_{\text{eff}} \equiv \omega_{\text{eff}}/H$ is the effective gyromagnetic ratio at the mode crossing. I can express the temperature and field dependence of $\Lambda$ near the $J = 2^-$, $M = +1$ mode crossing in terms of scaled temperature and field variables, $\alpha = s\,(T/T_+ - 1)$ and $\beta = H/B_+$,

$$\frac{\Lambda}{\lambda_f} = \frac{2}{\sqrt{\alpha + \beta} - \sqrt{\alpha}} = \begin{cases} 4\sqrt{\alpha}/\beta & , \quad \beta \ll \alpha \\ 2\left(1 + \sqrt{\alpha/\beta}\right)/\sqrt{\beta} & , \quad \beta \gg \alpha\,, \end{cases} \tag{46}$$

where $\lambda_f = 2\pi v_f/\Omega_+$. For $T_+/T_c \simeq 0.4$, $\omega/2\pi = 80\,\text{MHz}$ and $F_1^s \simeq 15$ the field scale is $B_+ \simeq 1\,\text{Tesla}$. Thus, except for a narrow temperature range near resonance, i.e. $|T - T_+| < 10^{-2}T_+$ at $H = 100\,\text{G}$, the Faraday rotation period varies as

$$\Lambda \simeq 4\frac{B_+}{H}\sqrt{s(T-T_+)} \simeq \frac{\xi_{\Omega_+}\pi v_f}{\gamma_{\text{eff}}g_{2-}}\sqrt{\frac{2}{75}F_1^s\rho_s}\,\frac{\sqrt{T-T_+}}{H}\,, \tag{47}$$

with $\xi_\Omega = 8|\Omega'_+|T_+/\Omega_+ \simeq 1$ at $T \simeq 0.44T_c$. In magneto-optics the field-independent prefactor, $V = \Lambda H/2\pi$, is called the Verdet constant.

*Broken Spin-Orbit Symmetry*

The acoustic Faraday effect provides a clear demonstration of *spontaneously broken spin-orbit symmetry*. The direction of mass flow is altered by the application of a magnetic field along the propagation direction. The role of broken spin-orbit symmetry is evident in the the contribution to the stress fluctuations from the order parameter,

$$\delta\pi_{ij} \sim \sum_M \mathcal{D}^{(-)}_{2,M}\,\mathbf{t}^{2,M}_{ij}\,. \tag{48}$$

Inspection of Eqs. (18) shows that the transverse stress couples the transverse current fluctuations only to the $M = \pm 1$ modes with *total* angular momentum $J = 2$,

$$\mathbf{t}^{(2,\pm 1)}_{ij} \sim (\hat{\mathbf{q}}_i\mathbf{e}^{\pm}_j + \mathbf{e}^{\pm}_i\hat{\mathbf{q}}_j)\,. \tag{49}$$

Broken time-reversal symmetry is also important for producing circular birefringence. The Zeeman splitting of the $M = \pm 1$ modes in a field aligned along $\mathbf{q}$ produces a difference in the stress induced by the $M = +1$ and $M = -1$ modes, and thus, different propagation velocities.

These general symmetry considerations also imply that superfluid $^3$He-A, which spontaneously breaks time-reversal symmetry as well as relative spin-orbit rotation symmetry, should exhibit spontaneous acoustic circular birefringence or dichroism for transverse excitations propagating along the $\boldsymbol{\ell}$ vector. In fact observation of this effect for a uniform texture would provide a direct evidence of broken time-reversal symmetry in $^3$He-A [38].

## 5. Observation of Transverse Current Waves in $^3$He-B

Propagating transverse current waves were recently observed in $^3$He-B at low pressure [19]. The excitation and detection of these waves was made in an acoustic cell with a transducer operating at a frequency of $\omega/2\pi = 82.26\,\text{MHz}$ that *generates and detects* transverse currents with a fixed linear polarization. In a high attenuation regime the acoustic response measures the transverse acoustic impedance of bulk $^3$He. However, if a propagating mode develops with low attenuation then current waves propagate across the cell, are reflected from the boundary opposite the transducer and are detected by the same transducer. In this regime standing waves develop and are observable in the acoustic impedance because of interference between the source and reflected wave at the detector. Oscillations of the impedance occur as the phase velocity and attenuation change with temperature, pressure or field. A linearly polarized wave with $\mathbf{q} \parallel \mathbf{H}$ undergoes Faraday rotation of its polarization as it propagates. Upon reflection from the opposite wall of the cavity the reflected wave propagates with its polarization rotating with the same handedness relative to the direction of the field, i.e. the rotation of the polarization accumulates after reflection from a surface. The rotation of the polarization produces an oscillatory modulation of the impedance as a function of magnetic field, with the period of the impedance oscillations being determined by, $\Lambda/2 \propto 1/H$.

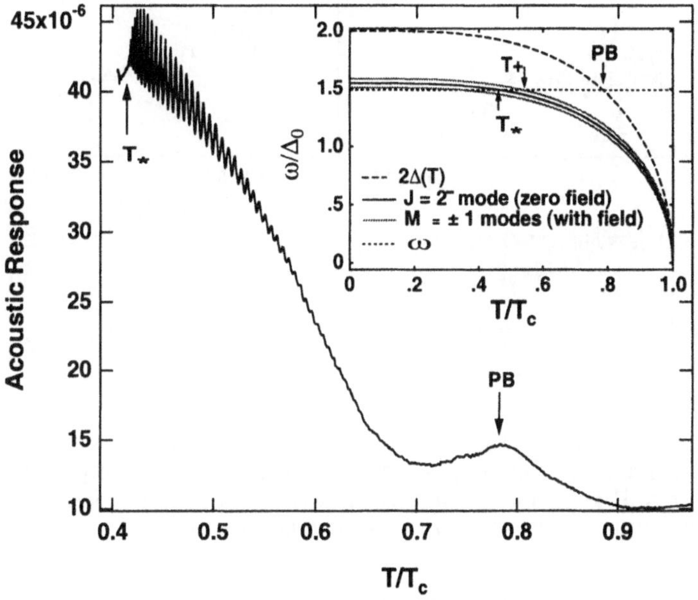


*Figure 4.* Transverse impedance oscillations in $^3$He-B at $\omega/2\pi = 82.26\,$MHz, $P = 4.31\,$bar and zero field. The oscillations develop in the intermediate frequency range $\Omega_{2^-} < \omega < 2\Delta(T)$, between the pairbreaking edge, labeled by PB ($\omega = 2\Delta(T_{\rm PB})$), and the extinction point determined by the resonance frequency of the $J = 2^-$, $M = \pm 1$ modes, i.e. $\omega = \Omega_{2^-}(T_*)$. The pair-breaking and resonance conditions for fixed frequency are shown in the inset.

The transverse acoustic response measured by [19] is shown for $^3$He at a pressure of 4.31 bar in Fig. 4. The impedance varies smoothly as the temperature drops below $T_c$ and shows a peak at the pairbreaking (PB) edge, $\omega = 2\Delta(T_{pb})$. Oscillations develop below a temperature of $T \simeq 0.65T_c$, indicating the presence of a propagating mode. The amplitude of the impedance oscillations increases as the temperature decreases indicating that the attenuation of the propagating mode decreases with decreasing temperature. The oscillations disappear at the temperature $T_* \simeq 0.42\,T_c$ corresponding to the resonance condition, $\omega = \Omega_{2^-}(T_*)$; i.e. transverse waves are extinguished below this temperature. All of these features are in agreement with theoretical predictions [25].

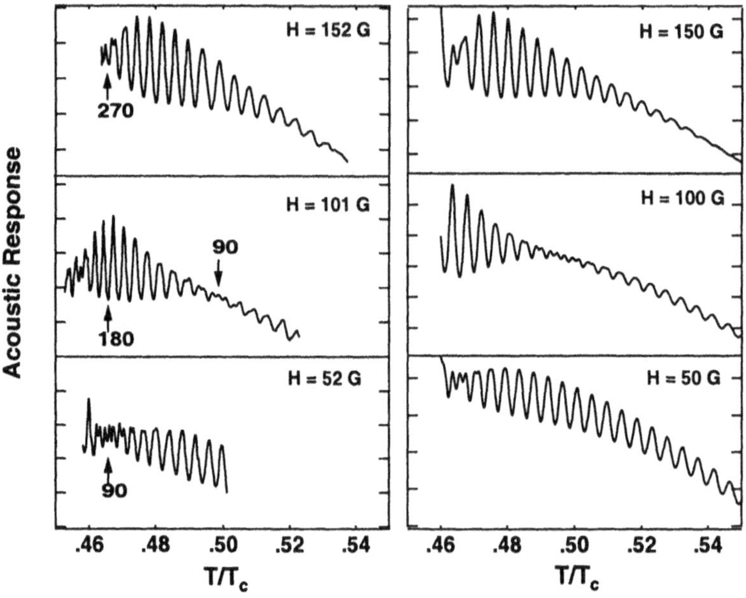


*Figure 5.* Left panels: Magnetic field dependence of the acoustic cavity response measured at 4.42 bar. The angles indicate the rotation of the polarization after one round-trip. Right panels: Simulation of the acoustic impedance for the same path length, temperature, pressure and magnetic fields.

Figure 5 (reproduced from [19]) shows the impedance oscillations at a pressure of 4.42 bar for magnetic fields of 52 G, 101 G and 152 G. The magnetic field modulates the zero field impedance oscillations. The transducer detects only transverse waves having the same linear polarization as the source wave. A field of 52 G suppresses the impedance oscillations near $T = 0.465\,T_c$ that were present in zero field, corresponding to a 90° rotation of the polarization of the reflected wave; i.e. the polarization of the reflected wave is orthogonal to the detection direction. Doubling the field restores the impedance oscillations near $T = 0.465\,T_c$. The oscillations are suppressed again at a field to 152 G. Near the 90° and 270° points, there are small amplitude impedance oscillations, with shorter period than the primary oscillations. These oscillations are due to interference of the source and waves that traverse the cavity twice. This interpretation was verified by comparing the impedance oscillations with the calculated cavity response, shown in the right panels of Fig. 5. The calculation is based on Eqs. (40) and (47) and takes as input the measured attenuation and phase velocity in zero field. The Verdet constant is obtained from the measurement at 52 G. The simulation reproduces the observed features of the impedance as a function of temperature including the maximum in the modulation at $T/T_c = 0.465$, $H = 101\,\mathrm{G}$ and the minimum at $T/T_c = 0.465$, $H = 152\,\mathrm{G}$, as well as the fine structure oscillations in the impedance near the points labeled 90° and 270°. The fine structure is observed when the polarization rotates by an odd multiple of 90° upon a single round trip in the cell. Waves that traverse the cell twice are then 180° out of phase relative to the source wave, and consequently the period of the impedance oscillations is halved. The amplitude of the oscillations is reduced because of attenuation over the longer pathlength. These observations of the rotation of the polarization of the current by a magnetic field prove that the impedance oscillations result from the interference of propagating *transverse* current waves [19].

### *The g-factor for the $J = 2^-$ Modes*

The temperature and field variations of the impedance oscillations provide a strong test of the theoretical predictions for transverse wave propagation in $^3$He-B. Here I compare the theoretical result for $\Lambda$ in Eq. (46) with the experimental data for magneto-acoustic rotation period for low fields, $H \ll 1\,\mathrm{kG}$, and temperatures above, but near, the extinction point $T_*$. Note that the temperature, $T_+$, is the extinction point in a field is slightly higher than the zero-field extinction point, $T_*$; at $H = 100\,\mathrm{G}$, $T_+ - T_* \approx 1\,\mu K$. The magnitude of the Faraday rotation period depends on well known properties of $^3$He-B, except for the Landé g-factor, $g_{2^-}$. For the calculations shown in Fig. 6 I used the measured Fermi-liquid data tabulated for $T_c$, $v_f$, $F_1^s$, $F_0^a$, and $F_2^a$ as a function of pressure in [13]. The mode data are $T_+/T_c$ and $\Omega_+(T_+) = \omega$, and $\gamma = 3.2434\,\mathrm{MHz/kG}$. The only undetermined parameter that enters Eq. (46) for $\Lambda$ is the Landé g-factor, which scales the magnitude of the Faraday rotation angle. The data exhibit the temperature dependence and pressure dependence predicted by the theory and provide a new determination of the Landé g-factor the $J = 2^-$ modes at low fields. The calculated results are in excellent agreement with the experimental data at three nearby pressures for $g_{2^-} = 0.02 \pm 0.002$.

This result differs from the value of $g_{2^-} \simeq 0.042$ reported by [26] for

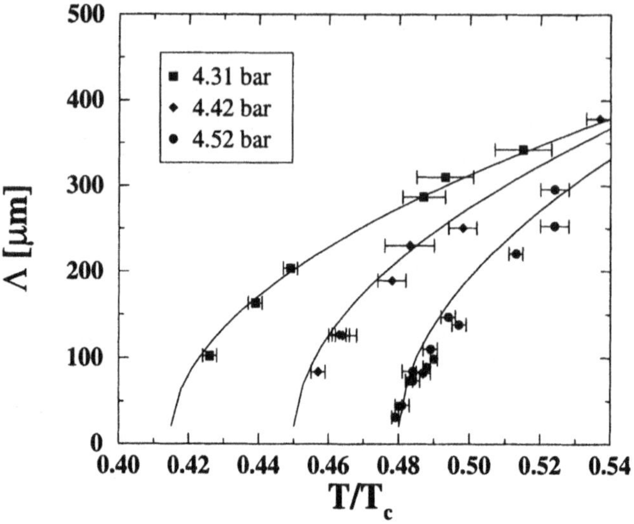


*Figure 6.* Comparison between the theoretical and experimental results for the temperature dependence of the Faraday rotation period for a field of $H = 100$ Gauss at pressures, $p = 4.31$ bar (square), $p = 4.42$ bar (diamond), and $p = 4.52$ bar (circle). The data for the corresponding pressures are shown as the solid squares with error bars (Y. Lee, et al. (1999)).

$p = 19$ bar, which was obtained from the splitting of the $J = 2^-$ multiplet in the absorption spectrum of longitudinal sound for fields above 2 kG. At these high fields the non-linear field contribution to the level splittings is comparable to the linear Zeeman splitting, which makes and accurate determination of the Landé g-factor from the absorption spectrum difficult. This complication is not present in the analysis of the Faraday rotation period at $H \approx 100$ G.

Earlier theoretical calculations for the order parameter collective mode frequencies showed that that the g-factors for these excitations are sensitive to Fermi liquid interaction effects and higher angular momentum pairing correlations [30]. Figure 7 shows the g-factor for the $J = 2^-$ modes as a function of temperature for several values of the f-wave pairing interaction, expressed in terms of the f-wave and p-wave transition temperatures, $x_3 = \ln(T_{\text{f-wave}}/T_c)$. Negative values correspond an attractive f-wave pairng interaction. The g-factor, $g_{2^-}$, also depends on the Landau interaction parameter, $F_2^s$, which is known from measurements of the normal-state zero sound velocity. The calculated values of $g_{2^-}$ are shown for a pressure of 4.31 bar, corresponding to $F_2^s \simeq -0.05$. The calculated g-factor is strongly reduced by attractive f-wave correlations.

The analysis of the Faraday rotation period yields a measurement of $g_{2^-}$. The fact that the $J = 2^-$ g-factor is intrinsically small (in contrast to the

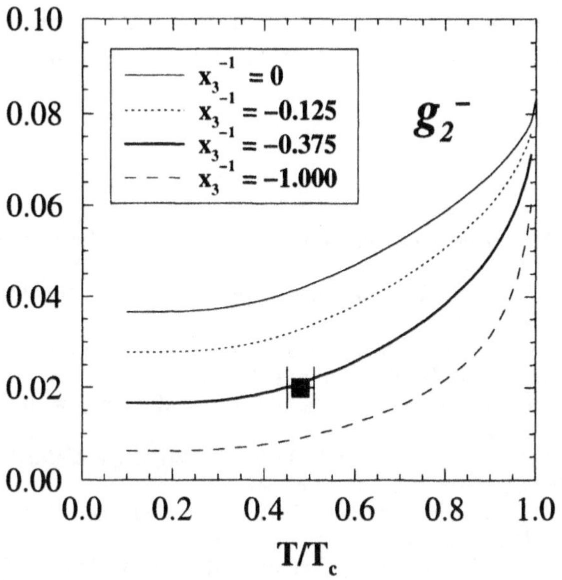


*Figure 7.* Comparison between the theoretical and experimental results for the Landé g-factor of the $J = 2^-$ modes. The theoretical curves are include the effects of f-wave pairing correlations. The experimental result for $g_{2^-}$ at $T/T_c = 0.46$ and $P = 4.32$ bar is fit with an attractive f-wave interaction, $x_3^{-1} \simeq -0.33$.

$J = 2^+$ g-factor), makes this parameter a sensitive test of many-body correlation effects in superfluid $^3$He. The implications of this result are that either substantial attractive f-wave pairing correlations or strong-coupling effects reduce the g-factor by a factor of two compared to the weak-coupling result. The value of $g_{2^-}$ obtained from Faraday rotation measurements suggests that f-wave pairing correlations at 4 bar are attractive with $T_{\text{f-wave}} \simeq 0.07\,T_c$. If the reduction is due to f-wave pairing correlations then there should be a collective mode with total angular momentum $J = 4$ [29] lying just below the pair breaking edge which might be observable in the dispersion of TZS at high frequencies near, but below the pair breaking edge. In summary, the discovery of propagating transverse currents and magneto-acoustic rotation of the polarization of mass currents in superfluid $^3$He-B opens a new spectroscopy of collective excitations in superfliud $^3$He-B. The precision of this spectroscopy offers the possibility of measuring strong coupling effects on the collective mode frequencies and Zeeman levels, as well as effects of higher angular momentum pairing correlations at ultra-low temperatures.

## Acknowledgements

I thank my colleagues and collaborators, Geneva Moores, Yoonseok Lee,

Tom Haard, and Bill Halperin for their many contributions included in these lecture notes, and for many important discussions on acoustics and superfluid $^3$He.

## References

1. Abel, W. R., A. C. Anderson, and J. C. Wheatley (1966) Propagation of Zero Sound in Liquid $^3$He at Low Temperatures. *Phys. Rev. Lett.* **17**, 74.
2. Anderson, P. W. (1958) Random-Phase Approximation in the Theory of Superconductivity. *Phys. Rev.* **112**, 1900.
3. Avenel, O., E. Varoquaux, and H. Ebisawa (1980) Field Splitting of the New Sound Attenuation Peak in $^3$He-B. *Phys. Rev. Lett.* **45**, 1952.
4. Balian, R. and N. R. Werthamer (1963) Superconductivity with Pairs in a Relative P-state. *Phys. Rev.* **131**, 1553.
5. Bogoliubov, N. N., Tolmachev, and Shirkov (1958) *New Methods in the Theory of Superconductivity.* Moscow: Academy of Science.
6. Combescot, M. and R. Combescot (1976) Transverse Zero Sound Propagation in Superfluid $^3$He. *Phys. Lett.* **58A**, 181.
7. Corruccini, L., J. S. Clark, N. D. Mermin, and J. W. Wilkins (1969) Attenuation of Transverse Zero Sound in $^3$He. *Phys. Rev.* **180**(1), 225.
8. Fishman, R. S. and J. A. Sauls (1986) Response Functions and Collective Modes of Superfluid $^3He$-B in Strong Magnetic Fields. *Phys. Rev.* **B33**, 6068.
9. Flowers, E. G., R. W. Richardson, and S. J. Williamson (1976) Transverse Sound in Normal $^3$He. *Phys. Rev. Lett.* **37**, 309.
10. Fomin, I. A. (1968) Excitation of Transverse Zero Sound in Liquid $^3$He. *Sov. Phys. JETP* **27**, 1010.
11. Giannetta, R. W., A. Ahonen, E. Polturak, J. Saunders, and E. K. Zeise (1980) Observation of a New Sound Attenuation Peak in Superfliud $^3$He-B. *Phys. Rev. Lett.* **45**, 262.
12. Halperin, W. P. (1982) Acoustic Order Parameter Mode Spectroscopy in Superfluid $^3$He-B. *Physica* **109-110B**, 1596.
13. Halperin, W. P. and E. Varoquaux (1990) Order Parameter Collective Modes in Superfluid $^3He$. In: W. P. Halperin and L. P. Pitaevskii (eds.): *Helium Three.* Amsterdam: Elsevier Science Publishers, p. 353.
14. Ho, T. and V. B. Shenoy (1996) Local Spin-Gauge Symmetry of the Bose-Einstein Condensates in Atomic Gases. *Phys. Rev. Lett.* **77**, 2595.
15. Koch, V. E. and P. Wölfle (1981) Coupling of New Order Parameter Collective Modes to Sound Waves in Superfluid $^3He$. *Phys. Rev. Lett.* **46**, 486.
16. Kurkijärvi, J. and D. Rainer (1990), Andreev Scattering in Superfluid $^3He$. In: edited by W. P. Halperin and L. P. Pitaevskii (eds.): *Helium Three.* (Elsevier Science Publishers, Amsterdam, p. 313.
17. Landau, L. D. (1956) The Theory of a Fermi Liquid. *Sov. Phys. JETP* **30**, 1058.
18. Landau, L. D. (1957) Oscillations in a Fermi Liquid. *Sov. Phys. JETP* **32**, 59.
19. Lee, Y., T. Haard, W. Halperin, and J. A. Sauls (1999) Discovery of an Acoustic Faraday Effect in Superfluid $^3$He-B. *Nature* **400**, 431.
20. Leggett, A. J. (1975) A Theoretical Description of the New Phases of Liquid $^3He$. *Rev. Mod. Phys.* **47**, 331.
21. Maki, K. (1974) Propagation of Zero Sound in the Balian-Werthamer State. *J. Low Temp. Phys.* **16**, 465.
22. Maki, K. and H. Ebisawa (1977) Transverse Zero Sound in Superfluid $^3$He-B. *J. Low Temp. Phys.* **26**, 627.
23. Mast, D. B., B. K. Sarma, J. R. Owers-Bradley, I. D. Calder, J. B. Ketterson, and W. P. Halperin (1980) Measurements of High Frequency Sound Propagation in $^3$He-B.

*Phys. Rev. Lett.* **45**, 266.

24. McKenzie, R. H. and J. A. Sauls (1990) Collective Modes and Nonlinear Acoustics in Superfluid $^3He$-B. In: edited by W. P. Halperin and L. P. Pitaevskii (eds.): *Helium Three.* Amsterdam: Elsevier Science Publishers, p. 255.
25. Moores, G. F. and J. A. Sauls (1993) Transverse Waves in Superfluid $^3$He-B. *J. Low Temp. Phys.* **91**, 13.
26. Movshovich, R., E. Varoquaux, N. Kim, and D. M. Lee (1988) Splitting of the Squashing Collective Mode of Superfluid $^3$He-B. *Phys. Rev. Lett.* **61**, 1732.
27. Nozières, P. (1964) *Theory of Interacting Fermi Systems.* New York, NY: Benjamin.
28. Roach, P. R. and J. B. Ketterson (1976) Observation of Transverse Sound in Normal $^3$He. *Phys. Rev. Lett.* **36**, 736.
29. Sauls, J. A. and J. W. Serene (1981) Coupling of Order-Parameter Modes with $l > 1$ to Zero Sound in $^3$He-B. *Phys. Rev.* **B23**, 4798.
30. Sauls, J. A. and J. W. Serene (1982) Interaction Effects on the Zeeman Splitting of Collective Modes in Superfluid $^3He$-B. *Phys. Rev. Lett.* **49**, 1183.
31. Schopohl, N., M. Warnke, and L. Tewordt (1983) Effect of Gap Distortion on the Field Splitting of the Collective Modes in Superfluid $^3He$-B. *Phys. Rev. Lett.* **50**, 1066.
32. Serene, J. W. (1983) Order Parameter Modes, Zero Sound and Symmetries in Superfluid $^3$He. In: *Quantum Fluids and Solids -1983*, Vol. 103. p. p. 305.
33. Serene, J. W. and D. Rainer (1983) The Quasiclassical Approach to $^3He$. *Phys. Rep.* **101**, 221.
34. Shivaram, B. S., M. W. Meisel, B. K. Sarma, W. P. Halperin, and J. B. Ketterson (1983) Nonlinear Zeeman Shifts in the Collective Mode Spectrum of $^3$He-B. *Phys. Rev. Lett.* **50**, 1070.
35. Tewordt, L. and N. Schopohl (1979) Gap Parameters and Collective Modes for $^3He$-B in the Presence of a Strong Magnetic Field. *J. Low Temp. Phys.* **37**, 421.
36. Volovik, G. (1999) The Helium-Centric Universe. In: *Topological Defects and the Non-equilibrium Dynamics of Symmetry Breaking Phase Transitions.* Dordrecht.
37. Wölfle, P. (1973) Attenuation of Zero Sound in Liquid $^3$He: a Probe of Superfluid Pairing. *Phys. Rev. Lett.* **30**, 1169.
38. Yip, S. K. and J. A. Sauls (1992) Circular Dichroism and Birefringence in Unconventional Superconductors. *J. Low Temp. Phys.* **86**.

# RELATIVISTIC DYNAMICS OF VORTEX DEFECTS IN SUPERFLUIDS

BRANDON CARTER
*Département d'Astrophysique Relativiste et de Cosmologie,*
*Centre National de la Recherche Scientifique,*
*Observatoire de Paris, 92195 Meudon, France.*

Superfluid condensates are known to occur in contexts ranging from laboratory liquid helium to neutron stars, and are also likely to occur in cosmological phenomena such as axion fields. In the zero temperature limit, such condensates are describable at a mesoscopic level by irrotational configurations of simple relativistic perfect fluid models. The general mechanical properties of such models are presented here in an introductory review giving special attention to the dynamics of vorticity flux 2-surfaces and the action principles governing both individual flow trajetories and the evolution of the system as a whole. Macroscopic rotation of such a condensate requires the presence of a lattice of quantised vortex defects, whose averaged tension violates perfect fluid isotropy. It is shown that for any equation of state (relating the mass density $\rho$ to the pressure $P$) the mesoscopic perfect fluid model can be extended in a uniquely simple and natural manner to a corresponding macroscopic model (in a conformally covariant category) that represents the effects of the vortex fibration anisotropy. The limiting case of an individual vortex defect is shown to be describable by a ("global") string type model with a variable tension $\mathcal{T}$ (obtained as a function of the background fluid density) whose "vorton" (i.e. closed loop equilibrium) states are derived as an exercise.

## 1. Introduction

These lectures offer an introduction to the dynamics of vortices in simple superfluid models of a general category that includes the kind appropriate in the context of laboratory condensed matter for the representation of Helium-4 in the zero temperature limit (for which no "normal" entropy carrying constituent is present), as well as the ultrarelativistic "stiff" kind that has been widely used [1] for the representation of a massless axion field in cosmology. Between these extremes, this category also includes the kind of model suitable as a simple approximation for the description of the neutron-pair condensation superfluid that is generally believed (for both theoretical



*Y.M. Bunkov and H. Godfrin (eds.),*
*Topological Defects and the Non-Equilibrium Dynamics of Symmetry Breaking Phase Transitions,* 267–301.

and observational reasons [2]) to occur in the intermediate layers of neutron stars. (A more sophisticated treatment would include allowance for the presence of an interpetrating ionic crust lattice or of an independently superconducting proton pair condensate, not to mention complications such as spin.)

The declared purpose of this school for which these lectures are intended is to study analogies between low temperature laboratory phenomena and high energy cosmological phenomena. However it is desirable, when possible, to reinforce mere analogy by interpolation. I would therefore like to emphasize the particular interest for this purpose of the intermediate regime of neutron star interiors, about which a great deal of admittedly indirect information is available from analysis of pulsar frequency variations. In contrast with most other areas of astrophysics, but in common with typical laboratory applications of condensed matter physics, the temperatures in typical neutron star interiors can usually be considered to be very low compared with the relevant energy scales. On the other hand, in contrast with typical laboratory applications of condensed matter physics, but in common with high energy cosmological scenarios, the treatment of neutron star interiors requires allowance for significant relativistic effects.

While evidently indispensible for an accurate treatment of neutron star matter, a relativistic treatment is commonly – but wrongly – considered to be an unnecessary complication for low temperature laboratory applications. For a reasonably accurate treatment of laboratory liquid Helium a relativistic treatment is indeed unnecessary, but what is wrong is to suppose that it is more complicated. On the contrary, as I hope these lectures will make clear, the relativistic treatment, in so far as it is available, is mathematically simpler. (The essential reason for this is that the Lorentz group is, in the technical sense "semi-simple" whereas the Galilei group is not.)

The work, in collaboration with David Langlois [3], on which these lectures are based, was originally inspired by the observation by Rick Davis and Paul Shellard [4] (in precisely the spirit that is the official raison d'être for the present school) of a strong analogy between the behaviour of vortices in the "stiff" (ultrarelativistic) massless axion model at one extreme, and in the incompressible superfluid model that is commonly used for the description of zero temperature Helium-4 at the opposite extreme.

What is shown here is the way to carry out the analogous, but not quite so trivially simple, treatment that is needed for the intermediate category of generically compressible superfluid models. At a mesoscopic level these models will be of irrotational perfect fluid type, and will be characterised by a subluminal speed, $c_{\rm I}$ say, of ordinary "first" sound, that will be determined – by an equation of state specifying the pressure $P$ as a function of the mass density $\rho$ – according to the familiar formula

$$c_{\rm I} = (dP/d\rho)^{1/2} \, . \tag{1}$$

Rotation at a macroscopic level will entail the presence of an Abrikosov type lattice of quantised vortex defects whose averaged tension produces a deviation from perfect fluid isotropy. Analysis of individual vortex tubes [5] indicates that, in the limit for which their relative "drift" or "flight" velocity is highly subsonic, their averaged effect can be described by the inclusion of an extra isotropy violating term of a uniquely simple and natural kind [3] in the action for a corresponding macroscopic model. Despite their violation of isotropy, such models (for different equations of state) will be shown to belong to a category that is conserved by conformal transformations. More particularly, it will be shown that the specific model corresponding to the high pressure limit of a gas of negligibly interacting particles as characterised by $c_{\rm I}^2 = c^2/3$ (where $c$ is the speed of light) has the same kind of conformal invariance property as the well known example of Maxwell's equations.

The scope of the present review does not extend to the recently perfected relativistic analogue [6, 7] of Landau's original non-dissipative two constituent superfluid model (which has recently been shown to be expressible in a very elegant Galilean covariant form [8]) for the purpose of allowing for the dynamical effect of the "normal" entropy flux current that would be present at a non zero temperature). Such effects are not the ones that are most important in the context of neutron stars, the main application for which the work described here is intended, where the temperature will typically be so low (compared with the relevant MEV range energy scales) that thermal corrections can for most purposes be neglected. For application to neutron stars, a more important kind of generalisation is to a relativistic models capable of describing neutron superfluid penetration [9] of the solid material forming the crust, and of describing [10] the protonic superconductivity that is expected [2] within the neutron superfluid at a deeper level below the crust.

## 2. Canonical treatment of relativistic flow trajectories

Before proceeding it is desirable to recall some essentials the relativistic kinematics and dynamics, particularly in view of the regretable tradition in non-relativistic fluid theory – and most notably in non-relativistic superfluid theory – of obscuring the essential distinction between velocity (which formally belongs in a tangent bundle) and momentum (which formally belongs in a cotangent bundle) despite the fact that the distinction is generally respected in other branches of non-relativistic condensed matter theory, such as solid state physics, where the possibility of non-alignment between the 3-velocity $v^a$, and the effective 3-momentum $p_a$ of an electron travelling in a metallic lattice is well known.

In a non relativistic treatment it is only in strictly Cartesian (rather than e.g. cylindrical or comoving) that the distinction between contravariant entities such as the velocity $v^a$ and covariant entities such as the momentum $p_a$ can be ignored. In a relativistic treatment, even using coordinates $x^\mu \leftrightarrow \{t, x^a\}$ of Minkowski type, with a flat spacetime metric $g_{\mu\nu}$ whose components are of the fixed standard form diag$\{-c^2, 1, 1, 1\}$, (where $c$ is the speed of light) the necessity of distinguishing between raised and lowered indices is inescapable. Thus for a trajectory parametrised by proper time $\tau$, the correponding unit tangent vector

$$u^\mu = \frac{dx^\mu}{d\tau} \tag{2}$$

is automatically, by construction a contravariant vector: its space components, $u^a = \gamma v^a$ with $\gamma = (1-v^2/c^2)^{-1/2}$ will be unaffected by the index lowering operation $u^\mu \mapsto u_\mu = g_{\mu\nu}u^\nu$, but its time component $u^0 = dt/d\tau = \gamma$ will differ in sign from the corresponding component $u_0 = -\gamma c^2$ of the associated covector $u_\mu$. On the other hand the 3-momentum $p_a$ and energy $E$ determine a 4-momentum covector $\pi_\nu$ that is intrinsically covariant, with components $\pi_a = p_a$, $\pi_0 = -E$ . The covariant nature of the momentum can be seen from the way it is introduced by the defining equation,

$$\pi_\nu = \frac{\partial L}{\partial u^\nu}, \tag{3}$$

in terms of the relevant position and velocity dependent Lagrangian function $L$, from which the corresponding equation of motion is obtained in the well known form

$$\frac{d\pi_\nu}{d\tau} = \frac{\partial L}{\partial x^\nu}. \tag{4}$$

In the case of a free particle trajectory, and more generally for fluid flow trajectories in all the simple "barotropic" perfect fluid models with which the present lectures will be concerned, the function will have the familiar standard form

$$L = \tfrac{1}{2}\mu g_{\mu\nu}u^\mu u^\nu - \tfrac{1}{2}\mu c^2, \tag{5}$$

in which (unlike what is needed for more complicated chemically inhomogeneous models[11, 9]) it is the same scalar spacetime field $\mu$ that plays the role of mass in the first term and that provides the potential energy contribution in the second term. The momentum will thus be given by the simple proportionality relation

$$\pi_\nu = \mu u_\nu, \tag{6}$$

so that one obtains the expressions $E = \gamma\mu c^2$, $p_a = \mu\gamma v_a$, in which the field $\mu$ is interpretable as the relevant effective mass.

In the case of a free particle model, the effective mass $\mu$ will of course just be a constant, $\mu = m$. This means that if, as we have been supposing so far, the metric $g_{\mu\nu}$ is that of flat Minkowski type, the resulting free particle trajectories will be obtainable trivially as straight lines. However the covariant form of the equations (2) to (6) means that they will still be valid for less trivial cases for which, instead of being flat, the metric $g_{\mu\nu}$ is postulated to have a variable form in order to represent the effect of a gravitational field, such as that of a Kerr black hole (for which, as I showed in detail in a much earlier Les Houches school [12], the resulting non trivial geodesic equations still turn out to be exactly integrable).

In the case of the simple perfect fluid models with which we shall be concerned here, the effective mass field $\mu$ will be generically non-uniform. In these models the equation of state giving the pressure $P$ as a function of the mass density $\rho$ can most conveniently be specified by first giving $\rho$ in terms of the corresponding conserved number density $n$ by an expression that will be decomposible in the form

$$\rho = mn + \frac{\epsilon}{c^2}, \tag{7}$$

in which $m$ is a fixed "rest mass" characterising the kind of particle (e.g. a Cooper type neutron pair) under consideration, while $\epsilon$ represents an extra compression energy contribution. The pressure will then be obtainable using the well known formula

$$P = (n\mu - \rho)c^2, \tag{8}$$

in which the effective dynamical mass $\mu$ (or equivalently the "specific enthalpy" $\mu c^2$) is given by

$$\mu = \frac{d\rho}{dn} = m + \frac{1}{c^2}\frac{d\epsilon}{dn}. \tag{9}$$

It is this that is to be taken as the effective mass function appearing in the specification (5) of the relevant .

When one is dealing not just with a single particle trajectory but a spacefilling fluid flow, it is possible and for many purposes desirable to convert the Lagrangian dynamical equation (4) from particle evolution equation to equivalent field evolution equations [11, 9]. Since the momentum covector $\pi_\nu$ will be obtained as a field over spacetime, it will have a well defined gradient tensor $\nabla_\rho\pi_\nu$ that can be used to rewrite the left hand side of (4) in the form $d\pi_\nu/d\tau = u^\rho\nabla_\rho\pi_\nu$. Since the value of the Lagrangian will also be obtained as a scalar spacetime field $L$, it will also have a well defined gradient which will evidently be given by an expression of the form

$\nabla_\nu L = \partial L/\partial x^\nu + (\partial L/\partial \pi_\rho)\nabla_\nu \pi_\rho$. We can thereby rewrite the Lagrangian dynamical equation (4) as a field equation of the form

$$u^\rho \nabla_\rho \pi_\nu + \pi_\rho \nabla_\nu u^\rho = \nabla_\nu L\,. \tag{10}$$

An alternative approach is of course to start from the corresponding Hamiltonian function, as obtained in terms of the position and momentum variables (so that formally it should be considered as a function on the spacetime cotangent bundle) via the Legendre transformation

$$H = \pi_\nu u^\nu - L\,. \tag{11}$$

In this approach the velocity vector (2) and the dynamical equation (4) are recovered using the familiar formulae

$$\frac{dx^\mu}{d\tau} = \frac{\partial H}{\partial \pi_\nu}\,, \qquad \frac{d\pi_\nu}{d\tau} = -\frac{\partial H}{\partial x^\nu}\,. \tag{12}$$

The consideration that we are concerned not just with a single trajectory but with a spacefilling fluid means that, as in the case of the preceding equations, in a similar way this familiar Hamiltonian dynamical equation can also be converted to a field equation which takes the form

$$2u^\rho \nabla_{[\rho}\pi_{\nu]} = -\nabla_\nu H\,, \tag{13}$$

with the usual convention that square brackets are used to indicate index antisymmetrisation. On contraction with $u^\nu$, the left hand side will evidently go out, leaving the condition

$$u^\nu \nabla_\nu H = 0\,, \tag{14}$$

expressing the conservation of the value of the Hamiltonian along the flow lines.

The actual form of the Hamiltonian function that is obtained from the particularly simple kind of Lagrangian function (5) with which we are concerned will evidently be given by

$$H = \frac{1}{2\mu} g^{\nu\rho}\pi_\nu \pi_\rho + \frac{\mu c^2}{2}\,. \tag{15}$$

In order to ensure the proper time normalisation for the parameter $\tau$ the equations of motion (in whichever of the four equivalent forms (4), (10), (12), (13) may be preferred) are to be solved subject to the constraint that – in order for $u^\mu$ to be correctly normalised – the numerical value of the Hamiltonian should vanish,

$$H = 0 \qquad \Rightarrow \qquad u^\mu u_\mu = -c^2\,, \tag{16}$$

initially, and hence also by (14) at all other times. In the more general systems that are needed for some purposes the Hamiltonian may be constrained in a non uniform manner [11, 9] so that the term on the right of (13) will be non zero, but in the simpler systems that suffice for our present purpose the restraint (16) ensures that this final term will drop out, leaving a Hamiltonian equation of the very elegant and convenient form

$$u^\mu \nabla_{[\mu} \pi_{\nu]} = 0 \,. \tag{17}$$

## 3. Vorticity conservation and flux 2-surfaces

The preceding form (17) of the dynamical equations is particularly handy for the analysis of symmetries and the derivation of conservation laws according to which physically interesting quantities are preserved by various kinds of continuous displacement [11, 9]. The variation induced by an infinitesimal displacement generated by an arbitrary vector field $k^\mu$ say, will be given by the corresponding Lie derivative operator, whose effect on a scalar field, $\mu$ say, will evidently be given simply by

$$\vec{k}\mathcal{L}\mu = k^\nu \nabla_\nu \mu \,, \tag{18}$$

while its effect on the metric will be given by the well known though not quite so immediately obvious formula

$$\vec{k}\mathcal{L}g_{\mu\nu} = 2\nabla_{(\mu} k_{\nu)} \,, \tag{19}$$

using round brackets to denote index antisymmetrisation. If $k^\mu$ is a Killing vector field, i.e. if the displacement generated by $k^\mu$ is a symmetry of the spacetime metric, then the right hand side of (19) will vanish. In curved space such symmetries are rare, but in ordinary flat space time there is of course a ten parameter family of such Killing vector fields generating the Poincaré group, whose algebra has as its basis the four independent generators of uniform spacetime translations and the six independent generators of the Lorentz group. The effect of the Lie differentiation operation on another vector field, $u^\rho$, say will be given simply by their mutual commutator bracket,

$$\vec{k}\mathcal{L}u^\rho = [\vec{k}, \vec{u}]^\rho = -\vec{u}\mathcal{L}k^\rho \,, \qquad [\vec{k}, \vec{u}]^\rho = k^\nu \nabla_\nu u^\rho - k^\nu \nabla_\nu k^\rho \,. \tag{20}$$

Such Lie differentiation is an example of an operation for which the distinction between covectors and ordinary contravariant vectors is important: except when performed with respect to a Killing vector, Lie differentiation does not commute with index raising and lowering. The rule that applies to the covector $u_\mu = g_{\mu\nu}u^\nu$ can evidently be obtained by combining (19) and (20), so that for $\pi_\nu$ one obtains a formula that can be conveniently expressed in terms of exterior (antisymmetrised) derivatives in the form

$$\vec{k}\mathcal{L}\pi_\nu = 2k^\rho\nabla_{[\rho}\pi_{\nu]} + \nabla_\nu(\pi_\rho k^\rho)\,. \tag{21}$$

Although it does not commute with index raising and lowering, Lie differentiation does commute with exterior differentiation. Thus for the relativistic vorticity tensor, which is defined to be the antisymmetrised derivative of the momentum covector, i.e.

$$w_{\mu\nu} = 2\nabla_{[\mu}\pi_{\nu]}\,, \tag{22}$$

so that its own exterior derivative will automatically vanish, i.e.

$$\nabla_{[\mu}w_{\nu\rho]} = 0\,, \tag{23}$$

it follows that its Lie derivative will be obtainable just by taking the exterior derivative of (21) which gives

$$\vec{k}\mathcal{L}w_{\mu\nu} = -2\nabla_{[\mu}(w_{\nu]\rho}k^\rho)\,. \tag{24}$$

The preceding general formula (21) can immediately be used to rewrite the Lagrangian dynamical equation (10) in the expressive form [11, 9]

$$\vec{u}\mathcal{L}\pi_\nu = \nabla_\nu L\,, \tag{25}$$

from which by taking the exterior derivative, one can immediately derive the dynamical vorticity conservation law that is the key to superfluidity theory (and much else) in the form

$$\vec{u}\mathcal{L}w_{\mu\nu} = 0\,. \tag{26}$$

The interpretation of this crucially important result is that for any flow governed by the Lagrangian equations (4) or their Hamiltonian equivalent (12) (which have so far simply been postulated ex cathededra, but whose validity for any "barotropic" perfect fluid be made clear in the following section) the vorticity field (22) will simply be convected onto itself by the flow field $u^\mu$, with the implication that if it vanishes initially $w_{\mu\nu}$ will remain zero throughout the flow, which in this case will be describable as "irrotational".

The foregoing results can be considerably strengthenned in cases such as those of the simple "barotropic" perfect fluids considered here for which the proper time normalisation is ensured by the Hamiltonian constraint (16). This has the effect of reducing the dynamical equations to the particularly simple form (17), whose interpretation is that the flow vector $u^\mu$ must be an zero eigenvalue eigenvector of the vorticity tensor $w_{\mu\nu}$. The posession of a zero eigenvalue requires that $w_{\mu\nu}$ should satisfy the degeneracy condition

$$w_{\mu[\nu}w_{\rho\sigma]} = 0\,, \tag{27}$$

which excludes the possibility of it having matrix rank 4, with the implication that unless it actually vanishes it must have rank 2 (since an antisymmetric tensor can never have odd integer rank). This means that the flow vector $u^\mu$ is just a particular case within a whole 2-dimensional tangent subspace of eigenvectors $e^\mu$ satisfying

$$e^\mu w_{\mu\nu} = 0\,. \tag{28}$$

This subspace will be spanned by a unit worldsheet element tangent bivector $\mathcal{E}^{\mu\mu}$ of the kind whose use was developped by Stachel [13], and that is definable, wherever the vorticity magnitude

$$w = \left(\tfrac{1}{2}w_{\mu\nu}w^{\mu\nu}\right)^{1/2} \tag{29}$$

does not vanish, as being proportional to the dual vorticity tensor $W^{\mu\nu}$, i.e.

$$\mathcal{E}^{\mu\nu} = \frac{1}{w}W^{\mu\nu}\,, \qquad W^{\mu\nu} = \frac{1}{2}\varepsilon^{\mu\nu\rho\sigma}\,w_{\rho\sigma}\,, \tag{30}$$

(note that sign convention used for the worldsheet orientation here is the opposite of what was used in the preceding article [3]) where $\varepsilon^{\mu\nu\rho\sigma}$ is the totally antisymmetric tensor normalised by the convention that its non zero components are equal to 1 or -1 (depending on whether the index ordering is an even or odd permutation) with respect to locally Minkowskian coordinates with $g_{\mu\nu} \leftrightarrow \mathrm{diag}\{-c^2, 0, 0, 0\}$, so that with respect to an arbitrary coordinate system the non zero component values of $\varepsilon^{\mu\nu\rho\sigma}$ will be given by $\pm c||g||^{-1/2}$.

It can be seen that this worldsheet tangent bi-vector will satisfy

$$\mathcal{E}^{\mu\nu}\mathcal{E}_{\mu\nu} = -2c^2\,, \qquad \mathcal{E}^{\mu\nu}w_{\nu\rho} = 0\,. \tag{31}$$

It follows from this last equation that the contraction of any covector with $\mathcal{E}^{\mu\nu}$ will provide a solution of the vortex worldsheet tangentiality condition (28). A noteworthy example is the helicity $h^\mu$ as defined [11, 6] by

$$h^\mu = w\mathcal{E}^{\mu\nu}\pi_\nu\,, \tag{32}$$

which can be seen from (23) and (27) to satisfy the helicity current conservation law [11, 9]

$$\nabla_\nu h^\nu = 0\,. \tag{33}$$

Again using the degeneracy condition (27) and the condition that the vorticity also satisfies the Poincaré closure condition (23), it can be shown [9] that the tangent elements characterised by $\mathcal{E}^{\mu\nu}$ and generated by solutions of (28) (or more specifically by $u^\nu$ and $h^\nu$) will automatically satisfy the relevant Frobenius condition for integrability, meaning that they will mesh together to form well behaved timelike 2-dimensional worldsheets. This makes it possible to extend the flow line conservation laws resulting from any continuous symmetries that may be present.

The simplest example is the Bernouilli type theorem that applies – even if the Hamiltonian does not satisfy the constraint (16) – whenever $k^\rho$ is a symmetry generator of the system, so that in particular $\vec{k}\mathcal{L}H$ and $\vec{k}\mathcal{L}\pi_\nu$ both vanish: it can be seen from from (13) and (21) that $u^\nu\nabla_\nu(\pi_\rho k^\rho)$ will vanish, so $\pi_\rho k^\mu$ will be constant along each flow line. In the most obvious application, the Killing vector is just the generator of time translations in flat space with the Minkowski coordinates, $k^\rho \leftrightarrow \{1, 0, 0, 0\}$, so that the Bernouilli constant will simply be identifiable with the negative of the effective energy per particle, i.e we shall have $\pi_\rho k^\rho = -E$ with $E = \gamma\mu c^2$ which automatically includes allowance for both compression energy and kinetic energy contributions.

This conclusion can be greatly strengthenned when the Hamiltonian constraint (16) holds: one then gets $e^\nu\nabla_\nu(\pi_\rho k^\rho) = 0$ for any vector $e^\nu$ satisfying (28) which evidently means that $\pi_\rho k^\nu$ will not just be constant along each flow trajectory but that it will be constant throughout each of the 2-dimensional vorticity flux worldsheets. In the irrotational case

$$w_{\mu\nu} = 0\,, \tag{34}$$

for which (28) is satisfied trivially by any vector $e^\nu$ at all, one can draw the even stronger conclusion that $\pi_\rho k^\rho$ will be constant throughout the fluid:

$$\vec{k}\mathcal{L}\pi_\rho = 0 \quad \Rightarrow \quad \nabla_\nu(\pi_\rho k^\rho) = 0\,. \tag{35}$$

(A similar conclusion of global uniformity of $\pi_\rho k^\rho$ would also be obtainable immediately from (16) and (21) in the alternative, more widely familiar, case of motion that is rigid in the sense of having a flow vector that is aligned with the Killing vector, i.e. for which $u^{[\nu}k^{\rho]} = 0$.) It is to be emphasised that the applicability of this kind of generalised Bernouilli theorem is not limited to the case of ordinary stationarity, for which $k^\rho$ is a time translation generator, but is just as well applicable to cases of axisymmetry for which the Killing vector is a rotation generator, and it has recently been found to be very useful [15] for a hybrid case (involving a non-rigidly rotating binary pair of tidally deformed and thus non-axisymmetric neutron stars which, from the point of view of a distant observer, are not stationary but periodically evolving, but which are nevertheless stationary from the point of view of a local observer with respect to a suitably rotating frame).

An alternative way of obtaining the local vorticity conservation theorem (26) is as the differential limit of the global Kelvin-Helmoltz theorem to the effect that, as is manifest from the form (25) of the Lagrangian dynamical equation, the action integral

$$S = \int \pi_\nu dx^\nu \tag{36}$$

will be preserved if taken round a closed circuit that is convected by the flow field $u^\mu$. In the irrotational case (34), the Jacobi type action integral (36) between any two fixed endpoints will be unaffected by continuous displacements of the path between them, and hence can be used to construct a locally well defined field $S$ that is not only such that one has

$$\pi_\nu = \nabla_\nu S\,, \tag{37}$$

but that will also be a solution of the Hamilton Jacobi equation specified by setting the Hamiltonian function to zero with the gradient of S substituted in place of the momentum.

The special case of a (simple, zero temperature limit) superfluid is specified by the existence of a well defined mesoscopic phase factor $e^{i\varphi}$ (representing the phase factor of an underlying bosonic condensate that might consist of Helium-4 atoms or Cooper type neutron pairs) in which the phase angle $\varphi$ is given according to the usual correspondence principle by

$$\varphi = S/\hbar\,. \tag{38}$$

In a multiconnected configuration of a classical irrotational fluid the Jacobi action field $S$ obtained from (37) might have an arbitrary periodicity, but in a superfluid there will be a U(1) quantisation requirement that the periodicity of the phase angle $\varphi$ should be a multiple of $2\pi$, and thus that the periodicity of the Jacobi action $S$ should be a multiple of $2\pi\hbar$. The simplest configuration for any such superfluid is a uniform stationary state in a flat Minkowski background, for which the phase will have the standard plane wave form

$$S/\hbar = k_a x^a - \omega t\,, \tag{39}$$

from which one obtains the correspondence $\pi_\nu \leftrightarrow \{-\hbar\omega, \hbar k_a\}$, which means that the effective energy per particle will be given by $E = \gamma\mu c^2 = \hbar\omega$ and that the 3-momentum will be given by $p_a = \mu\gamma v_a = \hbar k_a$.

It is to be remarked that for ordinary timelike superfluid particle trajectories the corresponding phase speed $\omega/k$ of the wave characterised by (39) will always be superluminal, a fact which people working with liquid

Helium-4 in the laboratory can blithely ignore, since what matters for most practical purposes is not the phase speed but the group velocity of perturbation wave packets. Our present discussion will be limited to the strict zero temperature limit for which no such packets are excited, but it is easy to extend the relativistic analysis to low but non-zero temperatures for which the relevant excitations are phonons [7]. Although their phase speed and group velocity are the same, both being given by the formula (1) for the ordinary ("first") soundspeed which will of course be subluminal, phonons do nevertheless have a tachyonic aspect of their own: their 4-momentum covector is always spacelike, in contrast with that of an ordinary fluid or superfluid particle which is timelike. This means that whereas the effective energy $E$ of an ordinary fluid or superfluid particle is always positive, the effective energy $E$ of a phonon may be positive or negative, depending on whether the frame of reference with respect to which it is measured is moving subsonically or supersonically. The well known implication is that if the superfluid is in contact with a supersonically moving boundary there will inevitably be an instability giving rise to dissipative phonon creation.

## 4. Conventional formulation of perfect fluid and simple superfluid theory

Although sufficient for the derivation of many important properties of the flow, the dynamical equations on which the preceding sections have been based contain only part of the information needed for a complete determination of the perfect fluid evolution. The most usual way of presenting the complete set of equations of motion of a simple perfect fluid of the barotropic type we are considering – meaning one whose intrinsic local physical state is characterised just by a single independent scalar field – is in the form of a conservation law of the standard form

$$\nabla_\nu T^{\mu\nu} = 0 \,, \tag{40}$$

for a stress momentum energy density tensor that is specified as a function of the timelike unit flow tangent vector $u^\nu$ and a single independent scalar field variable, such as the conserved particle number density $n$, on which the other relevant quantities, such as the effective mass $\mu$ given by (9), will be functionally dependent.

For a system of this simple kind, the 4 independent components of the energy momentum conservation law (40) provide all that is needed to determine the evolution of the 4 independent components that characterise the local state of the system, which can be taken to be the scalar $n$ and the 3 space components $u^a$ of $u^\mu$ (since the remaining component $u^0$ is not dynamically independent but determined by the unit normalisation condition (16) as an algebraic function of the 3 other components).

In any perfect fluid model the mass density $\rho$ and pressure $P$ are physically characterised by their role in the specification of the stress momentum energy density tensor, for which the standard expression is

$$T^{\mu\nu} = (\rho + \frac{P}{c^2})u^\mu u^\nu + P g^{\mu\nu} . \tag{41}$$

In the simple barotropic case, the relation (8) between the dependences of $P$ and $\rho$ on $n$ can be seen to be necessitated by the requirement that the dynamical system (40) should ensure conservation of the number current

$$n^\mu = n u^\mu . \tag{42}$$

It is easy to check, using (8) that contraction of (42) with $u_\mu$ does indeed lead to the required result, namely

$$\nabla_\mu n^\mu = 0 . \tag{43}$$

One can also verify the not quite so well known result [14] that the remaining independent equations (40) can be reorganised in the canonical uniformly Hamiltonian form (17) on which the work [11, 9] of the preceding subsection was based, and which is expressible succinctly as

$$n^\mu w_{\mu\nu} = 0 . \tag{44}$$

Thus in addition to the formula (44) that has been used so far, the only additional information needed for the complete specification of the dynamics of a barotropic fluid system is the obvious particle conservation law (43).

Given a dynamical system, one of the first things any physicist is inclined to ask is whether it is derivable from a Lagrangian type variation principle. We have already seen in the previous sections that (44) by itself is obtainable from Lagrangian equations of motion for the individual trajectories, which are of course obtainable from a one dimensional action integral of the form $\int L\, d\tau$ with $L$ as given by (5). The question to be adressed now is how to obtain the complete set of dynamical equations (40), including (43) as well as (44), from an action integral over the 4-dimensional background manifold $\mathcal{S}^{(4)}$ of the form

$$\mathcal{I} = \int \mathcal{L}\, d\mathcal{S}^{(4)} , \qquad d\mathcal{S}^{(4)} = \frac{||g||^{1/2}}{c} d^4 x , \tag{45}$$

for some suitable scalar Lagrangian functional $\mathcal{L}$.

Several radically different procedures are available for doing this. Although ultimately equivalent "on shell", they involve variation over "off shell" bundles that differ not just in structure but even in dimension. The oldest and most economical from a dimensional point of view is the world-line variation procedure developed by Taub[16], followed Clebsch type variation procedure developed by Schutz[17], but for our present purpose it will be more convenient to employ the more recently developed Kalb-Ramond type method[18] that has been specifically designed for dealing with problems of macroscopic superfluidity.

The problem is greatly simplified if, to start off with, one restricts oneself to the purely irrotational case (34), which is all that is needed for the description of zero temperature superfluidity at a mesoscopic level. For this case independent variable can be taken to be just the Jacobi action $S$, or equivalently in a superfluid context, the phase $\varphi$ as given by (38), and the action is simply taken to be the pressure P expressed as a function of the effective mass $\mu$, with the latter constructed as proportional to the amplitude of the 4-momentum, according to the prescription

$$\mu^2 c^2 = -\pi_\nu \pi^\nu \tag{46}$$

with the 4-momentum itself given by the relation (37) that applies in the irrotational case, i.e.

$$\pi_\nu = \hbar \nabla_\nu \varphi \,. \tag{47}$$

Thus setting

$$\mathcal{L} = P \,, \tag{48}$$

and using the standard pressure variation formula $\delta P = c^2 n \delta\mu$ one sees that the required variation of the Lagrangian will be given by

$$\delta\mathcal{L} = -n^\nu \delta\pi_\nu = -\hbar n^\nu \nabla_\nu (\delta\varphi) \,. \tag{49}$$

Demanding that the action integral (45) be invariant with respect to infinitesimal variations of $\varphi$ then evidently leads to the required conservation law (43).

## 5. Introduction of the dilatonic amplitude field $\Phi$

For an equation of state such that $P \propto \mu^2$, the Lagrangian (48) will, as it stands, have the quadratic field gradient dependence that is typical of simple physical field theories. This occurs in the special limit case, relevant to the massless axion field in cosmology [1, 4], of the "stiff" model, characterised by (62) as discussed below, for which the "first" sound speed (1) is equal to that of light. Except in this idealised limit case, a feature of the preceding Lagrangian function (48) that is widely considered to be undesirable is its generically non quadratic dependence on $\nabla\varphi$, an apparent drawback that is not uncommonly dealt with by recourse to approximation [19].

What this secion will show however is that that, not just for $P \propto \mu^2$ but for a quite general equation of state, the Lagrangian (48) can be reformulated in the much more desirably fashionable form

$$\mathcal{L} = -\frac{\hbar^2}{2}\Phi^2(\nabla_\nu\varphi)\nabla^\nu\varphi - V\{\Phi\} \tag{50}$$

– in which the potential energy density term $V$ is some suitably chosen algebraic function of the amplitude $\Phi$ – by a transformation of variables that is absolutely exact [18], without any need for recourse to approximation provided one adopts the correct definition for the auxiliary field variable $\Phi$.

What would involve an approximation would be to to implement a further step whereby $\Phi$ is "promoted" from the status of an auxiliary variable to that of an extra dynamical variable by adding in a supplementary kinetic term,

$$\Delta\mathcal{L}_{\{a\}} = -a^2\hbar^2(\nabla^\nu\Phi)\nabla_\nu\Phi\,, \tag{51}$$

for some constant value of $a$. Various kinds of gradient term, of which this is the most obvious, might indeed be added for the purpose of improving the physical precision of the model in regimes of rapid amplitude variation, where deviations from the strictly isotropic perfect fluid form (41) might be expected to become significant. However for a strongly interacting liquid like Helium-4 (as opposed to a weakly interacting gas) I know of no reliable procedure for the quantitative evaluation of such a term on the basis of an underlying many-particle quantum theory. The most commonly used [19] ansatz, namely to take $a = 1$, is not automatically guaranteed to provide an improvement but might even bring about a deterioration of physical precision in some circumstances. The standard choice $a = 1$ does not have theoretical or empirical foundations but is merely based on the purely mathematical consideration that it provides an adjusted Lagrangian $\mathcal{L}_{\{1\}} = \mathcal{L} + \Delta_{\{1\}}\mathcal{L}$ of the "relativistic Ginzburg Landau" form

$$\mathcal{L}_{\{1\}} = -\frac{\hbar^2}{2}(\nabla_\nu\Psi)\nabla^\nu\bar{\Psi} - V\{|\Psi|\}\,, \tag{52}$$

in which the phase variable $\varphi$ and the amplitude variable $\Phi$ have been combined to form a complex variable

$$\Psi = \mathrm{e}^{i\varphi}\,, \qquad \bar{\Psi} = \mathrm{e}^{-i\varphi}\,, \qquad \Phi = |\Psi|\,. \tag{53}$$

This feature is useful for some purposes – notably for providing a smoothed out treatment [20, 21] of the vortex defects that arise where the phase $\varphi$ becomes indeterminant – but not so convenient for deriving exact results such as the uniformity of the Bernouilli constant $\pi_\nu k^\nu$ that was shown above to be valid for a model of the original kind whenever $k^\rho$ is a symmetry generator.

As far as physical accuracy (as opposed to mathematical convenience) is concerned it will usually not matter very much whether one uses the exact perfect fluid model given by (50) or the associated Ginzburg Landau model given by (52) since their difference, as given by the extra kinetic term (51) can be expected to be very small except in the immediate neighbourhood of a vortex core where neither model can be expected to be physically accurate. What does matter for physical accuracy is the choice of the functional form for the amplitude $\Phi$.

When it comes to the point, without even attempting to provide any serious microscopic derivation, many introductory presentations start by assuming the validitity of the Landau Ginzburg framework, and then resort to crude guesswork for the specification of $\Phi$, typically taking $\Phi \propto \sqrt{n}$, or even more commonly

$$\Phi \propto \sqrt{\rho}\,. \tag{54}$$

Such a choice just happens to provide a quantitatively acceptable result in the non-relativistic limit, but the flimsiness of its theoretical basis is exposed by the fact that for a generic relativistic model it gives the wrong answer in the regime of slow amplitude variation where the (correctly calibrated) Landau Ginzburg model (52) should agree with the relevant perfect fluid model (50). The correct choice – the only way to get exact agreement between (48) and (50) – is to take

$$\Phi = \frac{n}{\sqrt{\rho + P/c^2}} = \left(\frac{n}{\mu}\right)^{1/2}, \tag{55}$$

the corresponding characterisation for the potential energy density function $V$ being [18] that it should be given by

$$V = \frac{\rho c^2 - P}{2}\,. \tag{56}$$

When characterised in this manner, – with the correct identification (55) rather than (54) – the perfect fluid model given by the Lagrangian (50) provides a uniquely canonical model for the representation at zero temperature of a simple superfluid such as Helium-4, at least in the regime where the rate of variation of the amplitude $\Phi$ is not too rapid compared with that of the phase. It is to be emphasised that for this regime the justification for this perfect fluid model is actually on a sounder footing than that of the corresponding Ginzburg Landau type model (52), since appart from the postulate of the isotropic perfect fluid form (41) for the stress momentum enery density tensor, whose applicability in the slow amplitude variation regime is hard to doubt, and the invocation of the generally valid conservation laws (40) and (42), the only assumption on which this fluid model is based, and for which it relies on an underlying microscopic quantum analysis, is that of the rather well established existence of the mesoscopic phase $\varphi$.

Having evaluated $V$ as a function of $\Phi$ one can recover the effective mass $\mu$, number density $n$, mass density $\rho$ and pressure $P$ of the fluid using the formulae

$$\mu^2 = \frac{1}{c^2\Phi}\frac{dV}{d\Phi}\,, \qquad n = \Phi^2\mu\,, \tag{57}$$

and

$$\rho = \tfrac{1}{2}\,\Phi^2\mu^2 + \frac{V}{c^2}\,, \qquad P = \tfrac{1}{2}\,\Phi^2\mu^2c^2 - V\,, \tag{58}$$

which are derivable from (8) and (9).

Since the preceding relations entail the variation rule

$$\delta V = \mu^2\, c^2\, \Phi\, \delta\Phi\,, \tag{59}$$

it can be seen that the full variation of the reformulated perfect fluid Lagrangian (50) will be given by

$$\delta\mathcal{L} = -(\pi_\nu\pi^\nu + \mu^2c^2)\Phi\delta\Phi - \hbar\Phi^2\pi^\nu\nabla_\nu(\delta\varphi)\,, \tag{60}$$

with the momentum covector $\pi_\nu$ as specified before by (47). Thus in this new formulation, instead of imposing (46) as a defining relation, we obtain it as a field equation from the requirement that the action integral should be invariant with respect to local variations of the auxiliary field $\Phi$, while as before the analogous requirement for variations of the phase variable $\varphi$ gives back the particle conservation law (42).

### 5.1. SPECIAL EQUATIONS OF STATE

It is to be remarked that whereas for a generic compressible fluid equation of state giving $\rho$ as a function of $n$, and hence giving $P$ as a function of $\mu$, the formula (59) will provide a corresponding function $V$ that provides a reformulated action function of the form (50) from which the required perfect fluid dynamical equations are obtainable by treating $\Phi$ and $\varphi$ as independent. There is however an exceptional case that works somewhat differently, namely that of the "stiff" Zel'dovich model – pertaining to the massless axion field in cosmology [1, 4] – which is characterised by

$$P = \rho c^2\,, \tag{61}$$

corresponding by (1) to sound propagation at the speed of light. This case is obtained from a primary equation of state of the form $\rho \propto n^2$ or equivalently (as observed at the beginning of Section 5) $P \propto \mu^2$, which gives $V = 0$ and $\mu \propto n$. This last relation means that the amplitude $\Phi$ will simply be a constant, and so will not act as an auxiliary variable in the usual way. In this case what one has to do to obtain the relevant "stiff" dynamical equations is simply to treat $\varphi$ as the only independent variable in the Lagrangian (50).

There is another important special limit for which it is not the new formulation but the original variational formulation (48) in terms of the pressure function that fails to work, namely the extreme case – which often useful as an approximation – of a "dust type" model characterised by $\rho \propto n$ for, which there is no pressure, i.e. $P = 0$. This low pressure limit model (the only model for which the naive identification (54) is exactly valid) is immediately obtainable within the framework of the new formulation (50) simply by taking $V \propto \Phi^2$.

The simplest non trivial potential energy function that can be used in the new formulation (50) is provided by the "radiation gas" model (61) characterised by

$$c_{\mathrm{I}}^2 = \frac{c^2}{3} \qquad \Rightarrow \qquad V \propto \Phi^4 \,, \tag{62}$$

which is obtainable from a primary equation of state of the form $\rho \propto n^{4/3}$. This is the model that is appropriate for representing the cosmological black body radiation (which is to a good approximation irrotational, though of course it is not a superfluid). It also applies to a degenerate Fermi gas of massless (or due to high compression effectively massless) non interacting particles, and for that reason was used as a first crude approximation in some of the pioneering studies of neutrons stars, whose intermediate layers are indeed believed to be superfluid. As will be explained below, this particular model is characterised by conformal invariance of the same kind as is familiar in the well known case of Maxwellian electromagnetism.

For more general purposes, as an approximation that can be usually be expected to be reasonably accurate within a limited density range, one can of course use the obvious generalisation of (62) to the standard form

$$V = \frac{m^2 c^2}{2}\Phi^2 + a^2\Phi^4 \,, \tag{63}$$

for suitably adjusted constants $m$ and $a$. For such a model the fluid mass density and pressure will be given according to (58) by

$$\rho = m^2\Phi^2 + \frac{3a^2}{c^2}\Phi^4 \,, \qquad P = a^2\Phi^4 \,, \tag{64}$$

from which it can be seen that the sound speed (the quantity one would probably want to use in practice to fix the appropriate value of $a$) will be given according to (1) by

$$c_{\rm I}^2 = \Big(\frac{3}{c^2} + \frac{m^2}{2a^2\Phi^2}\Big)^{-1} \tag{65}$$

## 6. Introduction of the gauge field

The treatment given in the preceding section is ideal for the description of a superfluid at a mesoscopic (i.e. intervortex) scale, but for the treatment of an ordinary perfect fluid with rotation, or for the treatment of a superfluid on a macroscopic scale (allowing for the averaged effect of a large number of vortices) more general models are required.

A first step towards the kind of generalisation that is needed is to formulate the current in terms of an antisymmetric Kalb Ramond type tensor field $B_{\mu\nu} = -B_{\nu\mu}$ whose exterior derivative

$$N_{\mu\nu\rho} = 3\nabla_{[\mu}B_{\nu\rho]}\,, \tag{66}$$

whose (Hodge type) dual

$$n^\mu = \frac{1}{3!}\varepsilon^{\mu\nu\rho\sigma}N_{\nu\rho\sigma} \tag{67}$$

is to be identified with the particle number current (42), which will evidently be invariant under the effect of Kalb Ramond gauge transformations $B_{\mu\nu} \mapsto B_{\mu\nu} + 2\nabla_{[\mu}\chi_{\nu]}$. It can then be seen that the consequent closure condition,

$$\nabla_{[\mu}N_{\nu\rho\sigma]} = 0\,, \tag{68}$$

is equivalent to the usual form (43) of the particle conservation law.

The idea now is to perform a Legendre type transformation $\mathcal{L} \mapsto \Lambda$ whereby the independent scalar field $\varphi$ of the preceding formulation based on $\mathcal{L}$ is replaced by the antisymmetric gauge tensor $B_{\mu\nu}$ in a new formulation of the same model in terms of a different dually related Lagrangian function $\Lambda$ which takes the form

$$\Lambda = -\frac{c^2}{12\Phi^2}\,N^{\mu\nu\rho}N_{\mu\nu\rho} - V\{\Phi\}\,. \tag{69}$$

In the preceding formulation based on $\mathcal{L}$ as given by (50) the irrotationality property was kinematically imposed in advance while the particle current conservation was obtained from the variational principle as a dynamical equation. However in the reformulated version based on the dual Lagrangian (69) it is the particle current conservation law that is obtained in advance as we have seen via the kinematic identity (68), while on the other hand the irrotationality condition (34), i.e.

$$\nabla_{[\mu}\pi_{\nu]} = 0\,, \tag{70}$$

is obtained directly from (69) in the equivalent dual form

$$\nabla_\nu(\Phi^{-2}N^{\nu\rho\sigma}) = 0\,. \tag{71}$$

from the requirement of invariance with respect to independent variations of the gauge 2-form $B_{\mu\nu}$.

The purpose of replacing the simple scalar field $\varphi$ by the tensorial field $B_{\mu\nu}$ is to enable the extension of the model to the general perfect fluid case, in which the particle conservation law (68) is retained, but the irrotationality condition (70) is abandoned. The way to do this [18] is to introduce a complete Lagrangian of the form

$$\mathcal{L} = \Lambda - \tfrac{1}{4}\varepsilon^{\mu\nu\rho\sigma} B_{\mu\nu} w_{\mu\nu}\,, \tag{72}$$

where, in analogy with what has already been done for the current 3-form, the vorticity 2-form is constructed from independent gauge fields in such a way that its conservation property, (23) is automatically ensured in advance as a kinematic identity. For this purpose the requisite independent gauge fields can be taken to be a pair of independent scalars, $\chi^\pm$ say, in terms of which an identically conserved vorticity flux will be given by

$$w_{\nu\sigma} = 2(\nabla_{[\nu}\chi^+)\nabla_{\sigma]}\chi^-\,. \tag{73}$$

When the Lagrangian (72) is substituted in the action integral (45), the requirement of invariance with respect to local variations of the (non-physical) dynamical gauge fields $B_{\mu\nu}$ and $\chi^\pm$, and of the (physical) auxiliary amplitude $\Phi$, can be seen to lead back to our original dynamical momentum transport equation (44). Thus (since the particle conservation law (43) has been imposed kinematically in advance) it provides the complete system of generic perfect fluid equations of motion as given by the standard stress momentum energy conservation law (40).

## 7. Macroscopic allowance for vortex quantisation

All we have done so far is to reformulate ordinary barotropic perfect fluid theory in such a way that the $w_{\mu\nu}$ comes in as an independent dynamical variable determined by the pair of scalar gauge fields $\chi^\pm$. The model describing simple zero temperature superfluidity is obtained within this framework simply by taking the scalars $\chi^\pm$ to be constants (e.g. zero) so as to get $w_{\mu\nu} = 0$.

For a macroscopic treatment of a bulk superfluid fibred by a dense congruence of discrete vortex defects, the need for an extra term in the action to allow for the extra tension and energy in the vortices was recognised long ago [22] in the laboratory context of Helium-4, and has more recently been taken into account in the Newtonian mechanical analysis of neutron star matter [23]. In an analogous manner, following earlier work [24] on the corresponding relativistic formulation needed for a more accurate treatment of neutron star matter, the more complete treatment [3] to be described here takes account of the averaged effect of quantised vortices aligned in an Abrikosov type lattice by summing over contributions of individual vortex cells as estimated [5] using the usual (very good) approximation in which the hexagonal cells are treated as if they were cylindrically symmetric. What this analysis suggests is that the averaged effect of such vortices can be represented rather well – provided any relative flow is highly subsonic – by a remarkably simple and mathematically elegant modification of the generic potential function introduced by (56): all that seems to be necessary is to make an adjustment of the very simple form

$$V\{\Phi\} \mapsto V\{\Phi\} + \Phi^2 \Upsilon\{w\}\,, \tag{74}$$

for some function $\Upsilon$ depending just on the scalar magnitude $w$ of the vorticity as given by (29). Furthermore, as was observed in analogous investigations in a non relativistic framework [22], the dependence on the vorticity magnitude is approximately linear, having the form

$$\Phi^2 \Upsilon = \lambda w\,, \tag{75}$$

in which

$$\lambda = \mathcal{K}\Phi^2\,, \qquad \Upsilon = \mathcal{K} w\,, \tag{76}$$

with a coefficient $\mathcal{K}$ that can be taken to be a constant of the order of the Planck value, $\mathcal{K} \approx \hbar$.

A remarkable consequence of the ansatz (74) in conjunction with the linearity postulate (76) is that, as will be described below, the extended model retains the noteworthy, though little known, conformal convariance property of the simple perfect fluid model.

A more specific quantitative estimate of the value of the coefficient in (76) is given by

$$\mathcal{K} = \frac{\hbar\,\hat{l}}{4}\,, \qquad \hat{l} = \ln\{\frac{\Delta^2}{\delta^2}\}\,, \tag{77}$$

where $\delta$ is an inner cut off length representing the microscopic vortex core radius, and $\Delta$ is a long range cut of length typically representing the mean intervortex separation distance.

For a very precise treatment one would of course need to allow for a weak logarithmic dependence of $\mathcal{K}$ on $w$ since $\Delta^2$ will be inversely proportional to $w$, but so long as $\Delta$ is very large compared with $\delta$, as will be the case in typical macroscopic applications, the effect of such a refinement will in practice be negligible. To obtain very high precision when the Mach (flow to sound speed) ratio is non-negligible [25] one might also have to allow for some sort of tensorial, not just scalar dependence on the vorticity tensor.

Assuming that a sufficiently accurate treatment is obtainable without the need to take account of tensorial vorticity dependence, it follows that the complete macroscopic superfluid Lagrangian will be given [3] by

$$\mathcal{L} = -\frac{c^2}{12\Phi^2} N^{\mu\nu\rho} N_{\mu\nu\rho} - V\{\Phi\} + \tfrac{1}{2}(\frac{\lambda}{c^2}\mathcal{E}_{\mu\nu} - B_{\mu\nu}) W^{\mu\nu} \,, \tag{78}$$

where $W^{\mu\nu}$ is the dual vorticity vector as defined by (30).

In order to obtain the equations of motion for this system, one needs to evaluate the corresponding variation which will be expressible by

$$\delta L = \pi_\nu \delta n^\nu - \tfrac{1}{2} w_{\mu\nu} \delta b^{\mu\nu} - \tfrac{1}{2}(b^{\mu\nu} + \lambda^{\mu\nu}) \delta w_{\mu\nu} + (\frac{c^2 n^2}{\Phi^3} - \frac{dV}{d\Phi} - \frac{2\lambda w}{\Phi}) \delta\Phi \tag{79}$$

using the notation

$$b^{\mu\nu} = \tfrac{1}{2} \varepsilon^{\mu\nu\rho\sigma} B_{\rho\sigma} \,, \qquad \lambda^{\mu\nu} = \frac{\lambda}{w} w^{\mu\nu} \,. \tag{80}$$

Requiring invariance of the corresponding integral with respect to $\delta B_{\mu\nu}$, or equivalently to $\delta b^{\mu\nu}$, just leads back again to the usual relation (22) specifying the vorticity 2-form $w_{\mu\nu}$ as the exterior derivative of the momentum 1-form $\pi_\nu$, so there will be a conserved helicity vector given by exactly the same formula (32) as before. The equation of state relation (57) is however modified by the addition of an extra term proportional to the vorticity: requiring invariance with respect to $\delta\Phi$ gives

$$\Phi^2 \mu^2 c^2 = \frac{n^2 c^2}{\Phi^2} = \Phi \frac{dV}{d\Phi} + 2\lambda w \,. \tag{81}$$

The final requirement is invariance with respect to variations of the gauge scalars determining the vorticity field according to (73), which is equivalent to the gauge independent requirement of invariance with respect to Lie transportation, $\delta w_{\mu\nu} = \vec{k}\mathcal{L} w_{\mu\nu}$, as given by (24) for an arbitrary displacement vector field $k^\mu$. The resulting dynamical equation is expressible in the form

$$\natural n^\mu w_{\mu\nu} = 0 \,, \tag{82}$$

which differs from its analogue in the perfect fluid limit only by the replacement of the conserved particle current $n^\mu$ by an "augmented" current $^{\natural}n^\mu$ that is also automatically conserved

$$\nabla_\mu {}^{\natural}n^\mu = 0\,, \tag{83}$$

whose specification is given by

$$^{\natural}n^\mu = n^\mu + \nabla_\nu \lambda^{\mu\nu}\,, \tag{84}$$

which is equivalent, by (80) , to the condition that it be obtainable from a corresponding "augmented" Kalb-Ramond gauge 2-form given by

$$^{\natural}B_{\mu\nu} = B_{\mu\nu} - \frac{\lambda}{c^2}\mathcal{E}_{\mu\nu}\,. \tag{85}$$

The geometrical interpretation of the modified dynamical equation (82) is facilitated by the observation that it is the same as the single vortex equation of motion (128) that will be derived below.

## 8. The conformal covariance property

It is well known that – when expressed in the usual way in terms of an electromagnetic gauge potential $A_\nu$ and charge current $j^\mu = en^\mu$, where $e$ is a charge coupling constant and $n^\mu$ is an automatically conserved current vector such as is obtainable from a corresponding closed 3-form $N_{\mu\nu\rho}$ according to the Hodge type duality formula (67) – the equations of ordinary Maxwellian electromagnetism are preserved by any conformal transformation of the form

$$g_{\mu\nu} \mapsto \tilde{g}_{\mu\nu} = \mathrm{e}^{2\phi} g_{\mu\nu}\,, \tag{86}$$

for an arbitrary scalar field $\phi$, on the understanding that the transformation affects neither the gauge 1-form, which obeys $A_\nu \mapsto A_\nu$, nor the closed current 3-form, which obeys $N_{\mu\nu\rho} \mapsto N_{\mu\nu\rho}$. This last condition means that, since (86) implies

$$\varepsilon^{\mu\nu\rho\sigma} \mapsto \tilde{\varepsilon}^{\mu\nu\rho\sigma} = \mathrm{e}^{-4\phi}\varepsilon^{\mu\nu\rho\sigma}\,, \tag{87}$$

the current vector itself will undergo a conformal transformation of the form

$$n^\mu \mapsto \tilde{n}^\mu = \mathrm{e}^{-4\phi} n^\mu\,. \tag{88}$$

In the framework of the Kalb-Ramond representation (66) this is evidently equivalent to the requirement of preservation of the gauge 2-form,

$$B_{\mu\nu} \mapsto \tilde{B}_{\mu\nu} = B_{\mu\nu}\,. \tag{89}$$

What is not so well known (since the utility of the auxiliary field $\Phi$, as correctly defined by (55), is not yet widely appreciated) is that the ordinary barotropic perfect fluid equations (43) and (44) are also preserved by such a conformal transformation, subject to the understanding that the auxiliary amplitude field undergoes a corresponding conformal transformation of the form

$$\Phi \mapsto \tilde{\Phi} = \mathrm{e}^{-\phi}\Phi\,, \tag{90}$$

which is what is required to ensure the preservation,

$$\pi_\nu \mapsto \tilde{\pi}_\nu = \pi_\nu\,, \tag{91}$$

of the momentum 1-form $\pi_\mu$ (whose role in the perfect fluid case is analogous to that of the gauge 1-form $A_\mu$ in the electromagnetic case).

The preservation of the form of the dynamical equations (43) and (44) is in general not quite sufficient for preservation of the complete system, because it is also necessary to satisfy the algebraical equation of state relation (57) specified by the function $V\{\Phi\}$ which governs the relation between the amplitude $\Phi$ and the 4-momentum magnitude $\mu = (-\pi^\nu\pi_\nu)^{1/2}$. However provided the conformal scalar $\phi$ is chosen to depend only the amplitude $\Phi$, making the later a function of new variable $\tilde{\Phi}$, then the system will be formally covariant in the sense that the new system will also behave as an ordinary barotropic fluid but with a modified equation of state function $\tilde{V}\{\tilde{\Phi}\}$ in place of the orginal potental $V\{\Phi\}$.

My original discussion of the perfect fluid case [18] envisaged a scenario involving gravitational coupling in the framework of a Brans-Dicke-Jordan type generalisation of Einstein's theory, with the dimensionless field $\phi = \ln\{\tilde{\Phi}/\Phi\}$ acting as a dilatonic coupling scalar. Conformal covariance of this scheme was found to require that the transformation law for $V$ should be given by an algebraic relation of the form $\tilde{V}/\tilde{\Phi}^4 = V/\Phi^4$. (For readers interested in this scenario I should warn that it is necessary to correct the final sentence, concerning a special case for which there is a transformation to a form in which "the dilatonic field is genuinely absent": the term "genuinely" should be replaced by "apparently", since although it disappears from the fluid sector of the Lagrangian it effectively turns up again in the gravitational sector instead.)

In the present discussion I wish to describe a kind of conformal covariance like that of the well known Maxwellian example, having nothing to do with any particular kind of gravitational coupling theory, whether it be that of Einstein or anyone else. In order to preserve the formal structure of the fluid system by itself (without involving anything to do with active gravitational coupling) it can be seen from (57) that since we shall have

$$\mu \mapsto \tilde{\mu} = \mathrm{e}^{-\phi}\mu\,, \tag{92}$$

it is necessary and sufficient that the effective potential should transform in such a way as to satisfy the differential condition

$$\tilde{\Phi}^{-3}\frac{d\tilde{V}}{d\tilde{\Phi}} = \Phi^{-3}\frac{dV}{d\Phi}\,. \tag{93}$$

This means, for example, that if the new system is to be characterised by a pressure free "dust" type equation of state of the form $\tilde{V} = \frac{1}{2}\tilde{m}^2c^2\tilde{\Phi}^2$ for some constant $\tilde{m}$ (which has the convenient feature that the flow trajectories will simply be geodesics with respect to the new metric $\tilde{g}_{\mu\nu}$) then the desired transformation giving $\tilde{\Phi}$ as a function of $\Phi$ will be obtainable immediately by taking $\tilde{m}^2c^2\tilde{\Phi}^{-2} = \Phi^{-3}dV/d\Phi$ except in the exceptional case for which the right hand side of (93) is constant.

Whereas a perfect fluid solution for a generic equation of state can thus be conformally transformed to a solution for any other generic equation of state, there is an exceptional case for which such a procedure fails, namely the case with a constant value for right hand side of (93), which is that of the "radiation gas" model for which the dependence on $\Phi$ of the effective potential $V$ is of the homogeneously quartic form (62). It is apparent that the particular form

$$P = \frac{\rho c^2}{3} \qquad \Rightarrow \qquad \mu \propto \Phi \tag{94}$$

of the equation of state relation for this particular model will automatically be preserved by (90) and (92), not just when the conformal scalar $\phi$ is chosen as a function of $\Phi$ but for any field $\phi$ whatsoever.

In summary, just as solutions of Maxwell's equation are well known to be mapped conformally onto other solutions of Maxwell's equations, solutions of the equations of the "radiation gas" model are analogously mapped conformally onto other solutions of the "radiation gas" model, while generically solutions of the barotropic perfect fluid equations for any equation of state can be conformally mapped onto solutions for another different equation of state. In view of the intrinsic isotropy of a perfect fluid these conformal properties are not very startling. However we conclude this subsection by the rather more surprising observation that these properties will still hold, subject to the same condition (93), when the intrinsic isotropy is violated by the inclusion of the term (75) that allows for the energy and tension in quantised vortices. Since we shall have

$$\mathcal{E}_{\mu\nu} \mapsto \tilde{\mathcal{E}}_{\mu\nu} = \mathrm{e}^{2\phi}\mathcal{E}_{\mu\nu}\,, \tag{95}$$

and

$$w \mapsto \tilde{w} = \mathrm{e}^{-2\phi}w\,, \tag{96}$$

it can be seen that in order to have invariance of ${}^{\sharp}B_{\mu\nu}$ as given by (85) and of the extra term $\lambda w$ in the action integral (remembering that the measure $d\mathcal{S}^{(4)}$ will transform proportionally to $e^{4\phi}$) we must have

$$\lambda \mapsto \tilde{\lambda} = e^{-2\phi}\lambda\,. \tag{97}$$

Rather remarkably, this condition just happens to be satisfied by the formula (76), but it would fail for a more complicated vorticity dependent term such as might be needed for an accurate treatment of relative flow at a speed comparable with that of sound.

## 9. The thin vortex string limit

Let us now consider the limiting case for which instead of being 4-dimensionally extended, the vorticity distribution is concentrated in the neighbourhood of some particular vorticity flux 2-surface $\mathcal{S}^{(2)}$ say, which might conveniently be characterised by zero values for the scalar gauge fields $\chi^{\pm}$. Such a 2-surface will be describable in terms of internal coordinates $\sigma^0$ and $\sigma^1$, which might conveniently be specified by the values on the worldsheet of the spacetime coordinates $t$ and $x^1$.

However the internal coordinates may be chosen, the worldsheet embedding $\{\sigma^0, \sigma^1\} \mapsto x^\mu = \bar{x}^\mu\{\sigma\}$ will induce (in the technical sense "pull back") a two surface metric with components $\gamma_{ab}$ on the worldsheet that will be given by

$$\gamma_{ab} = g_{\mu\nu}\bar{x}^\mu_{\ ,a}\bar{x}^\nu_{\ ,b}\,, \tag{98}$$

using a comma to denote partial differentiation with respect to the internal coordinates $\sigma^a$ $(a = 0, 1)$, and this induced metric in turn will specify the worldsheet measure

$$d\mathcal{S}^{(2)} = \frac{\|\gamma\|^{1/2}}{c}\, d\sigma^0\, d\sigma^1\,, \tag{99}$$

of the timelike 2-surface element spanned by the coordinate variations $d\sigma^0$ and $d\sigma^1$.

Let us start by supposing that the vorticity distribution is confined within a small but finite range specified by displacements $\delta\chi^+$, $\delta\chi^-$ of the comoving scalar fields, and then let us take the thin string limit as the size of these displacements tends to zero. In this limit it can be seen that the dual vorticity (30) will take the form of a two dimensional Dirac distribution that will be expressible in Dirac's notation as an integral over the 2-dimenssional world sheet $\mathcal{S}^{(2)}$ by the formula

$$W^{\mu\nu} = \frac{c}{\|g\|^{1/2}} \int \overline{W}^{\mu\nu}\delta^4[x^\mu - \bar{x}^\mu\{\sigma\}]\, d\mathcal{S}^{(2)}\,, \tag{100}$$

in which $\overline{W}^{\mu\nu}$ is a well behaved antisymmetric tensor on the worldsheet representing the concentrated 2 surface vorticity flux.

In the case of the continuous vorticity distribution considered in the preceding section, the vorticity conservation law (23) is expressible in its dual version in the form

$$\nabla_\nu W^{\mu\nu} = 0 \,. \tag{101}$$

When one goes over to the singular limit in which the distribution $W^{\mu\nu}$ is concentrated in the form (100) in the infinitesimal neighbourhood of a single flux worldsheet, the conservaton law (101) will translate into a corresponding condition on the regular worldsheet supported 2-surface vorticity flux tensor $\overline{W}^{\mu\nu}$. Using the abbreviation

$$\overline{\nabla}_\mu = \eta^\nu{}_\mu \nabla_\nu \tag{102}$$

for the operator of worldsheet tangential covariant differentiation, where $\eta^\nu{}_\nu$ is the "first" fundamental tensor [26, 27] of the worldsheet (i.e. the rank-2 projection operator whose contraction with an abritrary vector at a point on the worldsheet projects it onto its tangential part within the worldsheet) which will be given by

$$\eta^\nu{}_\nu = c^{-2} \mathcal{E}^{\nu\rho} \mathcal{E}_{\rho\mu} \,, \tag{103}$$

the condition expressing the conservation of the vorticity flux distribution in the worldsheet will expressible simply as

$$\overline{\nabla}_\nu \overline{W}^{\mu\nu} = 0 \,. \tag{104}$$

Since the canonically normalised worldsheet tangential bivector $\mathcal{E}^{\mu\nu}$ automatically satifies a conservation condition of the same form

$$\overline{\nabla}_\nu \mathcal{E}^{\mu\nu} = 0 \tag{105}$$

as a kinematic identity, it follows that (104) is interpretable as implying that the worldsheet supported surface vorticity flux tensor $\overline{W}^{\mu\nu}$ must have the form

$$\overline{W}^{\mu\nu} = \kappa \mathcal{E}^{\mu\nu} \,, \tag{106}$$

where $\kappa$ is a constant on the worldsheet, i.e. $\overline{\nabla}_\nu \kappa = 0$.

It can be seen that the constant $\kappa$ defined by (106) will be interpretable as the value in the thin string limit of the 2-surface integral of the vorticity across any small spacelike section through the world tube. By Stoke's theorem, using the defining relation (22), this will be equal to the value of the Jacobi action around the boundary circuit of the section, which can be taken to be any closed curve encircling the vortex string. Thus for any such surrounding circuit we shall have

$$\kappa = \oint dS = \oint \pi_\nu dx^\nu = 2\pi\hbar\nu\,, \tag{107}$$

where $\nu$ is an integer representing the winding number of the phase $\varphi$, i.e. $\nu$ represents the number of individual quantised vortices carrying the flux under consideration. This means that if we are considering a string representing just a single such quantised vortex, with orientation chosen so that $\nu = +1$ we shall simply have

$$\kappa = 2\pi\hbar\,. \tag{108}$$

Substituting (78) in formula (45) it can be seen that one obtains the total action integral in the form

$$\mathcal{I} = \mathcal{I}_{(4)} + \mathcal{I}_{(2)}\,, \tag{109}$$

in which the first part is just the 4-dimensional contribution from the irrotational superfluid outside the vortex, which by (69) is

$$\mathcal{I}_{(4)} = \int \mathcal{L}_{(4)}\, d\mathcal{S}^{(4)}\,, \qquad \mathcal{L}_{(4)} = -\frac{1}{12\Phi^2} N^{\mu\nu\rho} N_{\mu\nu\rho} - V\{\Phi\}\,, \tag{110}$$

while by (100) and (106) the second part reduces to a 2-dimensional string worldsheet integral given by

$$\mathcal{I}_{(2)} = \int \mathcal{L}_{(2)}\, d\mathcal{S}^{(2)}\,, \qquad \mathcal{L}_{(2)} = -\frac{\pi\,\hat{l}}{2}\,\hbar^2\Phi^2 - \pi\hbar B_{\mu\nu}\mathcal{E}^{\mu\nu}\,. \tag{111}$$

The first term in (111) is like the corresponding action density for a simple Nambu Goto (internally structureless) string, as characterised by a string worldsheet stress-energy density tensor $\overline{T}^\mu{}_\nu$ that is proportional to the fundamental tangential projection tensor $\eta^\mu{}_\nu$ of the worldsheet as defined by (103), except that for an ordinary Nambu-Goto model the effective string tension $\mathcal{T}$ is fixed. In the special limit case of the "stiff" fluid model (61) which describes the massless axion case, it is well known [1] that the vortices will have an effective tension $\mathcal{T}$ will also be fixed, but for a generic superfluid model one obtains

$$\overline{T}^\mu{}_\nu = -\mathcal{T}\eta^\mu{}_\nu\,, \qquad \mathcal{T} = \frac{\pi\hat{l}}{2}\,\hbar^2\Phi^2\,, \tag{112}$$

which shows that the effective tension $\mathcal{T}$ will vary proportionally to the square of the auxiliary field amplitude, which according to (55) will be given on shell by

$$\Phi^2 = \frac{n}{\mu}\,. \tag{113}$$

The second term in (111) has the form of what is known in the context of superstring theory as a "Wess-Zumino" coupling. When such a term turned up in the "stiff" massless axion case (61) it was at first interpreted [1] as a "very unusual interaction". However it soon came to be recognised [4] as a manifestation of an ordinary aerodynamic "lift" type force (arising from the Magnus effect) of the kind first evaluated for an aerofoil in the long thin (i.e. string type) limit by the Russian theorist Joukowski in the pioneering days of subsonic flight a hundred years ago.

## 10. The vortex string dynamical equations

To work out the effects of the pair of terms in (111), one must obtain the condition for the string type action $\mathcal{I}^{(2)}$ to be invariant with respect to an infinitesimal displacement of the worldsheet generated by an arbitrary vector field $k^\mu$ say. Any such displacement will give rise to corresponding "Lagrangian" variations $\delta\Phi$, $\delta B_{\mu\nu}$, $\delta g_{\mu\nu}$ (meaning variations as measured with respect to local coordinates that are comoving with the displacement) of the relevant background fields. Such "Lagrangian" variations will be given by corresponding Lie differentiation formulae of the kind introduced above for the respective cases of a scalar (18), a closed antisymmetric covariant field (24), and last but in importance not least, the metric itself (19). Thus in the trivial case of the scalar $\Phi$ we shall simply have

$$\delta\Phi = k^\nu \nabla_\nu \Phi \,, \qquad (114)$$

while in the case of the metric we have the familiar expression

$$\delta g_{\mu\nu} = 2\nabla_{(\mu} k_{\nu)} \,, \qquad (115)$$

(which would of course vanish if $k^\mu$ were not arbitrary but restricted to be a Killing vector generating a background spacetime isometry, i.e. in flatspace if $k^\mu$ were restricted to be a generator of some Poincaré combination of translations and Lorentz transformations). Since the antisymmetric field $B_{\mu\nu}$ is not closed (i.e. since its exterior derivative, the current 3-form $N_{\mu\nu\rho}$ is non vavishing) the formula for its Lie derivative is not quite so simple as that of its analogue (24) for the vorticity but contains an extra term: the relevant formula is

$$\delta B_{\mu\nu} = k^\mu N_{\mu\nu\rho} - 2\nabla_{[\mu}(B_{\nu]\rho} k^\rho) \,. \qquad (116)$$

The corresponding variation of the string type action integral specified by (111) takes the standard form

$$\delta\mathcal{I}_{(2)} = \int (\tfrac{1}{2}\overline{T}^{\nu\mu}\delta g_{\mu\nu} + \tfrac{1}{2}\overline{W}^{\nu\mu}\delta B_{\mu\nu} + \overline{\mathcal{F}}\delta\Phi)\, d\mathcal{S}^{(2)} \,, \qquad (117)$$

in which it can be seen, taking account of the variation of the surface element (99) due to the induced variation

$$\delta\|\gamma\|^{1/2} = \tfrac{1}{2}\|\gamma\|\eta^{\mu\nu}\delta g_{\mu\nu}\,, \tag{118}$$

that the relevant surface stress energy momentum tensor $\overline{T}^{\mu\nu}$ can be read out in the form (112), and that the relevant surface vorticity flux bi-vector will evidently be as given by (106) and (108), while finally the dynamical dual of the amplitude $\Phi$ can be read out simply as

$$\overline{\mathcal{F}} = -\pi\,\hat{l}\,\hbar^2\,\Phi\,. \tag{119}$$

Since we are only concerned with local variations, there will be no boundary contribution when we perform the usual operation of integration by parts using Green's theorem so as to eliminate gradients of the arbitrary dispacement vector $k^\mu$. The result is thus obtained in the standard form

$$\delta\mathcal{I}_{(2)} = \int k^\mu(\overline{f}_\mu - \overline{\nabla}_\nu\overline{T}^\nu{}_\mu)\,d\mathcal{S}^{(2)} \tag{120}$$

in which, after taking account of the surface vorticity flux conservation law (104), the effective surface force density acting in the ensuing dynamical equation,

$$\overline{\nabla}_\nu\overline{T}^\nu{}_\mu = \overline{f}_\mu\,, \tag{121}$$

can be read out as

$$\overline{f}_\mu = \tfrac{1}{2}N_{\mu\nu\rho}\overline{W}^{\nu\rho} + \overline{\mathcal{F}}\nabla_\nu\Phi\,. \tag{122}$$

In a model of this particular type (as a consequence of the fact that the vortex string has no internal degrees of freedom of its own) the tangentially projected part of the force balance equation (121), i.e. the part that is obtained by contracting it with the fundamental tensor $\eta^\mu{}_\rho$ given by (103), will automatically be satisfied as a kinematic identity. The only dynamical information in (121) is the part obtained by contracting it with the complement of the tangential projection tensor $\eta^\mu{}_\rho$, namely the orthogonal projection tensor defined by

$$\perp^\mu_\rho = g^\mu{}_\rho - \eta^\mu{}_\rho\,. \tag{123}$$

It can be seen that the remaining, purely orthogonal part of the left hand side of (121) can be evaluated in terms of the extrinsic geometry of the imbedding, without knowledge of the gradient of $\overline{T}^{\mu\nu}$, using the fact that, since the surface stress momentum energy tensor $\overline{T}^{\mu\nu}$ is purely tangential, i.e. since its contraction with $\perp^\mu_\rho$ vanishes, an integration by parts leads to the identity

$$\perp^\rho_\mu\,\overline{\nabla}_\nu\overline{T}^{\nu\mu} = \overline{T}^{\mu\nu}K_{\mu\nu}{}^\rho\,, \tag{124}$$

where the $K_{\mu\nu}{}^{\rho}$ is the second fundamental tensor as defined [26] in terms of tangential differentiation of the first fundamental tensor $\eta^{\rho}{}_{\sigma}$ by

$$K_{\mu\nu}{}^{\rho} = \eta^{\sigma}{}_{\nu}\overline{\nabla}_{\mu}\eta^{\rho}{}_{\sigma} \,. \tag{125}$$

One thus obtains an expression of the standard form [27]

$$\overline{T}^{\mu\nu}K_{\mu\nu}{}^{\rho} = \perp^{\rho\mu}\overline{f}_{\mu}\,, \tag{126}$$

for the extrinsic part of the dynamical equation (121), i.e. the part that governs the evolution of the worldsheet itself, which is the only part there is in the present case.

Due to the Nambu-Goto like form (112) obtained for $\overline{T}^{\mu\nu}$ in the simple vortex model under consideration here, it will not be necessary to work out the complete second fundamental tensor $K_{\mu\nu}{}^{\rho}$ in the present case, but only its trace, the curvature vector

$$K^{\rho} = K^{\nu}{}_{\nu}{}^{\rho} = \overline{\nabla}_{\nu}\eta^{\nu\rho}\,, \tag{127}$$

in terms of which it can be seen that the equation of motion (126) will reduce to the form

$$\mathcal{T}K_{\rho} = \perp^{\nu}{}_{\rho}\,\nabla_{\nu}\mathcal{T} + \pi\hbar\mathcal{E}^{\mu\nu}N_{\mu\nu\rho}\,, \tag{128}$$

with the effective tension $\mathcal{T}$ given by (112), according to which it is proportional to the square of the auxiliary field $\Phi$ in the ambient superfluid background. The first term on the right of (128), allowing for the effect of any non-uniformity of this field, was not needed in the special limit of the "stiff" fluid model (61) characterised by a fixed value of $\Phi$, which describes the massless axion case to which earlier studies of relativistic vortex string dynamics [1, 4] were restricted.

## 11. Vorton equilibrium in stationary background

As a simple exercise, let us apply the foregoing theory to the problem of a vorton, i.e. an equilibrium state of a vortex ring in a uniform background. As in the well known non relativistic version of this problem, relative motion is needed to provide the Joukowsky type force (due to the Magnus effect) that supports the string against its own tension. A superfluid vorton differs from local cosmic vorton [28, 27] in that the latter is supported not by a Magnus effect, but by the centrifugal effect of circulating current.

By definition, an equilibrium state is invariant with respect to the action of a time displacement vector, $k^{\mu}$ say, which, to preserve the background metric must satisfy the Killing equation given according to (19) by $2\nabla_{(\mu}k_{\nu)} = 0$. This holds as a triviality for an ordinary flat space time displacement vector, which satisfies

$$\nabla_\mu k_\nu = 0 \,. \tag{129}$$

Invariance requires that the vortex worldsheet be tangent to this vector, whose components in a corresponding Minkowski coordinate system will be given by $k^\mu \leftrightarrow \{1,\, 0,\, 0,\, 0\}$ with normalisation $k^\mu k_\mu = -c^2$.

The condition that the superfluid background is not just stationary but uniform with velocity $v$ relative to such a frame means that for suitably aligned space axes its 4-velocity will be given by $u^\mu \leftrightarrow \gamma\{1,\, v,\, 0,\, 0\}$ with $\gamma = (1 - v^2/c^2)^{-1/2}$. The assumption that the ambient fluid is uniform means that the gradient term will drop out of the dynamical equation (128), with the consequence that the background number density $n$ will also cancel out, leaving an equilibrium condition that is expressible just in terms of the 4-momentum covector $\pi_\nu = \mu u_\nu$ in the form

$$\hbar \hat{l}\, K_\rho = 2\mathcal{E}^{\mu\nu}\varepsilon_{\mu\nu\rho\sigma}\pi^\sigma \,. \tag{130}$$

In such a stationary case it is easy to evaluate the tangent bivector $\mathcal{E}^{\mu\nu}$ and the worldsheet curvature vector $K^\mu$ in terms of the unit tangent vector, $e^\mu$ say, that is orthogonal to $k^\mu$ within the worldsheet (which to be uniquely specified requires a choice of orientation). In terms of this vector, as characterised by $e^\mu k_\mu = 0$, $e^\mu e_\mu = 1$, we shall simply have

$$\mathcal{E}^{\mu\nu} = 2k^{[\mu}e^{\nu]} \,, \tag{131}$$

and (independently of the choice of orientation) the corresponding fundamental tensor will be given by

$$\eta^{\mu\nu} = -\frac{1}{c^2}u^\mu u^\nu + e^\mu e^\nu \,. \tag{132}$$

It can be seen from the defining relations (102) and (127) that when (129) is satisfied we shall simply have

$$K^\mu = e^\nu \nabla_\nu e^\mu \,. \tag{133}$$

The vorton equilibrium states in which we are interested will be symmetric about an axis aligned with the relative flow, which we take to be the $x_1$ axis of our Minkowski coordinate system. The worldsheet will be circular, with fixed radius $r$ say, meaning that any point on it will be located relative to the axis by a radial displacment vector that can be parametrised in terms of the corresponding axial angle $\theta$ in the form $r^\mu \leftrightarrow r\{0,\, 0,\, ,\cos\theta,\, \sin\theta\}$ with $r^\mu r_\mu = r^2$. The corresponding unit tangent vector will be given by

$e^\mu \leftrightarrow \{0, 0, , -\sin\theta, \cos\theta\}$ and thus the corresponding curvature vector is easily seen to be given by $K^\rho = r^{-2}r^\mu \leftrightarrow r^{-1}\{0, 0, , \cos\theta, \sin\theta\}$.

¿From this familiar result, namely that the curvature vector $K^\rho$ of a circle of radius $r$ is inwardly directed with magnitude $r^{-1}$, it can be seen that the solution to the equilibrium condition (130) will be obtained for a vorton radius given by

$$r = \frac{\hbar\,\hat{l}}{4\mu v\gamma}\,, \tag{134}$$

in which it is to be recalled that, according to (77), $\hat{l}$ is an order of unity factor that can be estimated as the (natural) logarithm of the ratio of the vorton area to the sectional area of the vortex core.

The vagueness of the prescription for the $\hat{l}$ symptomises an inherent limitation of any attempt (whether in a relativistic or in a Newtonian framework) to treat a "global" vortex as if it were a locally confined string type phenomenon, despite the "infra-red divergence" of its energy and tension, which can only be made finite by a long range cut off. This problem does not arise for the vortex defects in an ordinary metallic type II superconductor, in which the local gauge coupling to the electromagnetic field ensures an exponential fall off ensuring convergence without any need for a cut off. In the cosmic string context, there are locally gauged examples for which a string type description is highly accurate there are also examples of non-local vortex defects, such as the axionic case [1] (and cases involving electromagnetic [29] or gravitational [30] coupling) for which the thin string limit description is less satisfactory. If the vortex core radius $\delta$ is relatively small, the problem of sensitivity to the long range cut off $\Delta$ is mitigated by the fact that the dependence is only logarithmic. Thus a formula such as (134) (or its well known non-relativistic anologue, in which the $\gamma$ factor is omitted, and $\mu$ is replaced by a fixed mass $m$) can be usefully applied to macroscopic vorton configurations, but becomes quantitatively unreliable for describing the microscopic vortons, known as "rotons", that are important in the analysis of thermal excitations.

Whereas the problem of specification of the cut off can limit the utility of the string type description (111) for application to the dynamics of individual vortices, this reservation does not apply to the macroscopically averaged description for an Abrikosov type lattice of aligned vortices as given by (78), which is on a much sounder footing: in this case the relevant cut off is unambiguously determined by the lattice spacing. Although its accuracy is more questionable, the single string picture provides useful insight into the intepretation of the – at first sight rather mysterious – dynamical equation (82) for the more robust macroscopically averaged model, which is in fact formally identical to the more intuitively interpretable string dynamical equations (128).

## References


1. Vilenkin, A. and Vachaspati, T. (1987) Radiation of massless Goldstone bosons from cosmic strings, *Phys. Rev.* **35**, pp. 1138-1140.
2. Sauls, J.A. (1989) Superfluidity in the interior of neutron stars, in *Timing Neutron Stars* (NATO ASI **C262**, Izmir 1988) ed. Ogelman, H. and van den Heuvel, E.J.P., Kluwer, Dordrecht, pp. 457–502.
3. Carter, B. and Langlois, D. (1995) Kalb-Ramond coupled Vortex fibration model for relativistic Superfluid dynamics, *Nuclear Physics* **B454**, pp. 402–424. [hep-th/9611082]
4. Davis, R.L. and Shellard, E.P.S. (1989) Global strings and superfluid vortices, *Phys. Rev. Letters* **63**, pp. 2021–2024.
5. Carter, B. and Langlois, D. (1995) Explicit Relativistic Vortex solutions for Cool two-constituent Superfluid, *Phys. Rev.* **D52**, pp. 4640–4651. [hep-th 9507059]
6. Carter, B., and Khalatnikov, I.M. (1992) Momentum, Vorticity, and Helicity in Covariant Superfluid Dynamics, *Ann. Phys.* **219**, pp. 243–265.
7. Carter, B. and Langlois, D. (1995) The Equation of state for Cool relativistic two-constituent Superfluid dynamics, *Phys. Rev.* **D51**, pp. 5855–5864.[hep-th/9507058]
8. Carter, B. and Khalatnikov, I.M. (1994) Canonically covariant formulation of Landau's Newtonian superfluid dynamics, *Rev. Math. Phys.* **6**, pp. 277–304.
9. Carter, B. (1989) Covariant Theory of Conductivity in Ideal Fluid or Solid Media, in *Relativistic Fluid Dynamics (C.I.M.E., Noto, May 1987)* ed. Anile, A.M. and Choquet-Bruhat, Y., Lecture Notes in Mathematics **1385**, Springer–Verlag, Heidelberg, pp. 1–64.
10. Carter, B. and Langlois, D. (1998) Relativistic models for superconducting superfluid mixtures, *Nuclear Phys.* **B531**, pp. 478–504. [gr-qc/9806024]
11. Carter, B. (1979) Perfect Fluid and Magnetic Field Conservation Laws in the Theory of Black Hole Accretion Rings, in *Active Galactic Nuclei*, ed. Hazard C. and Mitton, S., Cambridge University Press, pp. 273–300.
12. Carter, B. (1973) Black Hole Equilibrium States, in *Black Holes* (proc. 1972 Les Houches Summer School), ed. DeWitt, B and C. Gordon and Breach, New York, pp. 57–210.
13. Stachel, J. (1980) Thickenning the string: I: The string perfect dust, *Phys. Rev.* **D21**, pp. 2171–2184.
14. Lichnerowicz, A. (1967) *Relativistic Hydrodynamics and Magnetohydrodynamics*, Benjamin, New York.
15. Bonazzola, S., Gourgoulhon, E., Marck, J.A. (1999) Numerical models of irrotational binary neutron stars in general relativity, *Phys. Rev. Lett.* **82**, pp. 892–895. [gr-qc/9810072]
16. Taub, A.H. (1954) General Relativistic Variational Principle for Perfect Fluids, *Phys. Rev.* **94**, pp. 1469–1470.
17. Schutz, B. (1970) Perfect Fluids in General Relativity: Velocity Potentials and a Variational Principle, *Phys. Rev.* **D2**, pp. 2762–2773.
18. Carter, B. (1994) Axionic Vorticity Variational formulation for Relativistic Perfect Fluids, *Class. Quantum Grav.* **11**, pp. 2013–2030.
19. Davis, R.L. (1990) Ginzburg-Landau from field theory, *Mod. Phys. Letters* **A6**, pp. 73–81.
20. Ben-Ya'acov, U. (1991) Strings, Superfluid Vortices, and Relativity, *Phys. Rev.* **D44**, pp. 2452–2458.
21. Ben-Ya'acov, U. (1992) Unified Dynamics of Quantum Vortices, *Nucl. Phys.* **B382**, pp. 597–615.
22. Bekarevich, I.L. and Khalatnikov, I.M. (1961) Phenomenological derivation of the equations of vortex motion in HeII, *Sov. Phys. J.E.T.P.* **13**, pp. 643–646.
23. Mendell, G. (1991) Superfluid hydrodynamics in rotating neutron stars. I Nondissipative equations. *Astroph. J* **380**, pp. 515–529.

24. Lebedev, V.V. and Khalatnikov, I.M. (1982) Relativistic Hydrodynamics of a Superfluid, *Sov. Phys. J.E.T.P.* **56**, pp. 923–930.
25. Carter, B., Langlois, D. and Priou, D. (1997) Moving vortex in relativistic irrotational perfect fluid or superfluid, *Phys. Rev.* **D56**, pp. 6673–6683. [astro-ph/9709090]
26. Carter, B. (1989) Stability and Characteristic Propagation Speeds in Superconducting Cosmic and other String Models, *Phys. Letters.* **B 228**, pp. 466–470.
27. Carter, B. (1995) Dynamics of Cosmic Strings and other Brane models, in *Formation and Interactions of Topological Defects* (NATO ASI **B349**, Newton Institute, Cambridge 1994) ed. Brandenberger R. and Davis, A.C., Plenum, New York, pp. 303–348. [hep-th/9611054]
28. Davis, R.L. and Shellard, E.P.S. (1998) The physics of Vortex Superconductivity, 2, *Phys. Lett* **B209**, 485–490.
29. Carter, B. (1997) Electromagnetic self interaction in strings, *Phys. Lett.* **B404**, pp. 246–252. [hep-th/9704210]
30. Carter, B. and Battye, R.A. (1998) Non divergence of gravitational self-interactions for Goto-Nambu strings, *Phys. Lett.* **B430**, pp. 49-53 . [hep-th/9803012]

# NUCLEATION AND PINNING OF VORTICES IN $^4$HE

ÉRIC VAROQUAUX
*Laboratoire de Physique des Solides*
*Université Paris-Sud, Bât. 510, 91405 Orsay*
*and*
*Service de Physique de l'État Condensé*
*Centre d'Études de Saclay, 91191 Gif-sur-Yvette Cedex (France)*

A large body of information about superfluid $^4$He vortices has been inferred from phase slippage experiments. In particular, it is now recognised that $^4$He vortices can be nucleated both by thermal fluctuations and through quantum tunnelling. The nucleation event can be accounted for by a simple model involving vortex half-rings.

## 1. Introduction

The existence of vortices in superfluid $^4$He was postulated first by Onsager [1], and later, independently by Feynman [2] who also suggested that they could account for the phenomenon of critical velocity in superfluid flows. A large body of work has been devoted to their studies as they constitute the archetypal defect of fluid dynamics (see, *e.g.*, [3, 4, 5, 6, 7, 8, 9]).

Vortices form spontaneously at the superfluid transition temperature as discussed at length in this Volume, and also when the local velocity of the superfluid with respect to its container or to an object moving through it exceeds a critical limit. The understanding of critical superflows through a micro-orifice remains incomplete but significant progress, reported in part below, has been made in recent years.

Some of the basic properties of vortices are reviewed in Sec. 1. Phase slippage is outlined in Sec. 2. The case for vortex nucleation is presented in Sec. 3. A simple nucleation model involving vortex half-rings is described in Sec. 4. The use of $^3$He quasiparticle impurities as local probes of the nucleation velocity is dealt with in Sec. 5. Wall contamination, which brings about collapses of the superflow and vortex pinning, is briefly touched upon in Sec. 6.



*Y.M. Bunkov and H. Godfrin (eds.),*
*Topological Defects and the Non-Equilibrium Dynamics of Symmetry Breaking Phase Transitions*, 303–323.

## 2. Vortices

At milliKelvin temperatures, which is the temperature range mainly considered here, vortices in superfluid $^4$He experience negligible friction from the fraction of normal fluid which is vanishingly small. If they deform only a little and slowly, they constitute stable fluid eddies with a microscopic core of radius $a_0$ (a few Å in $^4$He). The circulation of the velocity $\oint \mathbf{v}_s.\mathbf{dl}$ on any path around the core is constant, a consequence of the Kelvin-Helmholtz theorem [10], and furthermore, is quantised to $\kappa_4 = 2\pi\hbar/m_4$ because the phase $\varphi$ of the wavefunction can change only by multiples of $2\pi$ along any contour in the superfluid. The flow velocity induced by a straight vortex filament at a distance $r$ taken in the plane perpendicular to its core is then expressed by

$$v = \frac{\kappa_4}{2\pi r} , \tag{1}$$

provided that $r$ is larger than $a_0$ and smaller than a far distance cut-off $r_m$ which depends on the shape of the vortex, or that of the container, or else, the presence of other vortices [11]. For $r \geq r_m$, the flow velocity induced by the vortex falls off with distance faster than $1/r$ and rapidly becomes negligible. For $r \lesssim a_0$, the detailed structure of the core becomes important [3, 11, 12].

The kinetic energy associated with the flow around the core, obtained by integration of $\int \frac{1}{2}\rho_s v_s^2 d^3x$ where $\rho_s$ is the superfluid density, is expressed for a unit length of vortex by:

$$\epsilon_v = \frac{\rho_s \kappa_4^2}{4\pi} \ln\left(\frac{r_m}{a_0}\right) . \tag{2}$$

A vortex moves in a compressible, inviscid fluid according to the Kelvin-Helmholtz theorem. This theorem states that vorticity is conserved when moving with the fluid particles. That is, any vortex element moves with the local fluid velocity. Pure superfluid $^4$He close to $T = 0$ behaves as such an ideal fluid, at least for moderate superfluid velocity gradients. The velocity of the core of the vortex at point $\mathbf{r}$ is the sum of the velocity of the externally applied potential flow $\mathbf{v}_s$ at $\mathbf{r}$ and of the velocity $\mathbf{v}_v$ induced at $\mathbf{r}$ by the other parts of the vortex.

For a straight vortex filament, the velocity induced by the vortex onto itself is zero: the filament simply floats along with $\mathbf{v}_s$. For a circular ring of radius $R$, the vortex self-velocity lies along the axis and is given by:

$$v_v = \frac{\kappa_4}{4\pi R}\left[\ln\left(\frac{8R}{a_0}\right) - \frac{1}{4}\right] . \tag{3}$$

The ring moves in the laboratory frame without change of shape with a velocity $\mathbf{v}_s + \mathbf{v}_v$.

For an arbitrarily curved vortex, the self-velocity of each curve element is approximated by eq. (3), $R$ being replaced by the local radius of curvature. The validity of this approximation has been discussed, among others, by Schwarz [13, 14] who has used it in extensive numerical simulations of 3D vortex motion. In particular, the problem of single [14, 15] or multiple vortex [16] pinning on hemispherical asperities has been addressed within this framework. To achieve a stable configuration, the vortex has to take on a shape which is such that the local radius of curvature at each of its points results in a self-velocity which opposes exactly the local value of the applied flow. This dynamic equilibrium is what is meant here by pinning. Vortex pinning plays an important role in superfluids, bulk $^4$He as discussed here, films [17], $^3$He [18, 19, 20], neutron stars [21], ...

The eigenmodes of a straight isolated vortex, the Kelvin modes, are helical waves with a dispersion relation expressed by [6, 11, 22]

$$\omega^{\pm} = \frac{\kappa_4}{\pi\, a_0^2}\left[1 \pm \left\{1 + ka_0\left[\frac{\mathrm{K}_0(ka_0)}{\mathrm{K}_1(ka_0)}\right]\right\}^{\frac{1}{2}}\right], \tag{4}$$

where $\mathrm{K}_0$ and $\mathrm{K}_1$ are the modified Bessel functions of zeroth and first orders. In the short wavelength limit, $k^{-1} \to 0$, the high frequency mode becomes:

$$\omega^{+} = \frac{\kappa_4}{\pi a_0^2}\,. \tag{5}$$

Such a frequency is that of the cycloïdal motion taken by a long hollow cylinder impulsively pulled sideways in an inviscid fluid [6]. The cylinder plays the role of the core of the vortex, assumed to be hollow and with radius $a_0$. By analogy with the 2D-motion of point charges subjected to a magnetic field [23], this frequency is sometimes called the “cyclotron” frequency. It sets the shortest time scale on which vortices can be expected to respond.

The displaced mass per unit length of such a cylinder is $\rho\pi a_0^2$, a standard result of classical fluid dynamics [24]: this quantity stands for the mass of the vortex. A large correction to this mass has been pointed out by Ji-Min Duan [25, 26]. Owing to the compressibility of the fluid, the effective radius of the equivalent displaced cylinder is multiplied by a factor $\ln(r_m/a_0)$. A last contribution to the vortex mass in impure $^4$He comes from the mass of the $^3$He quasiparticles which condense onto the core at low temperature.

Fluctuations, which may be thermally activated, or may correspond to the zero point motion of the vortex [22], will be seen to be important for nucleation and unpinning processes. For a nearly straight filament, the fluctuating modes are the Kelvin modes given by eq. (4). In the long wavelength limit, the lower frequency branch is given by:

$$\omega^{-} = -\frac{\kappa_4}{\pi} k^2 \left[ \ln\left(\frac{1}{k\, a_0}\right) + 0.116 \right] . \tag{6}$$

## 3. Phase slippage

We now give a brief introduction to experiments designed to measure the superflow anomalies linked to the variations of the phase of the superfluid order parameter such as Josephson effects or phase slippage due to vortex motion. These experiments are conducted in a flexible-wall Helmholtz resonator in which one of the resonant vents is a microscopic aperture. The use of these resonators in superfluids was pioneered by Hess [27, 28], and later used by Zimmermann and coworkers [29], at Orsay-Saclay [30], at Berkeley [31, 32, 33] and recently in Tübingen. They can be equipped with either one micro-aperture, or two orifices, the micro-aperture (the weak link) and a larger, longer channel which, among other things, makes it easier to fill and cool the resonator when the liquid is still warm and in the normal (viscous) state.

These devices detect the anomalies which take place in the superfluid flow through the microscopic aperture, namely, the departure from ideal fluid mechanics when the flow velocity exceeds a certain critical value in the micro-aperture. Their asset lies in the fact that kinetic energy is almost entirely stored in the fluid motion: the flexible diaphragm moves very little and contributes a negligible amount to the inertia of the complete system.

When, under the action of the external drive, the superfluid flow through the micro-aperture reaches the critical velocity, discrete dissipation events are observed in a two-hole resonator, as shown in fig.1. These events are interpreted as slips of the phase difference through the weak link by $2\pi$ [30, 34] on grounds of reproducibility and because they obey the Josephson *ac*-relation:

$$\hbar \frac{\partial \varphi}{\partial t} = -\mu , \tag{7}$$

$\mu$ being the chemical potential.

Also, as is hardly necessary to point out, the occurrence of such slips fulfils theoretical expectations [35] (for a very accessible review, see [36]). In brief, discrete dissipation events are quantised, being single or multiple amounts of energy given by

$$\Delta E = \kappa_4 J_c, \tag{8}$$

$J_c$ being the critical mass flow rate. Eq.(8) is derived from the Josephson *ac*-relation, eq.(7), and from the drop of the superflow velocity which corresponds to a change of the phase difference through the micro-aperture of $2\pi$. Eq.(8) is well verified experimentally [30, 32, 34, 37].

The knowledge of the slip mechanism, *i.e.* of the process of formation of the vortex and of its subsequent trajectory, leads to a more detailed derivation of the result expressed by eq.(8), in particular by showing explicitly how the change of the phase difference through the aperture changes by exactly $2\pi$. This mechanism is described in refs. [36, 37, 38, 39, 40, 41, 42]. As discussed below, it involves the nucleation of a nanoscopic vortex loop and its subsequent motion across the micro-aperture, in the course of which it increases its energy at the expense of that of the potential flow. Multiple slips are sometimes seen in the experiment, as shown in fig.1; they involve a cascade of single slips and cause the phase difference to change by $n \times 2\pi$. These catastrophic events, or collapses, were first observed by Hess [28], and studied later by a number of authors (see refs. [43, 44] and references therein).

## 4. Vortex nucleation

As discussed in ref. [45], there is good evidence that, in these phase-slippage experiments, vortices are being *nucleated*, as opposed to getting unpinned or being milled out by reconnection processes. By "nucleated" it is meant here that they appear spontaneously under the influence of local velocity in the liquid. It should be mentioned here that it was first thought [46] that, vortices being objects large compared to atomic scale, the probability that they would thermally nucleate would be extremely small. Similar doubts about the possibility of quantum nucleation of vortices still linger about [47]. As will appear below, vortices are not big to start with, and correspond to a preferred collective motion of the fluid: this is why they can nucleate.

One of the clearest experimental clues for vortex nucleation arose from the temperature dependence of the critical velocity $v_c$, shown in fig.2 in ultrapure $^4$He at $P = 0$ [48]; the critical velocity increases in a quasi-linear fashion as the temperature is lowered, going as $1 - T/T_0$ with $T_0 \sim 2.45$ K [34, 45, 49]. Below $\sim 0.15$ K, $v_c(T)$ levels off in quite abrupt a manner to reach a plateau. In ultrapure $^4$He, this plateau extends all the way to the lowest temperatures reached in the experiment. These features have also been observed at Berkeley [32, 50], but with a higher slope of $v_c(T)$ and a slightly higher crossover temperature $T_q$ toward the $T$-independent low temperature regime. The linear dependence of $v_c$ *vs* $T$ was found by Beecken and Zimmermann [29] to hold up to 1.9 K. More complex behaviours are sometimes observed [27, 32, 41, 51] but the temperature dependence of $v_c$ shown in fig.2 can nonetheless, when phase slips are resolved, be considered as a well-established experimental fact.

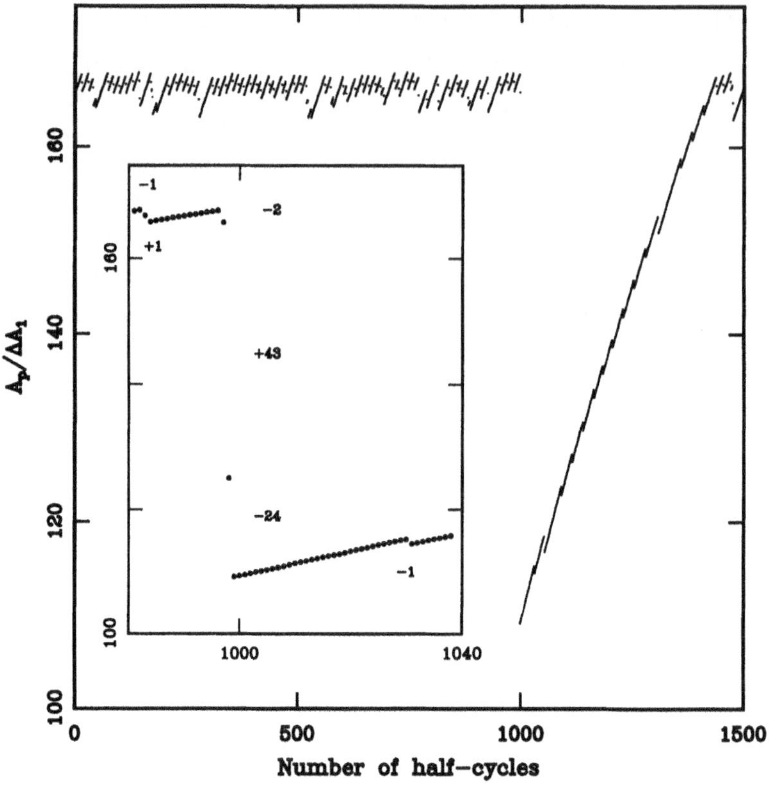


*Figure 1.* Phase slips are seen in the experiment as jumps in the peak amplitude of the resonator, plotted as a function of time for each upward and downward half-cycle of the membrane motion. The peak amplitude is normalised to the amplitude drop of a $2\pi$ phase slip. Occasionally, the succession of $2\pi$ phase slips is interrupted by larger amplitude jumps, or "collapses". The details of the large collapse, *i.e.* the multiplicity of the successive amplitude drops expressed as multiples of $2\pi$, appear in the inset.

The first conclusion that can be drawn from these observations is that the slip triggering event involves a very small energy, of the order of a few Kelvin. Such an energy is very small on a hydrodynamical scale, 1 Kelvin corresponding to the kinetic energy of a single (bare) $^4$He atom moving with a velocity of 64 m/s. This remark together with the conspicuous temperature dependence of $v_c$ below 1 K where the properties of $^4$He have settled close to their zero temperature values points toward thermally-assisted nucleation as the mechanism of vortex creation. Vortex nucleation was first considered by Vinen [52], and then as an explanation of the decay of persistent currents in porous media by Iordanskii, and by Langer and Fisher [53, 54, 55]. Such a mechanism has been invoked successfully by Muirhead,

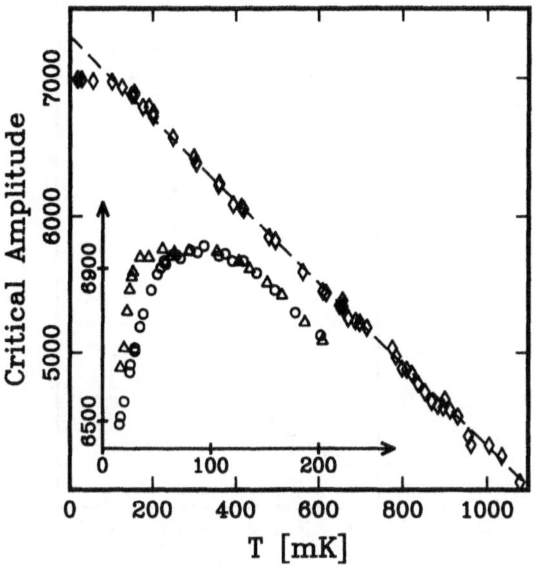


*Figure 2.* Critical velocity, in arbitrary units, *vs* temperature in ultrapure $^4$He. The effect of $^3$He impurities at two concentrations (△ 3 ppb, ○ 45 ppb) is shown in the inset.

Vinen and Donnelly [23, 56, 57] to account for the critical velocity of the negative ions. Another key finding of ion propagation work is that the emission of rotons at the Landau critical velocity, and vortex nucleation are distinct phenomena which are seen to co-exist. It can be expected that the underlying physical mechanisms for critical flow past obstacles and ion propagation have strong similarities. But, in orifice flows, the threshold for vortex formation is found to be much lower than the Landau critical velocity ($\sim$ 60 m/s at saturated vapour pressure).

If vortex nucleation really involves a thermally activated process over an energy barrier of a few Kelvin, then fluctuations in the critical velocity of individual slips should be seen. This is indeed so, although the experimental problem of showing that these fluctuations are intrinsic and meaningful is quite challenging [45, 48, 49, 58, 59, 60]. Consistent results for the statistical width of the critical transition have been obtained in two different setups in our laboratory [48, 58], and confirmed independently at Minneapolis [41]. The cumulative probability (*i.e.* the integral of the normalised histogram) for phase slips as a function of the flow velocity through the aperture is given in fig.(3). This probability shows the characteristic asymmetric "$S$"-shape profile which stems from a double exponential behaviour: the nucleation probability is a Poisson process, going as $p = 1 - e^{-\Gamma t}$, the nucleation

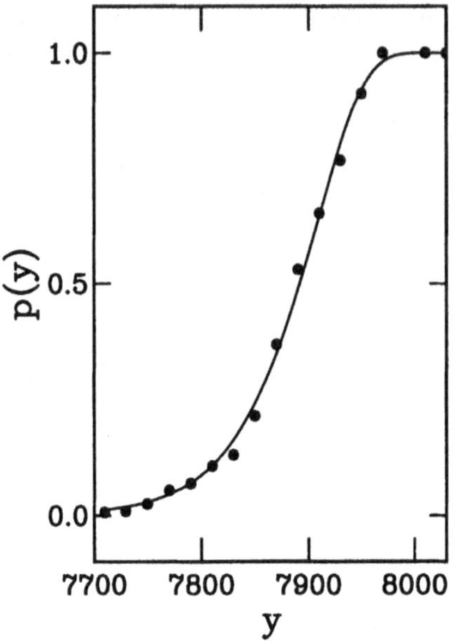


*Figure 3.* Cumulative probability $p(y)$ in terms of the resonance peak amplitude $y$ in arbitrary units in $^4$He containing 2000 ppm of $^3$He at low pressure and $T$-165 mK, from ref.[58].

rate $\Gamma$ depending itself exponentially on the energy barrier $E_a$. That the observed cumulative probability for phase slips in terms of velocity, shown in fig. 3, displays this characteristic shape is another experimental clue for vortex nucleation.

Below ~150 mK, the critical velocity as a function of temperature, in fig.2, reaches a plateau. The onset of this plateau is very sharp and is quite reproducible. A concomitant feature is also observed in fig.4 on the critical width of the critical transition, defined as $(\mathrm{d}p(y)/\mathrm{d}y)^{-1}$ at $p = 1/2$. This quantity is a measure of the "sharpness" of the critical velocity transition. These plateaus are interpreted as signalling the crossover from the thermally-assisted nucleation regime to a regime governed by quantum tunnelling. This interpretation is based on experimental grounds from two observations:

1. The curve of the width *vs* temperature displays no marked change at $T_q$, making a takeover by another mechanism such as a hydrodynamical instability [61] rather improbable.
2. The crossover region is very sharp, which is a feature of the destruction of quantum coherence by thermal fluctuations. If this sharp turning point were caused by a sudden decrease of the energy barrier, the width would

also considerably decrease.

Another reason supporting nucleation by thermal fluctuations and quantum tunnelling at low temperature is that a simple theoretical model, to which we turn in the next section, gives a consistent description of both regimes.

## 5. Vortex Half-Ring Model

This section describes the most rudimentary model for inhomogeneous vortex nucleation in which vortices appear as half-rings extending perpendicularly to the wall in the vicinity of an asperity where the local superfluid velocity $u_s$ is maximum. The asperity is assumed so small that its shape is unimportant: its only role is to break translational invariance. The healing of the wavefunction on the one hand and surface irregularities on the other are disregarded: the formulation of the problem is then quite simple.

The energy $E_0$ and the impulse $P_0$ of a vortex half-ring of radius $R$ are just half of those of a full ring in the bulk:

$$E_0 = \frac{1}{4}\rho_s\, \kappa_4{}^2\, R\, \eta\,, \quad P_0 = \frac{\pi}{2}\rho_s\, \kappa_4\, R^2\,,$$

with $\eta = \ln(8R/e^2 a_0) = \ln(R/b)$; $b = (e^2/8)a_0 = a_0/1.083$ is the core parameter and incorporates the core energy.

Although it will turn out that the vortices which are most relevant to the problem have radii larger than a few times $b$ and that the detailed behaviour of the energy and momentum at small $b$ is unimportant, it is nonetheless convenient to make the energy regular at small $R$ by redefining $\eta$ as $1/2\ln(1+R^2/b^2)$. The self-velocity of the vortex, taken as

$$v_v = \frac{\partial E_0}{\partial P_0} = \frac{\kappa_4}{4\pi R}\left\{\frac{1}{2}\ln(1+\frac{R^2}{b^2}) + \frac{R^2}{R^2+b^2}\right\}, \tag{9}$$

also becomes regular and tends to zero as $R \to 0$ (compare with eq. (3)).

Wall (inhomogeneous) nucleation is then treated in the same manner as bulk (homogeneous) nucleation, following the approach of Iordanski-Langer-Fischer (see Langer and Reppy [55]). When placed in a homogeneous superfluid velocity field $u_s$ with its impulse $P_0$ opposing the field, the vortex energy is reduced according to

$$E_v = E_0 - P_0 u_s = \frac{1}{4}\rho_s\kappa_4{}^2\left\{\frac{1}{2}R\ln(1+\frac{R^2}{b^2}) - \frac{2\pi}{\kappa_4}u_s R^2\right\}. \tag{10}$$

The activation energy $E_a$ is found as the difference between the maxima and minima of $E_v$ as a function of $R$. Once the activation energy is known, the rate at which thermal activation of vortices takes place is given by the Arrhenius formula [62, 63]:

$$\Gamma_{\rm cl} = \frac{\omega_0}{2\pi} \exp\left\{-\frac{E_{\rm a}(u_{\rm s})}{k_{\rm B}T}\right\}, \tag{11}$$

In principle, the attempt frequency $\omega_0$ can be obtained from the shape of the confining potential if the mass of the "particle" trapped in the well is known. It can be checked that if we take for this mass the mass of the nucleated vortex, we obtain a reasonable order of magnitude for $\omega_0/2\pi$, comparable to the "cyclotron" frequency. However, not much confidence can be placed in the accuracy of such an evaluation as neither the shape of $E_{\rm v}$ close to $R = 0$ nor the equivalent mass of the proto-vortex trapped at the wall is really well known.

A reliable (phenomenological) determination of $\omega_0$ is given by the following relation between the attempt frequency and the temperature $T_{\rm q}$ at which the thermal fluctuations take over quantum tunnelling [63, 64]:

$$\omega_0/2\pi = k_{\rm B}T_{\rm q}/\hbar, \tag{12}$$

The corresponding value of the attempt frequency, $\omega_0/2\pi = 1.9 \times 10^{10}$ Hz for $T_{\rm q} = 0.147$ K, is of a magnitude comparable to the "cyclotron" frequency, eq. (5), $\omega^+/2\pi = 8.1 \times 10^{10}$ Hz for $a_0 = 2.5$ Å .

Below $T_{\rm q}$, quantum fluctuations are abruptly restored and the $T = 0$ nucleation rate is given by [62, 63, 65]

$$\Gamma_{\rm q} = \frac{\omega_0}{2\pi}\left(864\pi\frac{E_{\rm a}}{\hbar\omega_0}\right)^{\frac{1}{2}} \exp\left\{-\frac{36}{5}\frac{E_{\rm a}}{\hbar\omega_0}\left[1+\frac{45\zeta(3)}{2\pi^3\omega_0\tau}\right]\right\}, \tag{13}$$

where the term proportional to $(\omega_0\tau)^{-1}$ expresses the contribution of damping to first order which is thought to be small in the present problem.

The nucleation rate in turn yields the probability that an event has taken place in the time interval $0 - t_{\rm f}$ :

$$p = 1 - \exp\left\{-\int_0^{t_{\rm f}} \Gamma(u_{\rm s})dt\right\}. \tag{14}$$

In the following, $t_{\rm f}$ will be taken as half the resonator period.

Close to the point where it vanishes, the energy barrier stemming from eq. (10) takes the form $E_{\rm a} = \frac{2}{3}E_{\rm J}\epsilon^3$ with $\epsilon = (1 - {u_{\rm s}}^2/{u_{\rm c0}}^2)^{\frac{1}{2}}$, $u_{\rm c0}$ being the value of the velocity $u_{\rm s}$ for which the energy barrier vanishes (the lability point). The calculation of $p$ can be carried out with asymptotic accuracy [66], with the following result:

$$p = 1 - \exp\left\{-\frac{\omega_0 t_{\rm f}}{\pi}\left[\frac{k_{\rm B}T}{6\pi E_{\rm a}}\frac{\epsilon^2}{(1-\epsilon^2)}\right]^{\frac{1}{2}} \exp\left\{-\frac{E_{\rm a}}{k_{\rm B}T}\right\}\right\}. \tag{15}$$

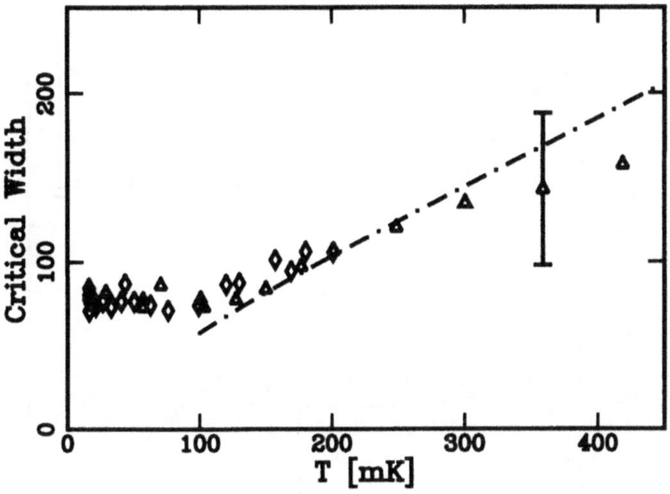


*Figure 4.* Statistical width of critical velocity transition, in arbitrary units, in terms of temperature. A $2\pi$ phase slip is about 42 instrumental units. The dash-dot line represents the limiting behaviour of eq.(17) at low $T$.

The critical value $u_c$ is defined as that for which $p = 1/2$ and is given by:

$$u_c = u_{c0} \left\{ 1 - \left[ \frac{3}{2} \frac{k_B T}{E_J} \gamma \right]^{\frac{2}{3}} \right\}^{\frac{1}{2}}, \tag{16}$$

$$\text{with } \gamma = \ln \left\{ \frac{1}{\ln 2} \frac{\omega_0 t_f}{\pi} \left[ \frac{k_B T}{4\pi E_J} \frac{1}{\epsilon(1-\epsilon^2)} \right]^{\frac{1}{2}} \right\}.$$

The probability $p$, eq.(14), is an abrupt function of $u_s$, being negligible nearly up to the critical value and rising very rapidly to unity over a very narrow span of velocity with the characteristic asymmetric "$S$" - shape mentioned above.

The width of the critical transition, defined [45, 58, 66] by $\Delta u_c = \partial u_p / \partial p$ evaluated at $p = 1/2$, is likewise found to be

$$\frac{\Delta u_c}{u_{c0}} = \frac{4}{\ln 2} \frac{(1-\epsilon^2)^{1/2} \epsilon^2}{1 + 6\gamma - 3\epsilon^2(1+2\gamma)}. \tag{17}$$

To express numerically the results, a value must be chosen for $b$, which is the only parameter in the model which can be varied to take into account the various phenomena that are expected to take place close to walls. Some of the corrections to the energy and impulse of the vortex half-ring

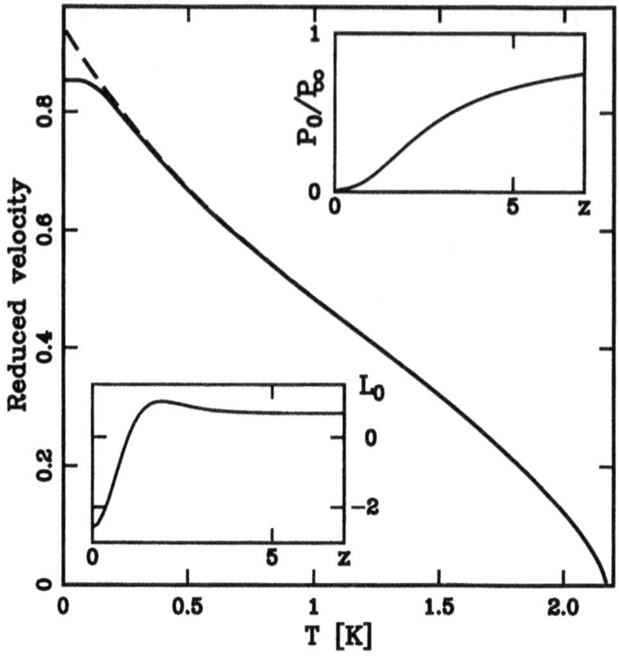


*Figure 5.* Critical velocity, in units of $\kappa_4/4\pi b$, computed as a function of $T$ for the model, for a value of $b$ of 4.5 Å, including estimates of the effect of the wall on the core contribution $L_0$ to the vortex energy $E_0$, in units of $\rho_s\kappa_4 b$, and on the vortex impulse $P_0$ with respect to its value in the bulk $P_\infty$. Quantum corrections are included in the plain curve which departs significantly from pure thermally-assisted case (dashed curve) below $\sim 0.3$ K.

which include the healing of the wavefunction close to a wall, evaluated in ref. [40], are shown in the insets of fig.5. With these corrections, a best fitting value for $b$ of 4.5 Å is found. The fit relies explicitly on the temperature dependences chosen for $\rho_s(T)$ [67] and $a_0$ [4]. The resulting curve for the temperature dependence of $v_c$ is shown in fig.5. Although not quite a straight line as a function of $T$, the computed critical velocity is seen to increase by a factor of 2 between 1 K and absolute zero and to extrapolate to zero at $T_0 \sim 2.45$ K as for the experimental data. Also, the statistical width computed with the same value of $b$ compares well with experimental data, as shown in fig.4.

Variants of this analysis have been performed by Shifflett and Hess [68], and by Zimmermann *et al.* [69] on their own critical velocity measurements: there is overall agreement over the main features of the half-ring nucleation model outlined here.

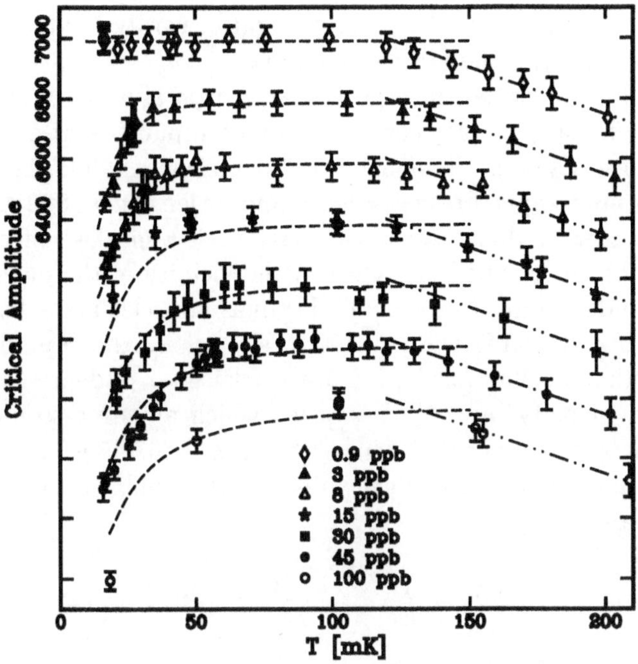


*Figure 6.* Effect of $^3$He impurities on the critical velocity expressed in arbitrary units. The curves have been shifted downward by 200 units for each concentration. The quantum plateau level and the high temperature behaviour are unaffected by $^3$He.

The half-ring model can be straightforwardly extended to the case of quantum tunnelling, following refs. [63, 65, 70] as done by Ihas *et al.*[48], by using $\Gamma_q$, eq. (13), instead of $\Gamma_{cl}$, eq. (11), in the probability, eq. (14).

Quantum nucleation can also be tackled directly in a Lagrangian formalism, following refs. [71, 72, 73, 74]. This approach is based on the fact that, in a number of cases - for vortex pairs, for half-ring vortices, *etc...* - the vortex equations of motion can be put in Hamiltonian form. The tunnelling rate is then proportional to $\exp -2Im\,\mathcal{S}/\hbar$, $\mathcal{S}$ being the action along the "bounce" trajectory which the vortex follows in the escape problem. The two methods lead to expressions which differ formally but which yields numerical values which are nonetheless very close: in the limited range of parameter values relevant to experiment, they are equivalent.

## 6. $^3$He Quasiparticles as a Local Velocity Probe

The analysis of the effect of $^3$He quasiparticles on the low temperature behaviour of the critical velocity which is shown in the inset of fig.2, and in more detail in fig.6, leads to an estimate of the superflow velocity at the nucleation site, $u_s$, a quantity not straightforwardly related to the mean flow velocity through the micro-aperture $v_s$.

It must first be noted that quite a large drop of $v_c$, by more than 15 %, is caused by a really minute concentration of impurities: at $x_3 = 3$ ppb, the mean distance between $^3$He quasiparticles is 2500 Å, while the width of the rectangular micro-aperture is $\sim$ 3000 Å and its length $\sim$ 2000 Å. It is hardly conceivable that so dilute a gas can have such a marked effect unless the quasiparticles are trapped or collect in the vicinity of the nucleation site.

This is indeed the case because $^3$He quasiparticles are attracted by the Bernoulli effect towards the nucleation site where the superfluid velocity is at a maximum. The local pressure in the fluid is depressed in regions of high flow velocity by $\Delta P = -\frac{1}{2}\rho_s u_s^2$, which amounts to one third of a bar at 20 m/s, the superfluid density being $\rho_s = 0.145$ g/cm$^3$. The change in the $^3$He quasiparticle chemical potential at equilibrium is related to the pressure change by

$$\Delta\mu_3 = \int_P^{P+\Delta P} \frac{\partial\mu_3}{\partial P}\, dP = v_3\,\Delta P,$$

where $v_3$ is the $^3$He volume of one atom in the bath.

There exists another contribution [75] to the quasiparticle energy change induced by the superfluid flow field $u_s$ which arises from the difference between the bare mass $m_3$ and the effective mass $m^*$ of the quasiparticle ($\sim 2.27 m_3$). The backflow around the quasiparticle responsible for this difference directly interacts with the applied superflow and yields a contribution proportional to $1 - m_3/m^*$. The total energy change of the quasiparticle is expressed as a sum of the chemical potential difference and the effect of backflow:

$$\delta\epsilon = \frac{1}{2}\left[\rho_s v_3 + \left(1 - \frac{m_3}{m^*}\right) m_3\right] u_s^2.$$

Thus the nucleation site, at which the local velocity $u_s$ is at a maximum, acts as an atom trap for the $^3$He quasiparticles into which they fall when the temperature becomes low enough. The quasiparticle density $n_3$ at the nucleation site is related to that in quiescent regions of the cell, $n_{30}$, by [76]

$$n_3 = n_{30}\exp\{\delta\epsilon/k_B T\},$$

assuming that the quasiparticle system is not very far from thermal equilibrium. The local density of quasiparticles at the nucleation site can thus become large enough so that the coincidence of a local fluctuation tending to form a nascent vortex and a quasiparticle is not an unlikely event.

The binding energy of $^3$He to vortex cores in the superfluid bulk is estimated [12, 23] to be about 2.7 K. The energy barrier opposing nucleation is lowered by a quantity $\mathcal{L}_3 - E_k$ when one $^3$He atom resides on the core, $E_k = \hbar^2/2m_3R^2$ being the zero point energy that the $^3$He atom acquires when it is confined to the vortex core of a small half-ring of length $R$. The quantity $\mathcal{L}_3$ is a phenomenological parameter the value of which can be expected to lie between the binding energy of a $^3$He atom to a vortex in the bulk (2.7 K) and that at the wall, which is taken to be zero.

The vortex half-ring nucleation rate for pure $^4$He in the quantum regime, $\Gamma_q$, is given by eq.(13). With one $^3$He quasiparticle present, the basic process of nucleation is the same, with the same attempt frequency, except that the energy barrier is now lowered in such a manner that the nucleation rate reads:

$$\Gamma = \Gamma_q \exp\{36(\mathcal{L}_3 - E_k)/5\hbar\omega_0\}\,. \tag{18}$$

Eq. (18) can be extended to more than one $^3$He quasiparticle. The critical velocity is then calculated as in sec. 4 and fit to the data as discussed in ref. [77]. This procedure leads to $\delta\epsilon = 0.204$ K, to which corresponds a critical velocity at the nucleation site of $u_c = 22$ m/sec and $\mathcal{L}_3 = 1.68$ K, about two-third of the bulk value.

## 7. Pinning

Major qualitative changes in the Helmholtz resonator operation can be brought about by the formation of microscopic dust inside the resonator during the course of an experimental run. The micro-dust is formed by the sputtering of atoms adsorbed on the resonator walls, induced by an electric discharge. The presence of micro-dust in the cell causes a profound alteration in the peak amplitude charts, with an increase of the rate of occurrence of collapses by almost three orders of magnitude and no well-defined plateau for the critical velocity. However, in some instances, a simpler pattern with two reproducible critical velocity plateaus, as illustrated in fig.7, was observed. The upper plateau, with mostly single slips, corresponds to the ordinary behaviour found in the absence of contamination.

The existence of the lower plateau and the detailed experimental features associated with its onset and its disappearance can be explained as follows: The velocity at the nucleation site appears to be subjected to a local influence of substantial magnitude. Vortices are the only long-lived defects in superfluid $^4$He which can cause such a perturbing velocity, ions and electrons being short-lived. Therefore, a vortex line must, in this view, become trapped in the immediate vicinity of the vortex nucleation site. Such a situation is schematically illustrated in fig.8.

The shape of the pinned filament adjusts to the imposed superflow. Its orientation is such that, for one flow direction, it bends toward the wall and toward the nucleation site on which it induces a velocity $v_p^{\max}$. This

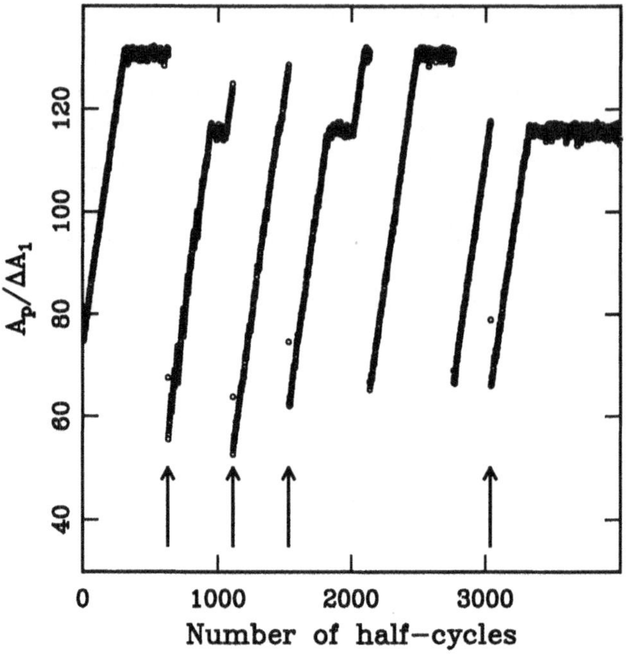


*Figure 7.* Chart of the peak amplitude of the resonator, in units of a single phase slip amplitude drop, *vs* time, expressed in number of half-cycles of the resonance, at 14 mK, 1.0 bar, in "nominally" pure $^4$He (100 ppb $^3$He. The upper plateau corresponds to the usual critical amplitude level of the resonator. The lower level constitutes one of the anomalous features due to vortex pinning. Collapses resulting in pinning are marked by arrows.

velocity adds to the membrane-driven flow. For the reverse flow direction, the vortex moves away from the nucleation site on which it induces a reduced velocity $v_\mathrm{p}^\mathrm{min}$ which subtracts from the membrane-driven flow. After a random amount of time, the upper plateau level is regained when the pinned vortex unpins and ceases to perturb the nucleation site. All the experimental features are fully accounted for by such a picture, as discussed in refs. [26, 78].

A detailed analysis on the unpinning process in the experimental temperature range, between 14 and 44 mK, limited at high temperature by the proliferation of collapses , leads to the following results:

• the unpinning velocity $v_\mathrm{u}$ is nearly independent of $T$, as shown in fig.9;
• the fluctuations of $v_\mathrm{u}$ about the mean value are also independent of $T$;
• the cumulative probability for unpinning in terms of velocity displays the asymmetric "$S$" - shape characteristic of nucleation processes.

All of these features tend to favour escape of the vortex from its pinned

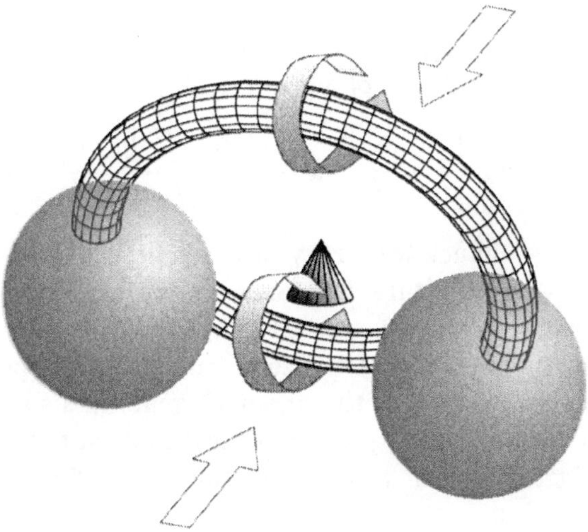

*Figure 8.* A possible arrangement of the pinned vortex close to the nucleation site, represented by a cone. The direction of the imposed flow is shown by the straight arrow closest to each vortex position. The curved arrow indicates the direction of the flow induced by the vortex.

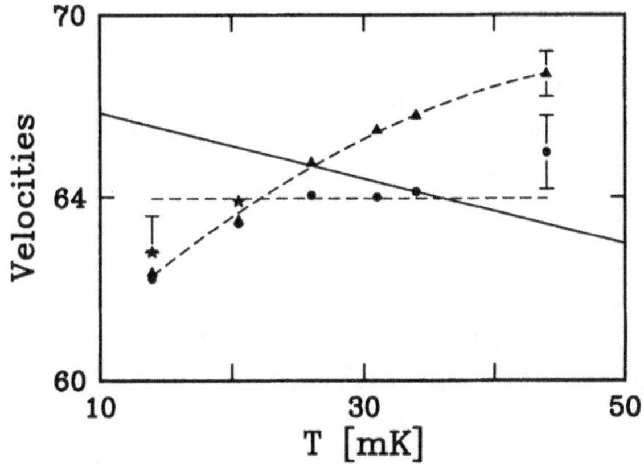


*Figure 9.* Nucleation velocity ($\Delta$) and unpinning velocity ($\bullet$, $\star$) as a function of temperature. The straight plain line shows the $T$-dependence of $v_u$ if the nucleation were thermally assisted at a temperature of 34 mK and below.

state by quantum tunnelling. This possibility is borne out by quantitative arguments as follows.

If unpinning were thermally-activated, the unpinning velocity $v_u$ would decrease when the temperature is raised and thermal fluctuations become more and more important. The temperature dependence of $v_u$ can be estimated from the knowledge of the energy barrier, represented by a Taylor expansion as $E_p(1 - v_u/v_{u0})$ following ref. [45] (compare with eq. (16)):

$$v_u = v_{u0}\left\{1 - \frac{k_B T}{E_p}\ln\left[\frac{\omega_0}{\pi}\frac{t_f}{\ln 2}\sqrt{\frac{2}{\pi}\frac{k_B T}{E_p}\frac{v_{u0}}{v_u}}\right]\right\} , \tag{19}$$

$\omega/2\pi$ being the resonance frequency of the cell (9.4 Hz). The straight line representing this temperature dependence with the value of $E_p/k_B$ found from the statistical width for unpinning at 34 mK is shown in Fig. 9. The data for $v_u$ clearly does not show such a temperature dependence: thermal activation seems ruled out as the underlying mechanism for the observed fluctuations of the unpinning process while quantum tunnelling is favoured by both the constancy with $T$ of $v_u$ and that of the fluctuations of both $v_u$ and $v_p^{max}$.

## 8. Concluding remarks

The experimental observations discussed above point towards a thermally assisted regime for vortex nucleation above 0.15 K. The sharp but continuous turnover to a $T$-independent regime below $\sim$ 0.15 K is most naturally interpreted as the onset of quantum tunnelling.

Both regimes can be explained in terms of a model which makes use of the classical fluid mechanical expressions for the energy and momentum of a vortex half-ring. The detailed geometry of the asperity is neglected as well as the perturbation of the order parameter due to the wall. Such a coarse semiclassical approach works because the energy barrier for nucleation extends over a length of *cir.* 5 to 10 Å which is larger than the coherence length.

A clear difference exists between vortex nucleation and unpinning. In some instances, the two phenomena have been observed concurrently with different critical velocities. Nucleation provides an inexhaustible supply of vortices and leads to reproducible phase slips. Unpinning gives rise to vagrant vorticity which is presumably short-lived unless the flow velocity is high enough to sustain a vortex tangle. This vagrant vorticity may be responsible for superflow collapses.

This work is supported by Centre National de la Recherche Scientifique and Commissariat à l'Énergie Atomique. The author ackowledges the fruitful collaboration with Olivier Avenel, Pertti Hakonen and Yuri Mukharsky.

## References

1. Onsager, L. (1949) Discussion on a paper by C.J. Gorter, *Nuovo Cimento Suppl.* **6**, p. 249

2. Feynman, R.P. (1955) *Prog. Low Temp. Phys.*, Vol. 1, p. 36, ed. C. J. Gorter, North-Holland - Amsterdam
3. Fetter, A.L. (1976) *The Physics of Liquid and Solid Helium - Part 1*, Chapt. 3, eds. K.H. Benneman, J. B. Ketterson, Wiley-Interscience, N.Y.
4. Glaberson, W.I and Donnelly, R.J. (1986) *Prog. Low Temp. Phys.*, Vol. IX, Ch. 1, ed. D. F. Brewer, North-Holland - Amsterdam
5. Vinen, W.F. (1994) *The Vortex State*, Vol. 438 of *NATO ASI Series*, Ch. 1, eds. N. Bontemps and Y. Bruynseraede, G. Deutscher, A. Kapitulnik, Kluwer Academic Press - Dordrecht
6. Donnelly, J. (1991) *Quantized Vortices in Helium*, C.U.P., Cambridge
7. Vinen, W.F. (1966) *Quantum Fluids*, p. 74, ed. D. F. Brewer, North-Holland - Amsterdam
8. Nozières, P. (1966) *Quantum Fluids*, p. 1, ed. D. F. Brewer, North-Holland - Amsterdam
9. Nozières, P., and Pines, D. (1959) *The Theory of Quantum Liquids, Vol. II*, Addison-Wesley, Redwood City, CA.
10. Landau, L.D. and Lifshitz, E.M. (1959) *Fluid Mechanics*, Pergamon Press - London
11. Sonin, E.B. (1987) Vortex oscillations and hydrodynamics of rotating superfluids, *Rev. Mod. Phys.* **59**, p. 87
12. Dalfovo, F. (1992) Structure of vortices in helium at zero temperature, *Phys. Rev.* **46**, p. 5482
13. Schwarz, K.W. (1978) Turbulence in superfluid helium: Steady homogeneous counterflow, *Phys. Rev.* **B18**, p. 245
14. Schwarz, K.W. (1985) Three-dimensional vortex dynamics in superfluid $^4$He: Line-line and line-boundary interactions, *Phys. Rev.* **31**, p. 5782
15. Schwarz, K.W. (1981) Vortex Pinning in Superfluid Helium, *Phys. Rev. Lett.* **47**, p. 251
16. Tsubota, M. and Maekawa, S. (1994) Dynamics of two quantized vortices on a pinning site, *Physica B* **194-196**, 721
17. Ellis, F.M., and Li, L. (1993) Quantum Swirling of Superfluid Helium Films, *Phys. Rev. Lett.* **71**, p. 1577
18. Hakonen, P.J., Nummila, K.K., Simola, J.T., Skrbek, L., and Mamniashvili, G. (1987) Surface Spin Waves in $^3$He-A, a Probe for Vortex Phenomena in Narrow Gap, *Phys. Rev. Lett.* **58**, p. 678
19. Krusius, M., Korhonen, J.S., Kondo, Y., and Sonin, E.B. (1993) Collective motion of quantized vortex lines in rotating superfluid $^3$He-B, *Phys. Rev.* **B47**, p. 15113
20. Zieve, R.J., Mukharsky, Yu, Close, J.D., Davis, J.C., and Packard, R.E. (1992) Precession of a Single Vortex Line in Superfluid $^3$He-B, *Phys. Rev. Lett.* **68**, p. 1327
21. Alpar, M.A., Nandkumar, R., and Pines, D. (1980) Vortex creep and the internal temperature of neutron stars: the Crab pulsar and PSR 0525+21, *Astrophys. J.* **288**, p. 191
22. Fetter, A. (1965) Vortices in an Imperfect Bose Gas. I. The Condensate, *Phys. Rev.* **138A**, p. 429
23. Muirhead, C.M., Vinen, W.F., and Donnelly, R.J. (1985) The nucleation of vorticity by ions in superfluid $^4$He II. Theory of the effect of dissolved $^3$He, *Proc. R. Soc. London A* **402**, p. 225
24. Lamb, Sir Horace (1945) *Hydrodynamics*, Cambridge University Press
25. Duan, Ji-Min (1994) Mass of a vortex line in superfluid $^4$He: Effects of gauge-symmetry breaking, *Phys. Rev.* **B49**, p. 12381
26. Varoquaux, E., Avenel, O., Hakonen, P., and Mukharsky, Yu (1998) Observation of single-vortex pinning in superfluid $^4$He, *Physica B* **255**, p. 55
27. Hess, G.B. (1971) Critical Velocities in Superfluid Helium Flow Through 10-$\mu$m-Diameter Pinholes, *Phys. Rev. Lett.* **27**, p. 977
28. Hess, G.B. (1977) Vortex generation in modulated superfluid $^4$He flow through a pinhole, *Phys. Rev.* **15**, p. 5204

29. Beecken, B.P., and Zimmermann, W. Jr. (1987) Variation of the critical order-parameter phase difference with temperature from 0.4 to 1.9 K in the flow of superfluid $^4$He through a tiny orifice, *Phys. Rev.* **35**, p. 1630
30. Avenel, O., and Varoquaux, E. (1985) Observation of singly quantized dissipation events obeying the Josephson frequency relation in the critical flow of superfluid $^4$He through an aperture, *Phys. Rev. Lett.* **55**, p. 2704
31. Amar, Ajay, Davis, J.C., Packard, R.E., and Lozes, R.L. ( 1990) Phase-slips in the flow of superfluid $^4$He through a submicron orifice, *Physica B* **165&166**, p. 753
32. Amar, Ajay, Sasaki, Y., Lozes, R., Davis, J.C., and Packard, R.E. (1992) Quantized Phase Slippage in Superfluid $^4$He, *Phys.Rev. Lett.* **68**, p. 2624
33. Davis, J.C., Steinhauer, J., Schwab, K., Mukharsky, Yu, Amar, Ajay, Sasaki, Y., and Packard, R.E. (1992) Evidence for Quantum Tunneling of Phase-Slip Vortices in Superfluid $^4$He, *Phys. Rev. Lett.* **69**, p. 323
34. Varoquaux, E., Avenel, O., and Meisel, M. (1987) Phase slippage and vortex nucleation in the critical flow of superfluid $^4$He through an orifice, *Can. J. Phys.* **65**, p. 1377
35. Anderson, P.W. (1966) Considerations on the Flow of Superfluid Helium, *Rev. Mod. Phys.* **38**, p. 298
36. Zimmermann, W., Jr. (1996) The flow of superfluid $^4$He through submicron apertures: phase slip and critical velocities due to quantum vortex motion, *Contemp. Phys.* **37**, p. 219
37. Varoquaux, E., and Avenel. O. (1987) Quantum phase slippage in superfluid $^4$He, *Physica Scripta* **T19B**, p. 445
38. Varoquaux, E., and Avenel, O. (1994) Phase slip phenomena in superfluid helium, *Physica B* **197**, p. 306
39. Bernard, M, Burkhart, S., Avenel, O., and Varoquaux, E. (1994) Mechanism for Quantum Phase-slips in $^4$He, *Physica B* **194-196**, p. 499
40. Burkhart, S., Bernard, M., Avenel, O., and Varoquaux, E., (1994) Scenario for a Quantum Phase Slip, *Phys. Rev. Lett.* **72**, p. 380
41. Zimmermann, W., Jr. (1993) The Critical Velocity Behavior of Superfluid $^4$He in a Micron-Size Aperture at Two Different Frequencies of Oscillatory Flow, *J. Low Temp. Phys.* **91**, p. 219
42. Schwarz, K.W. (1993) Fluid Dynamics of a Quantized Vortex Filament in a Hole, *J. Low Temp. Phys.* **93**, p. 1019
43. Avenel, O., Bernard, M., Burkhart, S., and Varoquaux, E. (1995) Autopsy of superflow collapses and multiple phase slips, *Physica B* **210**, p. 215
44. Varoquaux, E., Avenel, O., Bernard, M., and Burkhart, S. (1995) Multiple Quantum Phase Slips in Superfluid $^4$He, *J. Low Temp. Phys.* **101**, p. 821
45. Varoquaux, E., Zimmermann, W., Jr, and Avenel, O. (1991) *Excitations in Two-Dimensional and Three-Dimensional Quantum Fluids*, p. 343. eds. A. F.G. Wyatt and H. J. Lauter, Plenum press, New-York
46. Vinen, W.F. (1963) *Liquid Helium*, p. 336, ed. G. Careri, Academic Press - New-York
47. Volovik, G.E. (1997) Comment on vortex mass and quantum tunneling of vortices, *JETP Lett.* **65**, p. 217
48. Ihas, G.G., Avenel, O., Aarts, R., Salmelin, R., and Varoquaux, E. (1992) Quantum Nucleation of Vortices in the Flow of Superfluid $^4$He through an Orifice, *Phys. Rev. Lett.* **69**, p. 327
49. Varoquaux, E., Meisel, M.W., and Avenel, O. (1986) Onset of the critical velocity regime in superfluid $^4$He at low temperature, *Phys. Rev. Lett.* **57**, p. 2291
50. Davis, J.C., Close, J.D., Zieve, R., and Packard, R.E. (1991) Observation of Quantized Circulation in Superfluid $^3$He-B, *Phys. Rev. Lett.* **66**, p. 329
51. Hulin, J.P., D'Humières, D., Perrin, B., and Libchaber, A. (1974) Critical velocities for superfluid helium flow through a small hole, *Phys. Rev.* **A9**, p. 885
52. Vinen, W.F. (1968) Macroscopic quantum effects in superfluids, *Rep. Prog. Phys.*

**XXI**, p. 61
53. Iordanskii, S.V. (1965) Vortex ring formation in a superfluid, *Sov. Phys. JETP* **21**, p. 467
54. Langer, J.S., and Fisher, M.E. (1967) Intrinsic Critical Velocity of a Superfluid, *Phys. Rev. Lett.* **19**, p. 560
55. Langer, J.S., and Reppy, J.D. (1970) *Prog. Low Temp. Phys.*, Vol. 6, Ch. 1. ed. C. J. Gorter, North-Holland - Amsterdam
56. Muirhead, C.M., Vinen, W.F., and Donnelly, R.J. (1984) The nucleation of vorticity by ions in superfluid $^4$He I. Basic Theory, *Phil. Trans. Roy. Soc. A* **311**, p. 433
57. McClintock, P.V.E., and Bowley, R.M. (1991) *Excitations in Two-dimensional and Three-dimensional Quantum Fluids*, p. 567. eds. A. G. F. Wyatt and H. J. Lauter, Plenum Press, New York
58. Zimmermann, W., Jr., Avenel, O., and Varoquaux, E. (1990) Critical flow of superfluid helium-4 through a submicron aperture: width of the critical transition, *Physica B* **165&166**, p. 749
59. Avenel, O., Ihas G.G., and Varoquaux, E. (1993) The Nucleation of Vortices in Superfluid 4He: Answers and Questions, *J. Low Temp. Phys.* **93**, p. 1031
60. Ihas, G.G., Avenel, O., Aarts, R., and Varoquaux, E. (1994) $^3$He effect on Flow Dissipation of Superfluid $^4$He through a Microaperture, *Physica B* **194-196**, p. 583
61. Josserand, C., and Pomeau, Y. (1995) Generation of vortices by the Kadomtsev-Petviashvili instability, *Europhys. Lett.* **30**, p. 43
62. Larkin, A., Likharev, K.K., and Ovchinnikov, Yu.N. (1984) Secondary Quantum Macroscopic Effects in Weak Superconductivity, *Physica B* **126**, p. 414
63. Mel'nikov, V.I. (1991) The Kramers problem: Fifty years of development, *Phys. Reports* **209**, p. 1
64. Affleck, I. (1981) Quantum-Statistical Metastability, *Phys. Rev. Lett.* **46**, p. 388
65. Caldeira, A.O., and Leggett, A.J. (1983) Quantum Tunnelling in a Dissipative System, *Ann. Phys. (N.Y.)* **149**, p. 374
66. Varoquaux, E., Avenel, O., Ihas, G.G., and Salmelin, R. (1992) Phase Slippage in Superfluid $^3$He-B, *Physica B* **178**, p. 309
67. Maynard, J. (1976) Determination of the thermodynamics of He-II from sound-velocity data, *Phys. Rev.* **B14**, p. 3868
68. Shifflett, G.M., and Hess, G.B. (1995) Intrinsic critical velocities in superfluid $^4$He flow through 12-$\mu$m diam. orifices near $T_\lambda$: Experiments on the effect of geometry, *J. Low Temp. Phys.* **98**, p. 591
69. Zimmermann, W., Jr., Lindensmith, C.A., Flaten, J.A. (1998) The Interpretation of Phase-Slip and Critical-Velocity Data , *J. Low Temp. Phys.* **110**, p. 497
70. Hanggi, P., Grabert, H., Ingold, G.-L., and Weiss, U. (1985) Quantum Theory of Activated Events in Presence of Long-Time Memory, *Phys. Rev. Lett.* **55**, p. 761
71. Volovik, G.E. (1972) Quantum-Mechanical Formation of Vortices in a Superfluid Liquid, *Sov. Phys. JETP Lett.* **15**, p. 81
72. Sonin, E.B. (1995) Nucleation and creep of vortices in superfluids and clean superconductors, *Physica B* **210**, p. 234
73. Fischer, U.R. (1998) Geometric Laws of Vortex Quantum Tunneling, *Phys. Rev.* **B58**, p. 105
74. Fischer, U.R. (1998) On the theory of vortex quantum tunnelling in the dense Bose superfluid helium II, *Physica B* **255**, p. 41
75. Baym, G. (1969) *Mathematical Methods in Solid State and Superfluid Theory*, p. 134, eds. R.C. Clark and G.H. Derrick, Oliver and Boyd Ltd, Edinburgh
76. Kennard, E.H. (1938) p. 77, *Kinetic Theory of Gases*, McGraw-Hill, N.Y.
77. Varoquaux, E., Ihas, G.G., Avenel, O., and Aarts, R. (1993) Vortex Nucleation in Superfluid $^4$He Probed by $^3$He Impurities, *Phys. Rev. Lett.* **70**, p. 2114
78. Hakonen, P., Avenel, O., and Varoquaux, E. (1998) Evidence for Single-Vortex Pinning and Unpinning Events in Superfluid $^4$He, *Phys. Rev. Lett.* **81**, p. 3451

# TOPOLOGICAL DEFECTS IN $^3$HE SUPERFLUIDS

V. B. ELTSOV[1] AND M. KRUSIUS
*Low Temperature Laboratory,*
*Helsinki University of Technology,*
*Box 2200, FIN-02015 HUT, Finland*

Defects in the multi-dimensional macroscopic quantum field of the $^3$He superfluids are localized objects with a topological charge and are topologically stable. They include point-like objects, vortex lines, planar domain-wall-like structures, and 3-dimensional textures, which may or may not include singular points or lines. An inventory of illustrations is presented which concisely lists the different experimentally confirmed defect structures in the A and B phases of superfluid $^3$He.

## 1. Quantum Fields in $^3$He superfluids

Experimental information on defect structures in quantum fields is fragmentary and restricted by the properties and constraints of each particular degenerate many-body system in which they are studied. A notable exception are the p-wave-paired $^3$He superfluids. They provide a versatile laboratory system, with a multidimensional order parameter field, in which one can study objects of different dimensionality – point defects, quantized vortex lines, domain-wall-like topological solitons, and 3-dim textures. Some of these, like the point defects, have not yet been directly mapped, and their role remains elusive. Others on the other hand, have helped to illustrate general principles such as composite structure and topological confinement, nucleation, and interactions between objects of different topologies.

In addition to the diversity in structure, two more features have become important attributes of the current $^3$He work. First, bulk superfluid $^3$He is nearly devoid of extrinsic influence. The only weak heterogeneity is introduced by the surfaces of the containing vessel. Secondly, detailed theoretical understanding of the $^3$He order parameter field exists [1] and can be effectively correlated with new experimental results. The latter originate to

1Also at Kapitza Institute for Physical Problems, Russian Academy of Sciences, ul. Kosygina 2, 117334 Moscow.



*Y.M. Bunkov and H. Godfrin (eds.),*
*Topological Defects and the Non-Equilibrium Dynamics of Symmetry Breaking Phase Transitions,* 325–344.

a large extent from noninvasive nuclear magnetic resonance measurement on the superfluid contained in a rotating cylinder.

Similar to the application of a magnetic field on a superconductor, rotation of a superfluid is the most effective means for modifying existing structure in the order parameter field or for generating new structure, especially quantized vorticity. In the multi-dimensional order parameter field of the $^3$He superfluids quantized vorticity can have different topology and structure. For instance, in the vortex core the order parameter may become singular or it may have a continuous singularity-free distribution. Altogether eight different types of vorticity have been discovered and described: one continuous structure in $^3$He-A1, three continuous structures and one singular in $^3$He-A2, and three singular ones in $^3$He-B. Theoretically many more structures have been proposed [2]. In practice however, the question how to form or nucleate a particular new defect structure often becomes the threshold which precludes experimental verification.

The investigation of the $^3$He order parameter field can be characterized as "field theory in the laboratory". Conceptual similarities exist between the symmetry-breaking phase transitions of $^3$He and various other field theoretical models. This similarity extends far in the mathematical description of different phenomena [3]. Clearly the detailed properties of any one of these systems can only be worked out by studying each of them individually, with specific experimental input. However, in the very least comparative work on a model system, like the $^3$He superfluids, will provide general guide lines. Besides such comparative studies will teach us what kind of new questions should be answered by an experimental $^3$He program – questions, which perhaps normally would not be standard practice within condensed matter physics.

A comprehensive review on topological defects in the $^3$He superfluids does not exist at present time [4]. Here an abbreviated "pocket book" is provided, consisting of illustrations, extended figure descriptions, and the relevant references for further information.

## 2. Archive of defects in $^3$He superfluids

**Figure 1.** The phase diagram of the fermionic liquid $^3$He phases is shown here as a function of temperature $T$ and pressure $P$ at zero external magnetic field $H$. The normal phase exists at temperatures larger than the superfluid transition temperature $T_c$ and is the best example of a system described by the Landau fermi liquid theory. Superfluid $^3$He was the first system identified to exhibit Cooper pairing in the framework of the Bardeen-Cooper-Schrieffer theory in non-s-wave states with a total spin $S = 1$ and orbital momentum $L = 1$. The two major superfluid phases are

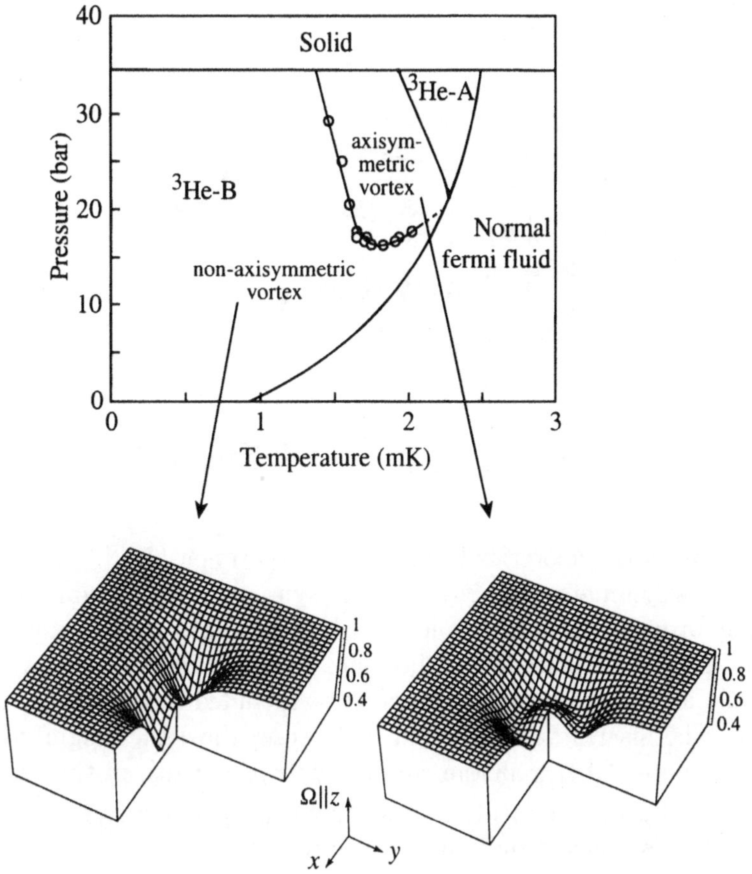


*Figure 1.* Phase transition in the singular core of B-phase vortices.

the axially anisotropic A phase at high pressures and the quasi-isotropic B phase at low temperatures.

In the B phase, with Cooper pairing in a state with total angular momentum $\mathbf{J} = \mathbf{L} + \mathbf{S} = 0$, vortex lines have similar topology as in the bosonic $^4$He superfluid and s-wave superconductors: a singular core which traps a phase winding by $2\pi$ around itself. The physical consequence from the trapped phase winding is a persistent superfluid current flowing around the core with a one-quantum circulation of $\kappa_0 = h/(2m_3) = 0.067$ mm$^2$/s. Already the very first rotating experiments displayed a first order phase transition [5], which only could be explained as a change in the vortex-core structure. This transition is shown in the phase diagram as a line with data points within the B phase region. This was the first example ever of a phase transition in a topologically stable quantized defect.

Quite surprisingly for an object associated with rotation, the core was

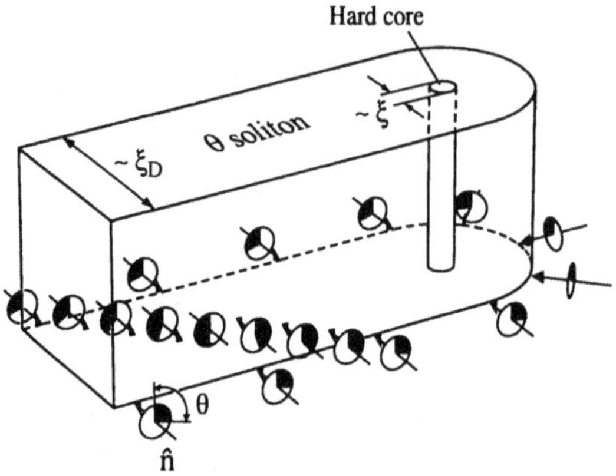


*Figure 2.* Spin-mass vortex in $^3$He-B.

later shown, both theoretically [6, 2] and experimentally [7], to undergo a spontaneous symmetry break from an axisymmetric to a double-core structure. The order parameter amplitude $|A_{ij}|$ does not vanish in the center of the core, but remains finite and becomes A-phase like in the axisymmetric case while additional components must be included in the case of the double core. The inserts below the phase diagram show the magnitude of the order parameter $\Sigma|A_{ij}|^2$ in the core [6], with the width of the area shown being about 1 $\mu$m. The rotation axis $\vec{\Omega} \parallel \hat{\mathbf{z}}$ is perpendicular to the $xy$ plane and vorticity is nonzero only in the core region.

**Figure 2**. The combined spin and mass current vortex in $^3$He-B [8] is a composite object, consisting of a linear and a planar object. The singular vortex has here two functions: It is simultaneously a quantized vortex lines with a trapped supercurrent and an endline of a planar domain-wall-like topological soliton of the bulk texture. The two possible configurations are shown in Fig. 3, in which the spin-mass vortex can exist in the rotating container at finite external magnetic field.

The soliton is a defect in the spin-orbit interaction, ie. a wall within which the spin-orbit interaction is not minimized and which separates two degenerate regions with oppositely oriented B-phase textures ($H \neq 0$). In $^3$He-B the relative orientation of the orbital and spin coordinate axes remains undefined as long as the minute spin-orbit interaction is not taken into account. To minimize the dipolar spin-orbit interaction the orbital and spin axes have to be rotated with respect to each other by an angle $\theta_L = \arccos\left(-\frac{1}{4}\right) \approx 104°$. The rotation is generally specified in terms of a rotation matrix $R_{\alpha i}(\hat{\mathbf{n}}, \theta)$, where the unit vector $\hat{\mathbf{n}}$ gives the orientation of the rotation axis around which the rotation by $\theta_L$ is performed. Superfluid coherence, ie. the requirement to minimize the textural gradient energy,

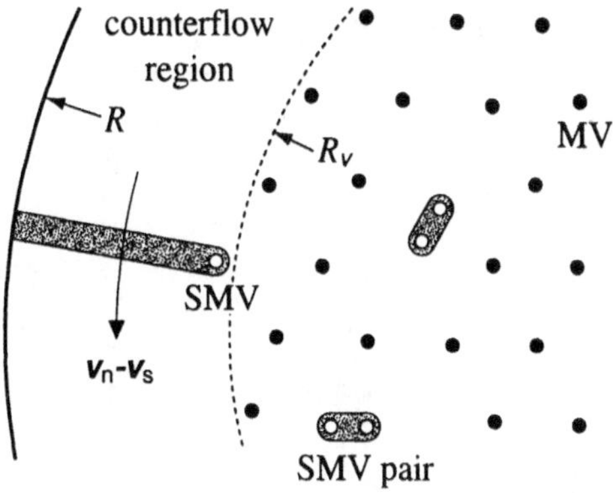


*Figure 3.* Spin-mass vortices in the rotating container with $^3$He-B.

implies that the $\hat{\mathbf{n}}$ vector field becomes a smoothly varying texture in 3-dim space. The healing length for a defect in the $\hat{\mathbf{n}}$ texture is $\xi_D \sim 10\ \mu$m, which thus is approximately the width of a textural soliton wall.

As depicted with arrows in the figure, $\hat{\mathbf{n}}$ is oriented perpendicular to the soliton wall in the bulk liquid and antiparallel on the opposite sides of the wall. Across the soliton $\theta$ traverses from $\theta_L$ to $\pi$ in the center, where the orientation of $\hat{\mathbf{n}}$ reverses, and back to $\theta_L$. On a path around the end of the soliton sheet $\hat{\mathbf{n}}$ smoothly changes orientation while $\theta$ remains fixed at $\theta_L$.

The singular hard core has a radius comparable to the superfluid coherence length $\xi$ which is typically 3 orders of magnitude smaller than the dipolar healing length $\xi_D$ and the width of the $\theta$ soliton sheet. The core of the spin-mass vortex is a combination of two defects: 1) The winding of the superfluid phase factor by $2\pi$ around the core corresponds to a trapped superfluid circulation of one quantum $\kappa_0$. This is the superfluid mass current. 2) The core is also a disclination line for the spin-orbit interaction with an accompanying trapped spin current. This is the spin current vortex.

**Figure 3.** The spin-mass vortex may be incorporated in a vortex cluster in the rotating container in two different configurations. A schematic cross-sectional cut transverse to the symmetry axis of the rotating cylinder is shown, with 5 spin-mass vortices in a vortex cluster. The usual mass current vortices are depicted with black dots while the spin-mass vortices correspond to the open circles. When the total number of vortex lines is less than the equilibrium value, then the existing lines are confined to a central vortex cluster which here is shown to have a radius $R_v$. The radius of the rotating container is $R$. The cluster is isolated from the cylinder wall by vortex-free counterflow, which as a function of the radial distance $r$ has the velocity $v = v_s - v_n = -\Omega r(1 - R_v/r)$ within the counterflow annulus $R > r > R_v$ [9].

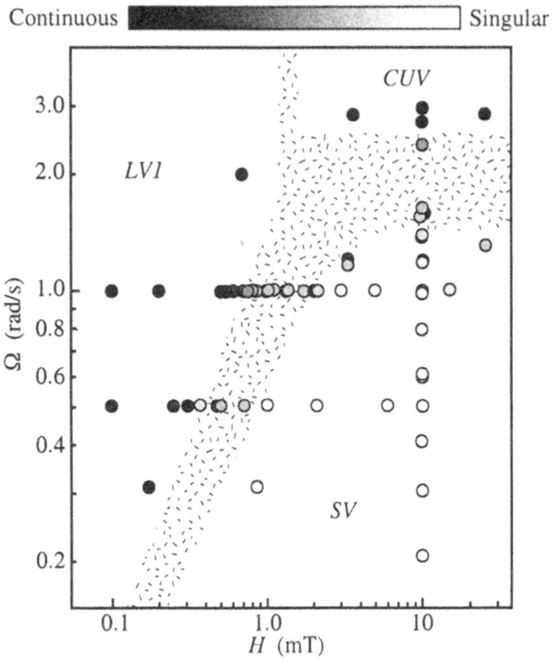


*Figure 4.* Measured phase diagram of different vortex structures in $^3$He-A at $T_c$.

The spin-mass vortices are the edge lines for the $\theta$ solitons. The latter are marked in grey. A $\theta$ soliton can terminate either in a spin-mass vortex or on the cylinder wall. The two possible configurations for a spin-mass vortex are: 1) A soliton connects pairwise two spin-mass vortex lines which move to a distance $d$ of each other where the soliton's surface tension $\sigma$ balances the inter-vortex repulsion $F_M = \kappa_0^2 \rho_s/(2\pi d)$. This occurs at a distance $d = \kappa_0^2 \rho_s/(2\pi\sigma) \sim 6\xi_D$ which is generally somewhat less than the inter-vortex distance $2r_v \approx \sqrt{2\kappa_0/(\pi\Omega)}$ between two mass-current vortices in a vortex array at usual experimentally accessible rotation velocities ($\Omega < 4$ rad/s). 2) A spin-mass vortex at the edge of the vortex cluster may have a soliton tail which connects to the cylinder wall. Compared to the Magnus force from the counterflow, $F_M = \kappa_0 \rho_s (v_s - v_n)$, the surface tension $\sigma$ is small and therefore the spin-mass vortex lies close to the outer boundary of the vortex cluster. In this configuration the length of the soliton is maximized, it gives rise to larger changes in the NMR response, and can be more readily identified.

**Figure 4.** At $T_c$ critical velocities vanish and the phase diagram of different vortex structures can be measured as a function of rotation velocity ($\Omega$) and applied magnetic field ($H$) [10]. Each data point represents a rotational state which was obtained by cooling slowly from the normal phase to

$^3$He-A at a steady rotation velocity $\Omega$ in an applied magnetic field $H$. During the slow second order transition the equlibrium state at $T_c$ is formed, with the equilibrium amount and the equilibrium structure of quantized vorticity.

Below $T_c$ in the A phase the type of vortex lines was analyzed from the nuclear magnetic resonance spectra: White circles denote a state with only singly-quantized vortex lines which have a singular core. Their structure is shown in Fig. 9. Black circles represent singularity-free vortex structures which at zero field are of the type shown in Fig. 7 and at high field as in Fig. 8. The degree of shading from white to black measures the relative amount of vorticity in the singularity-free structures.

The vertical phase boundary is the dipole locking $\leftrightarrow$ unlocking transition, shown in more detail in Fig. 5. The horizontal boundary is the transition in the dipole-unlocked regime from the singly-quantized singular-core vortex (Fig. 9) at low $\Omega < 0.7$ rad/s to the singularity-free doubly-quantized vortex (Fig. 8) at large $\Omega > 1$ rad/s. The phase boundaries agree semi-quantitatively with the calculated phase diagram without adjustable parameters [11].

**Figure 5.** In $^3$He-A a topological transition occurs in the vortex texture as a function of the applied magnetic field [12]. This dipole-locking $\leftrightarrow$ unlocking transition takes place in the spatial orientational distribution of the spin quantization axis. The orientation of the spin axis is usually specified in terms of a unit vector $\hat{\mathbf{d}}$ which is oriented perpendicular to the spin $\mathbf{S}$. The orientation of the orbital quantization axis, in turn, is denoted with a unit vector $\hat{\ell}$ which points in the direction of the local orbital momentum $\mathbf{L}$. At low fields the dipolar spin-orbit interaction wins and orients $\hat{\mathbf{d}} \parallel \hat{\ell}$. This is referred to as a *dipole-locked texture.* At high fields the coupling to the external field wins and orients $\hat{\mathbf{d}}$ uniformly perpendicular to $\mathbf{H}$ while the orbital part is left to produce the vorticity. High-field textures are called *dipole unlocked.* The critical field for the dipole-locking $\leftrightarrow$ unlocking transition is on the order of the dipolar field $H_D \sim 1$ mT at which the spin-orbit interaction equals the magnetic field anisotropy energy.

As illustrated schematically on the top, at high fields the dipole-unlocked texture is localized in the central part of the vortex, the so-scalled *soft vortex core.* Outside the soft core $\hat{\ell}$ is oriented uniformly parallel to $\hat{\mathbf{d}}$ (ie. $\hat{\ell} \parallel \hat{\mathbf{d}} \perp \mathbf{H}$). At low fields below the critical value $H_c$ the uniform $\hat{\ell}$ texture is reduced in area and the central part with the inhomogenous $\hat{\ell}$ texture increases, but now $\hat{\mathbf{d}}$ remains dipole locked to $\hat{\ell}$ even in this central region. Thus the dipole-locking $\leftrightarrow$ unlocking transition separates two different topologies of the $\hat{\mathbf{d}}$ texture in the central part of the vortex. It is the dipole-unlocked part of the texture which leaves its distinct signature in the NMR spectrum and makes possible the NMR spectroscopy of order

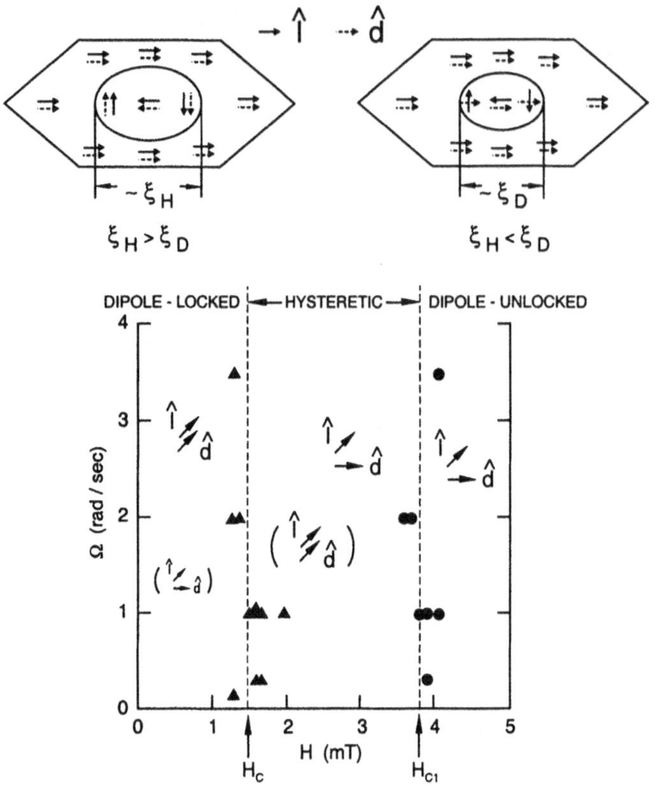


*Figure 5.* Topological dipole-locking ↔ unlocking transition in the A-phase vortex texture.

parameter defects in the A phase [13].

The low-field inhomogenous $\hat{\mathbf{d}}$ texture is thus not stable as a function of the external field, but undergoes a first order phase transition. A measurement of this transition is shown in the main part of the figure in the rotation velocity ($\Omega$) – applied magnetic field ($H$) plane. The different field regimes are denoted by the relative alignments of the $\hat{\mathbf{d}}$ and $\hat{\ell}$ vectors. Relative alignments in parenthesis denote a metastable field regime while without parenthesis the respective type of relative orientation is stable. The measured critical field $H_c(\Omega)$ is marked with triangles (▲), while filled circles (•) denote the catastrophe line $H_{c1}(\Omega)$ at which the low-field dipole-locked soft cores finally lose stability at the temperature $0.9\,T_c$ of the measurements. At lower temperatures a distorted form of the dipole-locked vortex remains metastable to much higher fields: Below $0.6\,T_c$ $H_{c1}$ has been found to be more than 10 mT [13].

**Figure 6.** In $^3$He-A the vortex texture in a rotating container is formed by the spatial variation in the orientation of the orbital quantization axis,

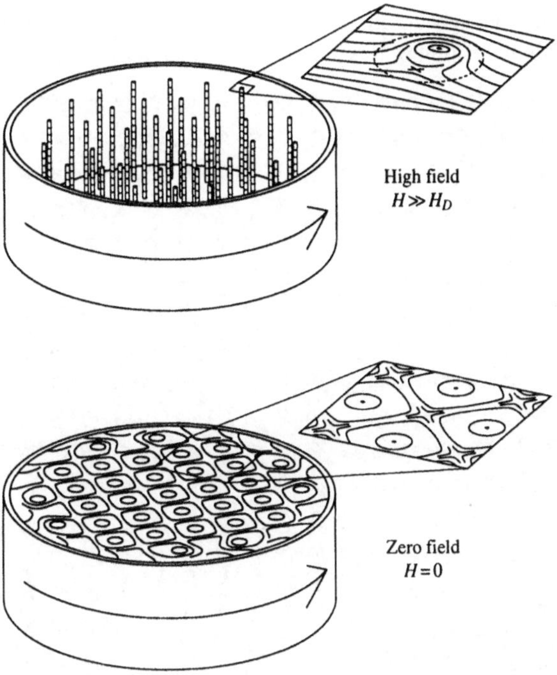


*Figure 6.* Rotating container with singularity-free vortex textures in $^3$He-A.

the $\hat{\ell}$ vector field. Vorticity is nonzero, $\vec{\nabla} \times \mathbf{v}_s \neq 0$, in the regions where the $\hat{\ell}$ texture is inhomogeneous. The textures are continuous or singularity-free, if rotation is started in the A phase, because they have typically an order of magnitude lower critical velocity than a vortex with a singular core [14].

The extreme cases as a function of the applied magnetic field are: *(Bottom)* Zero-field periodic dipole-locked vortex texture. Here the nonuniform part of the $\hat{\ell}$ texture fills the entire vortex lattice cell. The dominant structure has a square lattice cell with 4 circulation quanta, as shown in Fig. 7. *(Top)* High-field ($H \gg H_D$) dipole-unlocked texture. Here the soft vortex core is formed and superficially the triangular vortex lattice starts to resemble the more conventional case with singular vortex cores. Inside the soft core the $\hat{\ell}$ texture has the nonuniform and nonaxisymmetric structure shown in Fig. 8. The velocity $v_s$ of the of the suppercurrent increases smoothly from zero at the center of the soft core to a maximum at the edge. Outside the soft core the $\hat{\ell}$ texture is uniform, vorticity is zero, and the velocity of the persistent superflow decays with distance as $v_s = \kappa/(2\pi r)$, where $\kappa = 2\,\kappa_0 = h/m_3$ is the circulation trapped around the soft core.

**Figure 7**. The zero-field dipole-locked vortex in $^3$He-A has a square elementary lattice cell with 4 circulation quanta [15, 11]. The figure shows in the transverse $xy$ plane (when $\hat{\mathbf{z}} \parallel \vec{\Omega}$) the orientation of the orbital unit

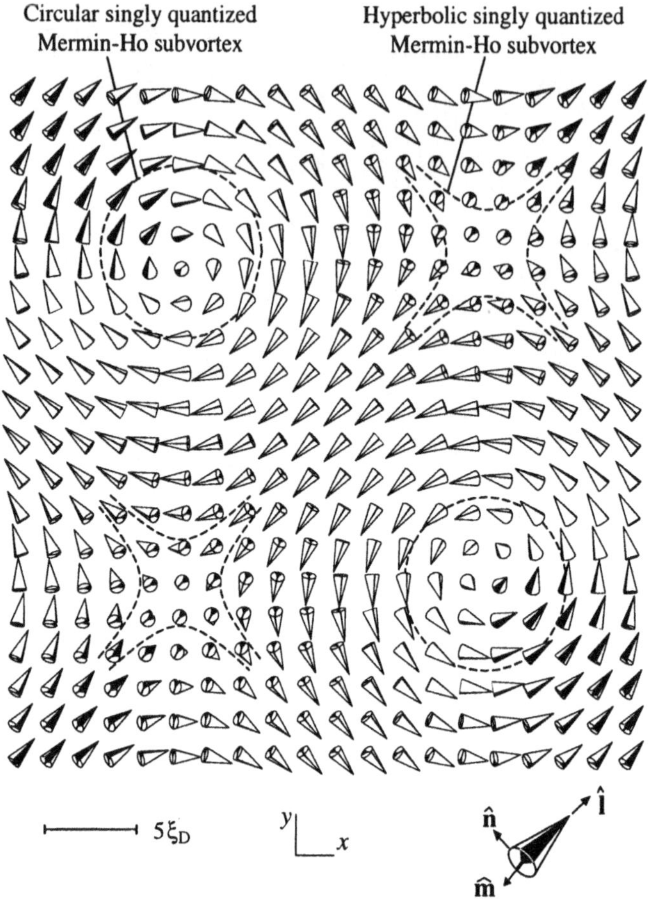


*Figure 7.* Orbital $\hat{\ell}$ texture of the dipole-locked vortex at zero external magnetic field in $^3$He-A.

vector triad $\hat{\ell} = \hat{\mathbf{m}} \times \hat{\mathbf{n}}$ within one vortex lattice cell. The yardstick gives the healing length $\xi_D$ of the dipolar spin-orbit interaction relative to the gradient energy. Because of dipole locking $\hat{\ell}$ and $\hat{\mathbf{d}}$ have the same orientation everywhere. The viewing direction of this figure (and the following similar ones) is inclined by 14° from the $z$ axis.

The texture is made up of 4 so-called Mermin-Ho vortices, two vortices with circular distribution of $\hat{\ell}$ orientations and two with hyperbolic. The boundaries of each of these 4 subunits is defined by the contours where $\hat{\ell}_z = 0$. A circular Mermin-Ho vortex includes all orientations of the positive unit hemisphere, where $\ell_z > 0$, and the hyperbolic vortex those of the negative hemisphere, where $\ell_z < 0$. Each of the Mermin-Ho vortices contributes one quantum of circulation to the total circulation of 4 quanta which is reached along the edge of the lattice cell. A stable configuration is attained with the pairwise antisymmetric arrangement in the square lattice. Other possibilities are a pairwise molecularization as in the dipole-unlocked high-field vortex (Fig. 8) or an alternating linear chain as in the vortex sheet (Fig. 12).

The above vortex texture has been derived from numerical minimization of the full textural free energy expression for an infinite bulk fluid with periodic boundary conditions at temperatures close to $T_c$ [11]. At zero magnetic field the energy arises entirely from the gradient terms which are minimized when the orientational distribution over the lattice cell is as smooth as possible. When the magnetic field is switched on, the structure remains qualitatively unchanged, but the magnetic anisotropy energy attempts to reduce $|\hat{d}_z|$ and thus regions with large $|\hat{\ell}_z|$ values are squeezed closer to the center of the individual Mermin-Ho vortices at the expense of an increase in the gradient energy. The calculations show that at nonzero magnetic field also two other dipole-locked vortex textures have similar textural energies and may in fact become stable equilibrium states. One of them resembles a linear chain of circular and hyperbolic vortices, the dipole-locked variant of the vortex sheet, while the other is the dipole-locked molecule of a circular – hyperbolic pair in a triangular lattice [11]. Dipole-locking $\leftrightarrow$ unlocking transitions are of first order because of the discontinuous change in the topology of the $\hat{\mathbf{d}}$ texture. They are experimentally easy to spot because of the associated textural hysteresis. In contrast, transitions between different dipole-locked textures would be weaker and harder to discern in the measurement.

**Figure 8.** The doubly-quantized high-field vortex in $^3$He-A [16, 11] has a nonaxisymmetric soft core which consists of a circular – hyperbolic pair of Mermin-Ho vortices. Together these two include all $4\pi$ orientations of $\hat{\ell}$ within a unit sphere. This is seen by following the rotation of $\hat{\mathbf{m}}$ and $\hat{\mathbf{n}}$ around $\hat{\ell}$ when one makes one full circle along the outer edge of the figure: Two full $2\pi$ rotations are performed, which means that the superfluid circulation trapped around the soft core corresponds to 2 quanta. The spin anisotropy axis $\hat{\mathbf{d}}$ is oriented along $\hat{\mathbf{x}}$ outside the soft core and is deflected only little from this orientation within the core. Thus the magnetic anisotropy energy is almost entirely minimized. The radius of the soft core is seen to be approximately $3\,\xi_D$, when compared to the yardstick on the bottom of the figure. It is determined by the balance of the kinetic flow energy outside the core and the spin-orbit energy within the core, since the total gradient energy in the core is roughly independent of the core size. The nonaxisymmetric structure has the consequence that the triangular elementary lattice cell is not exactly of the ideal hexagonal form, but elongated by several percent along $\hat{\mathbf{y}}$ (ie. the direction connecting the centers of the Mermin-Ho vortices).

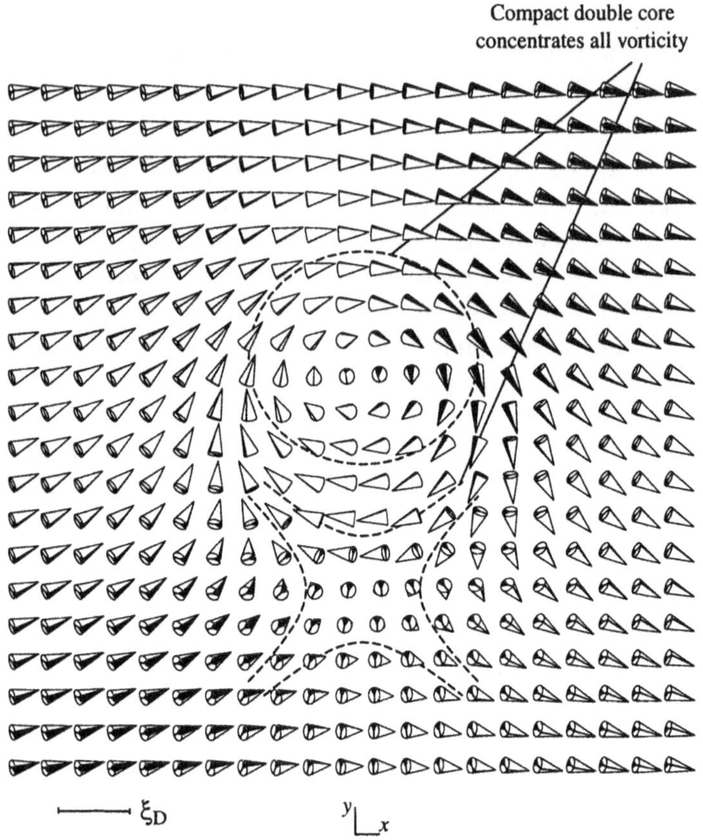


*Figure 8.* Dipole-unlocked orbital $\hat{\ell}$ texture in the soft core of the doubly-quantized high-field vortex in $^3$He-A.

The doubly-quantized vortex is generally always formed when rotation is started at slow angular acceleration in a magnetic field, if the liquid is already in the A phase state. Only the vortex sheet has a lower critical velocity, but it is not formed if a splay soliton sheet is not present and if $\Omega$ is slowly and monotonically increased.

**Figure 9.** The dipole-unlocked soft core of the singly quantized vortex in $^3$He-A [16, 11] includes also a singular hard core, where the order parameter amplitude deviates from the A phase form. As in Fig. 2, the radius of the hard core is on the order of the superfluid coherence length $\xi$ and orders of magnitude smaller than that of the soft core. A singly quantized vortex provides a smoother coverage of vorticity and thus a lower value in the hydrodynamic flow contribution. This property makes the singly quantized vortex preferable at low $\Omega$ (below 1 rad/s). However, as discussed in the context of Fig. 4, except very close to $T_c$, the critical velocity for creating

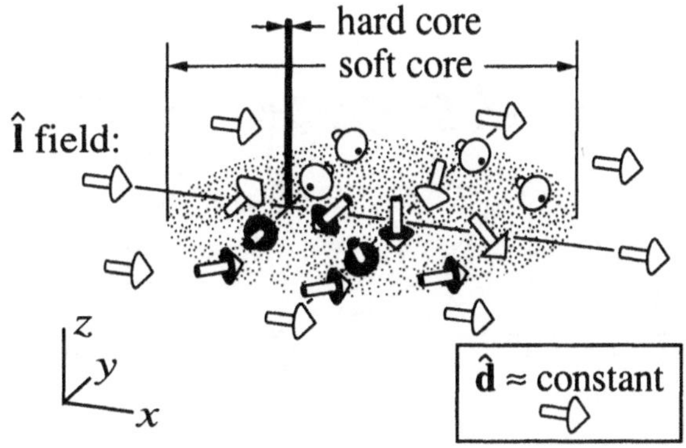


*Figure 9.* Sketch of the dipole-unlocked $\hat{\ell}$ texture in the soft core of the singly quantized vortex in $^3$He-A.

the singular core is much larger than that needed to form a singularity-free structure and thus singular vortices are not created at temperatures below $T_c$.

A singly quantized vortex in A phase could in principle have the structure of a simple phase vortex, like in $^4$He or $^3$He-B, with $\hat{\mathbf{d}} = \hat{\ell}$ in the transverse plane ($\perp \mathbf{H}, \vec{\Omega}$) everywhere else except at the hard core where the order parameter changes form. However, textural energy considerations show that a lower energy configuration is obtained by surrounding the hard core with a dipole-unlocked soft core. In this situation the vortex still looks like a phase vortex from outside the soft core, with $\hat{\ell} = \hat{\mathbf{d}} = \hat{\mathbf{x}}$, and $\hat{\mathbf{m}} + i\hat{\mathbf{n}}$ rotating about $\hat{\ell}$ by $2\pi$ on encircling the soft core once. This configuration thus has trapped one quantum of circulation around the soft core.

Within the soft core the orbital triad rotates around $\hat{\mathbf{n}}$ by about 90° on moving from the perimeter to the hard core. At the hard core one then finds that $\hat{\ell}$ rotates by $2\pi$ about $\hat{\mathbf{m}}$ when one moves around the core, or that the polar angle $\beta$ of $\hat{\ell}$ changes by $2\pi$. This means that there is no circulation about the hard core and the vorticity vanishes: $\vec{\nabla} \times \mathbf{v}_s = \sin\beta\,(\vec{\nabla}\beta \times \vec{\nabla}\alpha) = 0$, as the azimuthal angle $\alpha$ remains constant. The hard core is a pure disgyration line, all circulation arises from the soft core with its $2\pi$ distribution of $\hat{\ell}$ orientations, and $v_s$ goes smoothly to zero on approaching the hard core. Such a structure makes the soft core nonaxisymmetric, the hard core lies slightly displaced from the geometric center of the soft core, and the lattice is distorted from the ideal hcp structure.

**Figure 10**. In the $^3$He literature planar structures are called solitons. In a magnetic field ($H > H_D$) stationary solitons are localized and topologically stable dipole-unlocked structures [17]. Such a soliton sheet separates two regions with degenerate $\hat{\ell}$ textures: On one side of the sheet $\hat{\ell}$ is aligned parallel to $\hat{\mathbf{d}}$ while on the other side $\hat{\ell}$ and $\hat{\mathbf{d}}$ are antiparallel. The

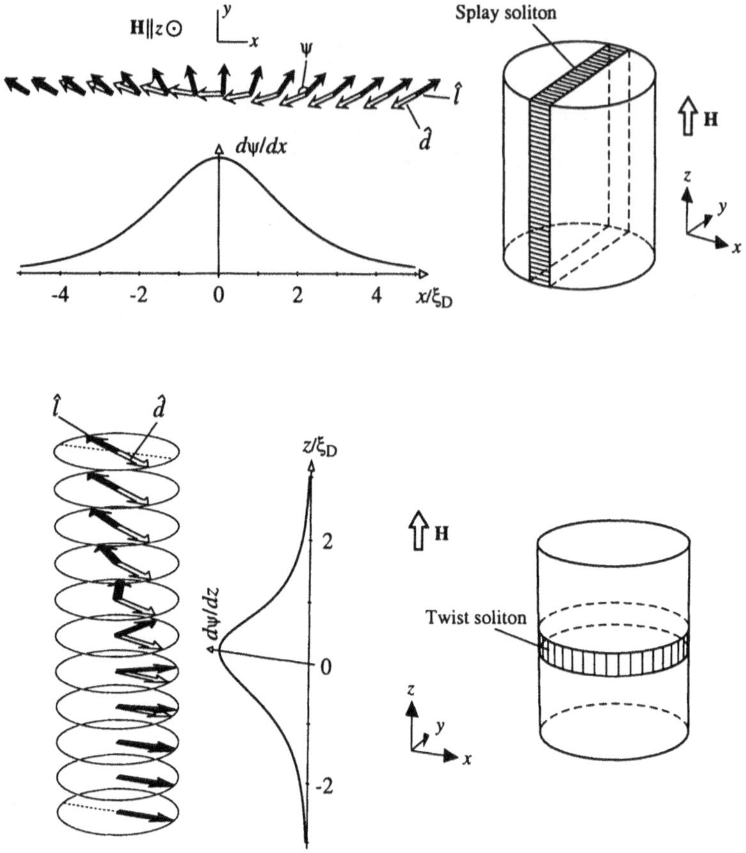


*Figure 10.* Topological soliton sheets in the $\hat{\ell}$ texture of $^3$He-A.

reorientation across the wall is primarily the responsibility of the $\hat{\ell}$ vector. Nevertheless, these sheets are called composite solitons since the orientation of $\hat{\mathbf{d}}$ also changes across the soliton: It participates in the reorientation process by performing approximately $\frac{1}{5}$ of the change. In the simple case of zero flow both $\hat{\mathbf{d}}$ and $\hat{\ell}$ are confined in the plane $\perp$ **H** also within the soliton.

In the cylindrical container two soliton structures are of particular interest: The transverse soliton appears as a plane perpendicular to the symmetry axis of the cylinder. In an axially oriented field it has *twist* structure. The longitudinal soliton is oriented along the cylinder axis and has *splay* structure in the axial field. Since $\hat{\mathbf{d}}$ is confined to the transverse plane independently of the soliton orientation, both structures preserve $\ell_z = 0$ also within the soliton sheet. In the twist soliton the $\hat{\ell}$ orientation is twisted around as a function of $z$ while traversing across the sheet along $\hat{\mathbf{z}}$. In the

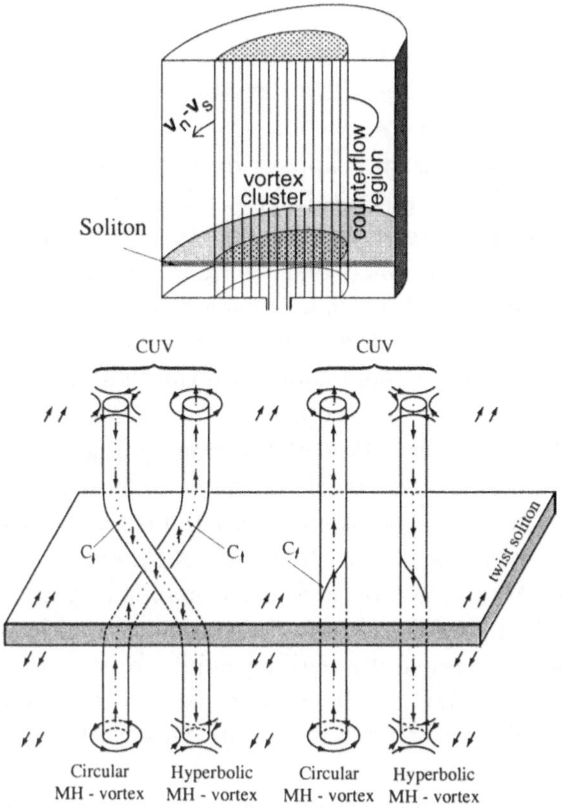


*Figure 11.* Intersection of the doubly-quantized singularity-free vortex line with a transverse soliton in $^3$He-A.

splay structure the reorientation of $\hat{\ell}$ occurs at constant $z$ within one and the same transverse plane. It is the magnetic field orientations which chooses between the two soliton structures. Thus in a transverse magnetic field the longitudinal and transverse solitons have the twist and splay structures, respectively.

Both structures are modified by counterflow, when rotation is started. Since $\hat{\ell}$ is aligned either parallel or antiparallel by the counterflow, initially the width of the transverse soliton is compressed. Furthermore, within the soliton the critical velocity for the creation of the doubly-quantized singularity-free vortex line (Fig. 8) drops typically by a factor of 3 from that in the bulk outside the soliton [14]. Thus vortex lines are created at low rotation velocity. These tend to destabilize the presence of a transverse soliton sheet within the cylinder such that it is pushed either to the top or bottom wall and annihilated there.

In contrast, the longitudinal soliton sheet is oriented perpendicular to the counterflow direction. It becomes unstable with respect to the creation of vorticity at essentially zero critical velocity and the vortex sheet (Fig. 12) starts to develop [18]. Thus the splay soliton, by transforming into the vortex sheet, is stabilized by rotation.

Transverse solitons usually appear during a rapid cool down through $T_c$, when the container is not rotating. In a long cylinder several transverse soliton sheets can then be stacked one on top of the other in a long-lived unstable state of metastability. During rapid rotational acceleration they can pairwise combine and annihilate each other or they can be swept to the top or bottom walls of the cylinder. The longitudinal soliton is only infrequently found after a rapid cool down through $T_c$ (at $\Omega = 0$), since its stability in the cylinder is most fragile.

**Figure 11**. The intersection of the doubly-quantized singularity-free vortex line with a transverse twist soliton is a continuous point-like $\hat{\ell}$ texture. It belongs to the $\pi_3$ homotopy group and is created when a line defect of $\pi_2$ topology crosses a planar $\pi_1$ defect. Although the $\hat{\ell}$ texture in this knot-like structure has not been directly mapped by the experiment, its existence must be inferred from the measurements and in this sense it represents the first experimentally verified point-like structure in the $^3$He superfluids. The intersection can be maintained in a state of unstable balance in the rotating cylinder at a relatively low density of vortex lines ($\Omega \lesssim 0.3$ rad/s).

There are two possible configurations in which the knot can appear: The circular and hyperbolic Mermin-Ho vortices, the two constituents of the doubly-quantized vortex (Fig. 8), may interchange places and form a tangled intersection within the soliton plane *(on the left)*. In this case the Mermin-Ho vortices remain continuous across the soliton plane. Since the $\hat{\ell}$ orientation is also changed by 180° across the soliton, this intersection leaves the $\hat{\ell}$ texture of the doubly-quantized vortex line in Fig. 8 unchanged.

In the second case *(on the right)* the Mermin-Ho vortices do not change places, but each of them is twisted separately while crossing the soliton. Here the doubly-quantized vortices on the two sides of the soliton plane become mirror images of each other. At present time it is not known which one of these two intersections represents the lower textural energy state.

**Figure 12**. The dipole-unlocked vortex sheet in $^3$He-A [18] is a combined object which is made up of a longitudinal planar soliton (Fig. 10) and the two linear Mermin-Ho vortices (Fig. 7). Within the sheet the two types of Mermin-Ho vortices, the circular and the hyperbolic vortex, are stacked one after the other as an alternating chain. At any reasonable value of $\Omega$ the vorticity becomes evenly distributed along the sheet: On moving in the $x$ direction along the centerline of the sheet, it is seen that $\hat{\ell}$ is winding around the $x$ axis at uniform pitch.

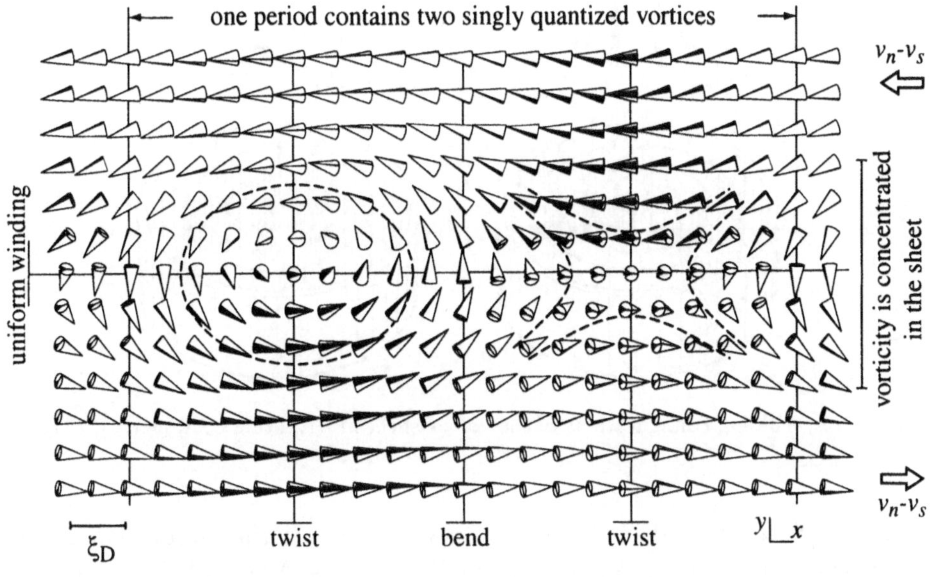


*Figure 12.* Dipole-unlocked orbital texture of the vortex sheet in $^3$He-A.

Another important characteristic is the alternating pattern of bend and twist sections. In the bend section $\hat{\ell}$ rotates by 180° while confined in the same transverse plane at constant $z$, whereas in the twist section $\hat{\ell}$ is rotated around the $y$ axis by 180°. Outside the sheet $\hat{\ell}$ and the local counterflow $\mathbf{v} = \mathbf{v}_s - \mathbf{v}_n$ are oriented parallel to the sheet, but on opposite sides this orientation is reversed by 180°. Across the sheet the velocity changes in magnitude due to the vorticity contained in the sheet. The overall configuration of the vortex sheet in the rotating container consists of parallel layers of vorticity (Fig. 13), separated by irrotational counterflow. This hydrodynamic structure was first discussed by Landau and Lifshitz [19].

**Figure 13.** In the equilibrium state the vortex sheet forms a continuous meander which has two connection lines with the lateral cylinder wall. Vorticity enters and leaves the sheet at these connection lines. The meandering sheet fills the rotating container evenly with equidistant folds. Their spacing is determined by the balance of the sheet's surface tension $\sigma$ and the kinetic energy of the counterflow: $b = [3\sigma/(\rho_{s\|}\,\Omega^2)]^{1/3}$. Thus with increasing rotation velocity the length of the sheet is extended $\propto \Omega^{2/3}$ and the number of folded layers increases.

The spacing $b$ of the folds and the periodicity $p$ within the sheet have to fulfill the requirement of solid-body rotation on a radial length scale $\gg b$, so that the large-scale average of the counterflow velocity becomes $< \mathbf{v} > = \vec{\Omega} \times \mathbf{r}$. This condition gives the usual equilibrium density of vorticity:

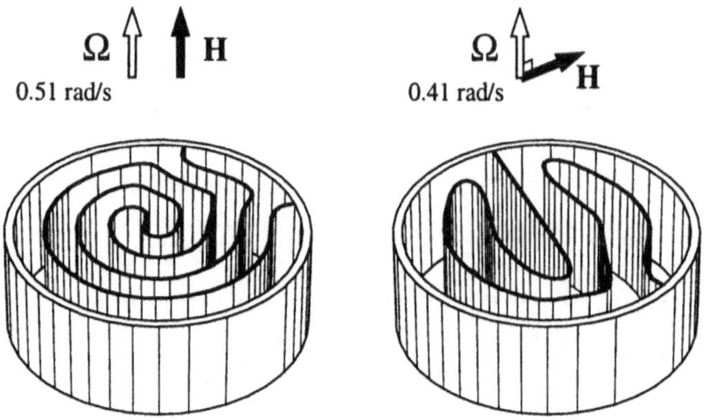


*Figure 13.* Folding configurations of the vortex sheet in a rotating container with $^3$He-A.

$\kappa/bp = |\vec{\nabla} \times \mathbf{v}| = 2\Omega$, where $\kappa = 2\kappa_0 = h/m_3$ is the circulation of one elementary unit with the two Mermin-Ho vortices. Although $b > p$, roughly speaking $b$ and $p$ are comparable to the intervortex distance in a vortex array consisting of doubly-quantized vortex lines. This means that $p$ is larger than the distance between the two Mermin-Ho constituents within one doubly quantized vortex. Due to this distributed structure, at higher rotation velocities the vortex sheet becomes an economic arrangement of quantized vorticity. Although not yet confirmed by measurement, numeric minimization of the textural energies suggests that at $\Omega > 3$ rad/s the vortex sheet becomes the equilibrium form of vorticity [11].

The arrangement of the folds depends on the direction of the applied magnetic field. In an axially oriented field a more symmetric configuration evolves, which resembles a double spiral of coaxial sheets. In the transverse field $\hat{\mathbf{d}}$ is confined in the transverse plane to the orientation perpendicular to the field. To minimize dipole coupling, $\hat{\ell}$ attempts to align itself parallel to $\hat{\mathbf{d}}$ in much of the container and therefore the folds prefer to form parallel walls perpendicular to $\mathbf{H}$. If the field is rotated to a new orientation after forming the vortex sheet, frustrated and distorted folding patterns are created.

An important property of the vortex sheet is its low critical velocity. If the rotation velocity is increased above the equilibrium value, a macroscopic counterflow forms which encircles the vortex sheet next to the container wall. It compresses the vorticity towards the center of the container and thereby leaves the two regions of the soliton sheet at the contact lines devoid of vorticity. In these two regions the flow crosses the soliton sheet and the critical velocity is substantially reduced.

The low critical velocity at the connection lines makes the vortex sheet the dominating form of vorticity as soon as the longitudinal soliton is present. In rapidly changing rotation the vortex sheet generally also dominates the creation and annihilation of vorticity. This is caused by the fact that the vortex sheet has the fastest dynamic response for minimizing the dominant energy contribution, the kinetic energy of the macroscopic counterflow $\sim \frac{1}{2}\rho_{s\|}(v_s - v_n)^2$, where $\rho_{s\|}$ is the density of the superfluid component. Consequently more often than not, the vortex sheet is the regnant form of vorticity in the rotating container.

## References


1. D. Vollhardt and P. Wölfle, *The superfluid phases of helium 3* (Taylor & Francis, London, 1990).
2. G.E. Volovik, *Exotic properties of superfluid* $^3$*He* (World Scientific Publishing Co., Singapore, 1992); M.M. Salomaa and G.E. Volovik, *Rev. Mod. Phys.* **59**, 533 (1988); A.L. Fetter, in *Prog. Low Temp. Phys.*, Vol X, p. 1, (Elsevier Science Publ., Amsterdam, 1986).
3. G.E. Volovik, *Physica* B **255**, 86 (1998); V.B. Eltsov, M. Krusius, and G.E. Volovik, *preprint* (http://xxx.lanl.gov/abs/cond-mat/9809125).
4. O.V. Lounasmaa and E.V. Thuneberg, *Proc. Natl. Acad. Sci. USA* **96**, 7760 (1999).
5. O.T. Ikkala, G.E. Volovik, P.J. Hakonen, Yu.M. Bunkov, S.T. Islander, and G.A. Kharadze, *Pis'ma Zh. Eksp. Teor. Fiz.* **35**, 338 (1982) [*JETP Lett.* **35**, 416 (1982)]; P.J. Hakonen, O.T. Ikkala, S.T. Islander, O.V. Lounasmaa, and G.E. Volovik, *J. Low Temp. Phys.* **53**, 425 (1983).
6. E.V. Thuneberg, *Phys. Rev. Lett.* **56**, 359 (1986); *Phys. Rev.* B **36**, 3583 (1987).
7. Y. Kondo, J.S. Korhonen, M. Krusius, V.V. Dmitriev, Yu.M. Mukharsky, E.B. Sonin, and G.E. Volovik, *Phys. Rev. Lett.* **67**, 81 (1991).
8. Y. Kondo, J.S. Korhonen, M. Krusius, V.V. Dmitriev, E.V. Thuneberg, and G.E. Volovik, *Phys. Rev. Lett.* **68**, 3331 (1992); J.S. Korhonen, Y. Kondo, M. Krusius, E.V. Thuneberg, and G.E. Volovik, *Phys. Rev.* B **47**, 8868 (1993-II).
9. M. Krusius, E.V. Thuneberg, and Ü. Parts, *Physica* B **197**, 376 (1994).
10. Ü. Parts, J.M. Karimäki, J.H. Koivuniemi, M. Krusius, V.M. Ruutu, E.V. Thuneberg, and G.E. Volovik, *Phys. Rev. Lett* **75**, 3320 (1995).
11. J.M. Karimäki and E.V. Thuneberg, *Phys. Rev.* B, in press (http://xxx.lanl.gov/abs/cond-mat/9902207).
12. J.P. Pekola, K. Torizuka, A.J. Manninen, J.M. Kyynäräinen, and G.E. Volovik, Phys. Rev. Lett. **65**, 3293 (1990).
13. V.M. Ruutu, Ü. Parts, and M. Krusius, *J. Low Temp. Phys.* **103**, 331 (1996).
14. V.M. Ruutu, J. Kopu, M. Krusius, Ü. Parts, B. Plaçais, E.V. Thuneberg, and W. Xu, *Phys. Rev. Lett.* **79**, 5058 (1997).
15. M. Nakahara, T. Ohmi, T. Tsuneto, and T. Fujita, *Prog. Theor. Phys.* **62**, 874 (1979); *ibid.* **60**, 671 (1978).
16. H.K. Seppälä and G.E. Volovik, J. Low Temp. Phys. **51**, 279 (1983); X. Zotos and K. Maki, Phys. Rev. B **30**, 145 (1984); V.Z. Vulovic, D.L. Stein, and A.L. Fetter, Phys. Rev. B **29**, 6090 (1984); T. Ohmi, J. Low Temp. Phys. **56**, 183 (1984); H.K. Seppälä, P.J. Hakonen, M. Krusius, T. Ohmi, M.M. Salomaa, J.T. Simola, and G.E. Volovik, Phys. Rev. Lett. **52**, 1802 (1984); P.J. Hakonen, M. Krusius, and H.K. Seppälä, J. Low Temp. Phys. **60**, 187 (1985).
17. K. Maki and P. Kumar, *Phys. Rev. Lett* **38**, 557 (1977); *Phys. Rev.* B **16**, 182 (1977); D. Vollhardt and K. Maki, *Phys. Rev.* B **20**, 963 (1979); R. Hänninen and E.V. Thuneberg, to be published.
18. Ü. Parts, E.V. Thuneberg, G.E. Volovik, J.H. Koivuniemi, V.H. Ruutu, M. Heinilä, J.M. Karimäki, and M. Krusius, *Phys. Rev. Lett* **72**, 3839 (1994); E.V. Thuneberg, *Physica B* **210**, 287 (1995); M.T. Heinilä and G.E. Volovik, *Physica B* **210**, 300

(1995); Ü. Parts, V.M. Ruutu, J.H. Koivuniemi, M. Krusius, E.V. Thuneberg, and G.E. Volovik, *Physica B* **210**, 311 (1995).
19. L. Landau and E. Lifshitz, *Dokl. Akad. Nauk.* **100**, 669 (1955).
20. V.M. Ruutu, Ü. Parts, J.H. Koivuniemi, M. Krusius, E.V. Thuneberg, and G.E. Volovik, *Pis'ma Zh. Eksp. Teor. Fiz.* **60**, 659 (1994).

# REGULAR VORTEX PRODUCTION BY A MOVING WIRE IN SUPERFLUID $^3$HE-B

D. I. BRADLEY AND G. R. PICKETT
*Department of Physics, Lancaster University, Lancaster, UK*

We report spectacular features on the velocity-force characteristics of a vibrating wire resonator in superfluid $^3$He-B at ultralow temperatures. We see steps and discontinuities in the characteristics. The steps display two separate critical velocities where firstly some 'event' occurs, causing the wire to lose energy and lose amplitude, followed by a second lower critical velocity where the 'event' decouples. We speculate that these events arise from vortex loop creation at asperities on the vibrating wire.

## 1. Introduction

The superfluid order parameter in superfluid $^3$He is rich in structure since it is characterised by both total spin angular momentum $S = 1$ and total orbital angular momentum $L = 1$. As a result there exist many topological defects. One such defect is the macroscopic quantum vortex.

Most studies of quantum vorticity in superfluid $^3$He have been carried out on rotating nuclear demagnetisation refrigerators where the rotating superfluid creates a regular lattice of vortices. These have mainly been studied by NMR[1, 2]. Vorticity is also believed to be created in the rapid cooling following local heating by neutron absorption[3, 4], the Kibble mechanism[5]. Finally, vortices can be cyclically created during the oscillatory flow which is set up when superfluid is forced to flow through an orifice (weak link)[6, 7].

We report here the inverse experiment to that of flow through a weak link, where an oscillating wire loop repetitively creates vorticity in superfluid $^3$He-B at temperatures approaching $100\,\mu$K at zero pressure. The experiments are carried out on $^3$He cooled by a standard Lancaster double cell style nuclear refrigerator[8].

The moving wire is made in the form of a standard vibrating wire resonator in which a thin superconducting filament, either $0.9\,\mu$m or $4.5\,\mu$m in diameter, is bent into an approximately semicircular loop with the ends secured by gluing through a non-conducting plate. The loop has a mechanical



*Y.M. Bunkov and H. Godfrin (eds.),*
*Topological Defects and the Non-Equilibrium Dynamics of Symmetry Breaking Phase Transitions*, 345–351.

resonance perpendicular to the loop plane. The frequency of the resonance, typically a few hundred Hertz, is determined by the stiffness, mass and geometry of the wire loops. A small magnetic field is applied in the plane of the loop. The resonance is driven by an AC current of the appropriate frequency passed through the loop. The Lorentz force on the current provides the drive. As the loop is superconducting the only voltage (ignoring the small inductive voltage) generated across the loop is that induced by the velocity in the field.

We observe features on the velocity-driving force characteristics which for a given wire are repeatable from run to run, even after cycling to room temperature, but which are different from wire to wire. These characteristics we measure by slowly increasing the amplitude of the driving current and monitoring the velocity response at resonance. These features are superimposed on top of the expected and well understood velocity-force characteristic[9].

The quasiparticle density in superfluid $^3$He-B, and hence the damping on a slowly moving object, is dominated by the rapidly varying Boltzmann factor $\exp(-\Delta/kT)$, where $\Delta$ is the superfluid energy gap. The unusual dispersion curve of the quasiparticle excitation gas and the Andreev scattering process gives rise to a force which becomes independent of the velocity $v$ as the velocity rises. The form of the force $F$ on a moving object in the superfluid is $F \propto kT\exp(-\Delta/kT)(1 - \exp(-p_F v/kT))$. In the low velocity regime ($p_F v \ll kT$) this gives a force linear in velocity, proportional to $\exp(-\Delta/kT)p_F v$. With increasing velocity the damping force falls below this linear variation and finally becomes independent of velocity when $p_F v \gg kT$. As the driving force increases, the peak velocity of the oscillating wire increases more and more steeply, due to the reduction in damping, until it reaches a velocity sufficient to cause pair breaking in the superfluid. At this point the damping increases extremely rapidly giving an almost drive independent velocity. This general form of the velocity-drive characteristic is shown inset in Fig. 1.

Fig. 1 shows the new features on velocity-force curves produced by a resonator made of the 0.9 $\mu$m diameter wire. A number of steps where the velocity of the wire is limited are clearly seen. At even higher velocities there also occur a number of discontinuities (which will not be discussed further) where the velocity suddenly drops with increasing driving force. Both these types of features are reproduced as the drive current is reduced. The different curves are taken at different temperatures. Since the background quasiparticle damping rapidly increases as the temperature rises, the curves are displaced. The steps are broadly independent of temperature and can be followed as the temperature rises until swamped by the rapidly increasing quasiparticle damping.

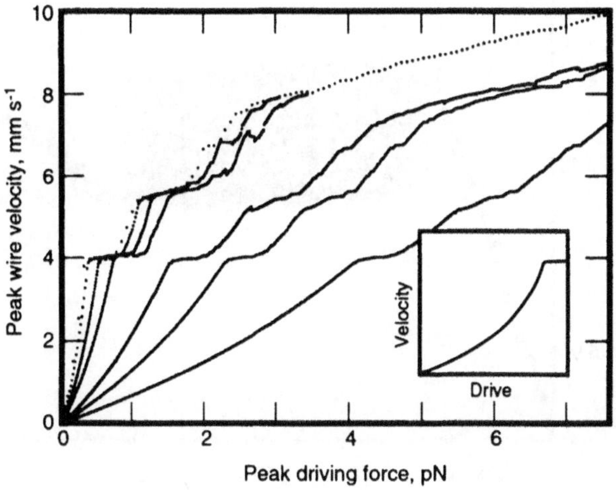


*Figure 1.* Force-velocity characteristic of a 0.9 $\mu$m vibrating wire resonator in superfluid $^3$He-B over a range of temperature from 110 $\mu$K to 150 $\mu$K. The inset shows the general form of the characteristic expected for a smooth cylinder moving in the liquid. The sharp change in slope occurs at the onset of pair-breaking.

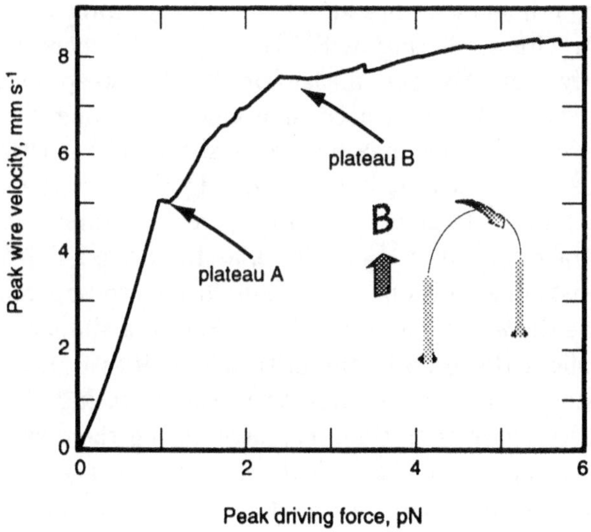


*Figure 2.* Force-velocity characteristics for a 4.5 $\mu$m vibrating wire resonator at a temperature of 114 $\mu$K. Several steps and discontinuities are clearly visible. The inset shows a schematic picture of the vibrating wire resonator.

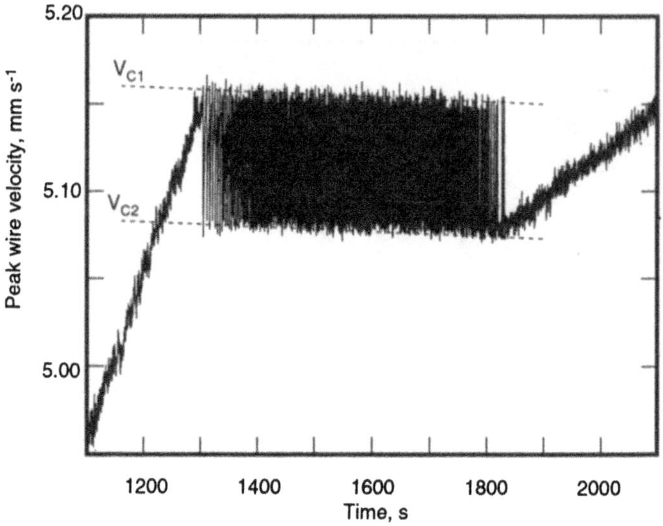


*Figure 3.* The peak wire velocity as a function of time during an increasing linear drive force sweep across the step labelled A in figure 2. The upper and lower critical velocity limits are obvious. The two critical velocities fall slightly as the driving force increases with time.

In later experiments we measured the velocity-force characteristic of a thicker $4.5\,\mu$m wire oscillating at 890 Hz, Fig. 2. Plateaux and discontinuities are clearly seen. For this measurement, the sweep lasting 1.5 hours, the detection electronics had a considerable smoothing time constant as was the case for the earlier measurements above. The startling feature occurs when this time constant is removed. Fig. 3 shows the step at a peak velocity of $5.1\,\mathrm{mm\,s^{-1}}$ (labelled A in Fig. 2) at high time resolution plotted as a function of *time*, rather than drive, as the drive is linearly increased. The step consists of a series of pulses limited by two apparent "critical velocities". If the drive is then swept down, Fig. 4, a similar series of pulses is seen. The slight decrease in the critical velocities during the up-sweep is not understood but it is not due to warming during the sweep as the critical velocities increase as time increases during the down-sweep. Apart from these small changes, these critical velocities are very stable, sufficiently so, that if the drive is halted on the step the pattern simply repeats with time for hour after hour. An expanded view of the time evolution of the beginning and end of step A during the up-sweep is shown in Fig. 5. In the centre of the step, the 'pulse' shape is approximately triangular. Similar pulsing behaviour is observed on the other clear step, at a peak velocity of $7.5\,\mathrm{mm\,s^{-1}}$ (labelled B in Fig. 2).

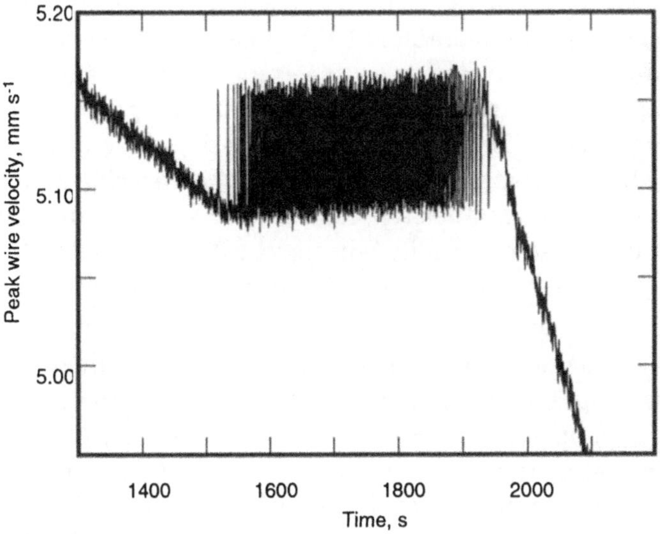


*Figure 4.* The peak wire velocity as a function of time during a down sweep across the same step shown in Fig. 3. The critical velocities increase across the step (as the driving force is reduced) mirroring the up-sweep indicating that this effect is not a result of either a time or temperature variation.

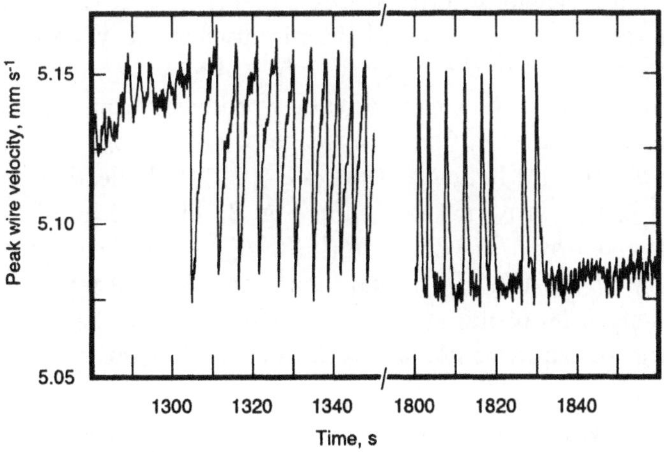


*Figure 5.* An expanded view of the pulses at the start and end of the step shown in figure 3. The change in shape of the pulses is clearly visible. In the centre of the step the shape is approximately triangular.

We see in Fig. 5 that as the drive is ramped the peak velocity smoothly increases until a critical velocity $v_{c1}$ is reached for the first time. The effective damping suddenly increases at this point and the velocity rapidly falls until at a second lower velocity $v_{c2}$ the extra damping vanishes as suddenly as it appeared. The velocity then rises again until $v_{c1}$ is reached and the cycle starts once more. This repeats with increasing drive until after an event at $v_{c1}$ the drive becomes high enough that the velocity is not allowed to fall to $v_{c2}$. At this point the step has been crossed, pulsing ceases, and the velocity again increases steadily with increasing drive but with a shallower slope than that below the step indicating the existence of an additional dissipation mechanism.

Assuming that each pulse corresponds to the production of a vortex loop, an assumption that is not proven, we present the following picture as to what is happening. At $v_{c1}$, the oscillation velocity of the wire is sufficiently high that a local critical velocity is exceeded resulting in the creation of a vortex loop or ring. Such a local critical velocity is likely as the cross section of the wires are anything but a simple cylinder. The structures visible on the velocity-force characteristics are different for different wires, and many steps are seen on an individual wire. We attribute the local critical velocities to the local increase in the superfluid velocity over asperities on the surface of wire, different asperities on a wire being responsible for each step. As the vortex loop or ring grows, the energy needed for its production is extracted from the kinetic energy of the wire and the wire velocity falls. As the peak velocity of the wire for the initial pulse in Fig. 5 takes $\sim 0.3$ seconds to fall to the lower critical value, it is clear that the production process continues over many cycles of the wire oscillation. Whether the ends of the vortex are pinned to the asperity or free to move is not known. In either case, we can estimate the length of vortex produced from the loss of energy from the resonating wire. Our knowledge of $v_{c1}$ and $v_{c2}$ together with the properties of the wire indicates that $\sim 8 \times 10^{-17}$ Joules is extracted from the motion of the wire for the step in Fig. 3. The energy of a vortex ring radius $R$ is given by $\rho/2(h/2m_3)^2 R\,(\ln(8R/\xi_0) - 1.615)$[10], where $\rho$ is the fluid density, $m_3$ is the $^3$He atomic mass and $\xi_0$ is the zero temperature coherence length ($\sim 65\,$nm at zero bar). This suggests that a vortex ring of radius $60\,\mu$m (or equivalently a length of order $300\,\mu$m) would account for the lost energy.

When the wire velocity reaches $v_{c2}$, the vortex is presumably released by the wire. The wire now travels without the extra damping and the velocity builds up again to the upper critical velocity, whereupon the process repeats.

This possible shedding of vortex loops or rings opens up several interesting potential experiments on vorticity. For example, this mechanism would provide the possibility of measuring vortex decay through the heat released by the decay without the need for neutron irradiation, and the unavoidable heating this produces, required in earlier experiments[4].

In conclusion, we have seen unusual features on the velocity-force characteristics of vibrating wires. We speculate that these features arise from the regular creation of vortex loops or rings which subsequently decouple from the wire. If true, this opens up the way to controlled creation of vorticity in superfluid $^3$He without the necessity for rotation or driven flows through narrow orifices.

We acknowledge many useful conversations with S. N. Fisher, A. M. Guénault and S. Iordanski together with the technical backup provided by I. E. Miller and M. G. Ward. This work was supported by the UK EPSRC under grant GR/L67172.

## References

1. Kondo, Y., Korhonen, J.S., Parts, U., Krusius, M., Lounasmaa, O.V. and Gongadze, A.D. (1992) Gap anisotrophy, vortex cluster, and nucleation of vortices in rotating $^3$He-B, Physica B **178**, 90–101.
2. Parts, Ü., Koivuniemi, J.H., Krusius, M., Ruutu, V.M.H. and Zakozov, S.R. (1994) Nucleation of single vortex lines in rotating $^3$He-B, Physica B **194–196**, 771–772.
3. Bäuerle, C., Bunkov, Yu.M., Fisher, S.N., Godfrin, H. and Pickett, G.R. (1996) Laboratory simulation of cosmic string formation in the early Universe using superfluid $^3$He, Nature **382**, 332–334.
4. Bradley, D.I., Fisher S.N. and Hayes, W.M. (1998) Measuring the decay of vorticity in superfluid $^3$He-B - A progress report, J. Low Temp. Phys. **113**, 687-692.
5. Kibble, T.W.B. (1976) Topology of cosmic strings and domains, J. Phys. **A9**, 1387-1398.
6. Packard, R.E., and Davis, J.C. (1994) Phase slip phenomena in superfluid helium, Physica B **197**, 315–323.
7. Varoquaux, E., and Avenel, O. (1994) Phase slip phenomena in superfluid helium, Physica B **197**, 306–314.
8. Bradley, D.I., Guénault, A.M., Keith, V., Kennedy, C.J., Miller, I.E., Mussett, S.G., Pickett, G.R. and Pratt Jr., W.P. (1984) New methods for nuclear cooling into the microkelvin region, J. Low Temp. Phys. **57**, 359–390.
9. Fisher, S.N., Guénault, A.M., Kennedy, C.J. and Pickett, G.R. (1989) *Phys. Rev. Lett.* **63**, 2566–2569.
10. Donnelly, R.J. (1991) *Quantum Vortices in Helium II*, Cambridge University Press, Cambridge.

# $^3$HE AND UNIVERSE PARALLELISM

G.E. VOLOVIK
*Low Temperature Laboratory, Helsinki University of Technology, FIN-02015 HUT, Espoo, Finland*
*and*
*Landau Institute for Theoretical Physics, Moscow, Russia*

We discuss topological properties of the ground state of a spatially homogeneous ensemble of fermions. There are several universality classes of topologically different fermionic vacua; in each case the momentum space topology of the vacuum determines the low-energy (infrared) properties of the fermionic energy spectrum. Among them there is class of the gapless systems which is characterized by the Fermi-hypersurface. This class contains conventional Landau Fermi-liquid and also Luttinger liquid. Another important universality class of gapless systems is characterized by the topologically stable point nodes (Fermi points). Superfluid $^3$He-A and electroweak vacuum belong to this universality class. The fermionic quasiparticles (particles) in this class are chiral: close to the Fermi points they are left-handed or right-handed massless relativistic particles. The low-energy dynamics acquires gauge invariance, Lorentz invariance and general covariance, which become better defined when the energy decreases. Interaction of the fermions near the Fermi point leads to collective bosonic modes, which look like effective gauge and gravitational fields. Since the vacuum of superfluid $^3$He-A and electroweak vacuum are topologically similar, we can use $^3$He-A for simulation of axial anomaly, event horizons, Hawking radiation, rotating vacuum, conical space, etc.

## 1. Introduction

The physical vacuum is a complicated condensed matter[1, 2, 3, 4]. At the moment we know only the low-energy properties of this substance, i.e. the properties at energies much smaller than the Planck energy scale, $E \ll E_{\rm P} = \sqrt{\hbar c^5/G}$, where $G$ is Newton's constant. We know that at the low-energy edge our vacuum has many different symmetries: $U(1)$ and $SU(3)$ gauge symmetries, Lorentz invariance, general coordinate invariance, and discrete $CPT$ symmetry. With increasing energy more elements of symmetry are added: $SU(2)$ symmetry of weak interactions, probably the GUT symmetry and even supersymmetry. When the temperature decreases such



*Y.M. Bunkov and H. Godfrin (eds.),*
*Topological Defects and the Non-Equilibrium Dynamics of Symmetry Breaking Phase Transitions*, 353–387.

symmetries become spontaneously broken. This is the traditional point of view, which is supported by the observation of the similar symmetry breaking in different (many-body) condensed matter systems: superfluids, superconductors, magnets, liquid crustals and ordinary crystals.

However there is another and actually opposite point of view: all symmetries known in the Universe spontaneously (but without any phase transition) appear at the low-energy corner[5]. They become more and more pronounced the lower the energy. The symmetries disappear at higher energies when the Planck energy is approached. This conjecture is also supported by the condensed matter analogy.

Here we discuss the topological origin of such enhanced symmetry in the infrared limit. It is caused by the so-called topologically stable Fermi points in the spectrum of the fermionic system. We show that if such points exist in quantum condensed matter, then at low enough temperature this system exhibits fully or partially the Lorentz invariance, general covariance, and gauge invariance. Moreover the collective modes which describe such a system at low $T$ all are represented by chiral relativistic fermions, gauge fields, and gravity. All are low-energy phenomena, which are absent at higher energies, where the condensed matter has only a very limited set of the global symmetries.

¿From the second point of view some directions in physics look artificial. In particular, since gravity exists only as a infrared phenomenon one should not quantize gravity: Only low-energy gravitons can be quantized [6]. The same concerns the multidimensional string theories, which also give rise to gravitation in the infrared limit: The Fermi point mechanism does not require a high dimensionality for the space-time. The topologically stable Fermi point is just a property of the conventional 3+1 dimensional space-time.

## 2. Manifolds of zeroes

The Fermi point is a particular case of the topologically stable manifolds of zeroes in the fermionic spectrum. Let us start with the simplest such manifold – the Fermi surface.

### 2.1. FERMI SURFACE AS TOPOLOGICAL OBJECT

#### 2.1.1. *Topological stability of Fermi surface*

The Fermi surface naturally appears in the noninteracting Fermi gas, where the energy spectrum of fermions is

$$E(p) = \frac{p^2}{2m} - \mu \ , \tag{1}$$

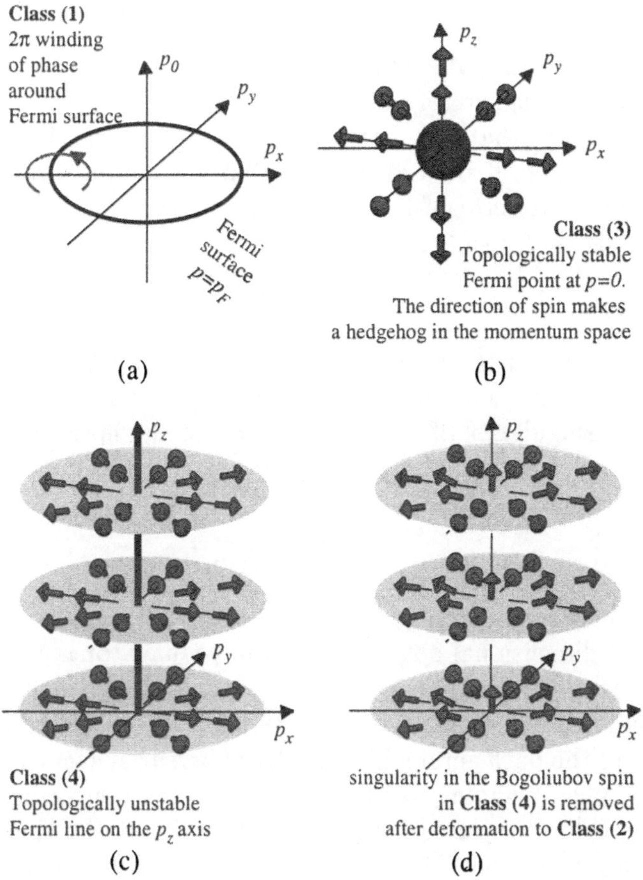


*Figure 1.* (a) The Fermi surface in the 2+1 momentum space. For simplicity the $p_z$ coordinate is hidden so that the Fermi surface is the line $(p_0 = 0, p = p_F)$ in the 2+1 space $(p_0, p_x, p_y)$. This line is a singularity, which is similar to a vortex in a real 3D-space: The phase of the propagator $G = (ip_0 - (p_x^2 + p_y^2 - p_F^2)/2m)^{-1}$ changes by $2\pi$ around the line in the momentum space in the same manner as the phase of the order parameter changes by $2\pi$ around a vortex in the real space. In general case the winding number $N_1$ – the momentum space topological invariant of the Fermi surface – is given by Eq.(4). (b) Fermi point at $\mathbf{p} = 0$ in the 3D momentum space $(p_x, p_y, p_z)$. At this point the particle energy $E = cp$ is zero. A right-handed particle is considered with its spin parallel to the momentum $\mathbf{p}$, *i.e.* $\mathbf{s}(\mathbf{p}) = (1/2)\mathbf{p}/p$. The spin makes a hedgehog in the momentum space, which is topologically stable. The topological invariant of the Fermi point is $N_3 = 1$, where $N_3$ is given by Eq.(13). (c) Fermi line – topologically unstable manifold of zeroes – is shown in the 3D momentum space $(p_x, p_y, p_z)$. The (Bogoliubov) spin (arrows) is confined into the $(p_x, p_y)$ plane and has a singularity on the $p_z$ axis. (d) This singularity can be removed by a continuous transformation. The spin escapes into a third dimension $(p_z)$ and becomes well defined on the $p_z$ axis. As a result, the quasiparticle spectrum becomes fully gapped (the "relativistic" fermion acquires the mass).

and $\mu > 0$ is the chemical potential. The Fermi surface bounds the volume in the momentum space where the energy is negative, $E(p) < 0$, and where the particle states are all occupied at $T = 0$. In this isotropic model the Fermi surface is a sphere of radius $p_F = \sqrt{2m\mu}$.

It is important that the Fermi surface survives even if interactions between particles are introduced. Such stability of Fermi surface comes from the topological property of the Feynman quantum mechanical propagator – the one-particle Green's function

$$\mathcal{G} = (z - \mathcal{H})^{-1} \; . \tag{2}$$

Let us write the propagator for a given momentum $\mathbf{p}$ and for the imaginary frequency, $z = ip_0$. The imaginary frequency is introduced to avoid the conventional singularity of the propagator "on the mass shell", i.e. at $z = E(p)$. For noninteracting particles the propagator has the form

$$G = \frac{1}{ip_0 - E(p)} \; . \tag{3}$$

Obviously there is still a singularity: On the 2D hypersurface $(p_0 = 0, p = p_F)$ in the 4-dimensional space $(p_0, \mathbf{p})$ the propagator is not well defined. This singularity is stable, i.e. cannot be eliminated by small perturbations. The reason is that the phase $\Phi$ of the Green's funtion $G = |G|e^{i\Phi}$ changes by $2\pi$ around the path embracing this 2D hypersurface in the 4D-space (see Fig. 1). The phase winding number $N = 1$ cannot change continuously, that is why it is robust towards any perturbation. Thus the singularity of the Green's funtion on the 2D-surface in the momentum space is preserved, even when interactions between particles are introduced.

Exactly the same topological conservation of the winding number leads to the stability of the quantized vortex in superfluids and superconductors, the only difference being that, in the case of vortices, the phase winding occurs in the real space, instead of the momentum space. The complex order parameter $\Psi = |\Psi|e^{i\Phi}$ changes by $2\pi N$ around the path embracing vortex line in 3D space or vortex sheet in 3+1 space-time. The connection between the space-time topology and the momentum space topology is, in fact, even deeper (see *e.g.* Ref.[7]). If the order parameter depends on space-time, the propagator in semiclassical aproximation depends both on 4-momentum and on space-time coordinates $G(p_0, \mathbf{p}, t, \mathbf{r})$. The topology in the 4+4 dimensional space describes: (i) The momentum space topology in the homogeneous system; (ii) The defects of the order parameter in space-time; and (iii) Topology of the energy spectrum within the topological defects [8].

In the more complicated cases, when the Green's function is the matrix with spin and band indices, the phase of the Green's function becomes

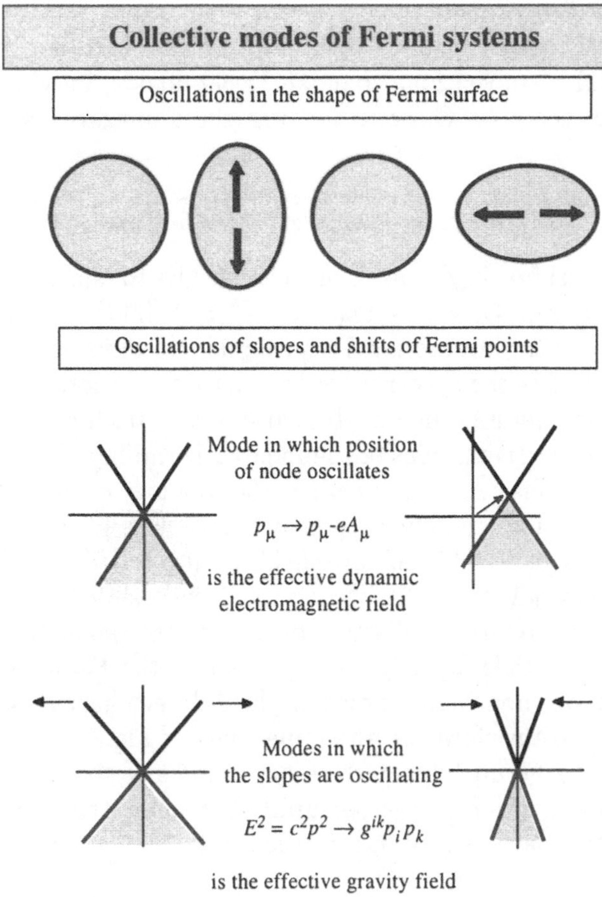


*Figure 2.* Collective modes of fermionic systems with Fermi surface and Fermi point.

meaningless. In this case one should use a general analytic expression for the integer momentum space topological invariant which is responsible for the stability of the Fermi surface:

$$N_1 = \mathbf{Tr} \oint_C \frac{dl}{2\pi i} \mathcal{G}(p_0, p) \partial_l \mathcal{G}^{-1}(p_0, p) \ . \tag{4}$$

Here the integral is taken over an arbitrary contour $C$ in the momentum space $(\mathbf{p}, p_0)$, which encloses the Fermi hypersurface; and $\mathbf{Tr}$ is the trace over the spin and band indices.

### 2.1.2. *Landau Fermi liquid*

The topological class of systems with Fermi surface is rather broad. In particular it contains conventional Landau Fermi-liquids, in which the propagator preserves the pole. Close to the pole the propagator is

$$G = \frac{Z}{ip_0 - v_F(p - p_F)} . \tag{5}$$

Evidently the residue $Z \neq 1$ does not change the topological invariant for the propagator, Eq.(4), which remains $N_1 = 1$. This is essential for the Landau theory of an interacting Fermi liquid; it confirms the assumption that in Fermi liquids the spectrum of quasiparticles at low energy is similar to that of particles in a Fermi gas. It is also important for the consideration of the bosonic collective modes of the Landau Fermi-liquid. The interaction between the fermions cannot not change the topology of the fermionic spectrum, but it produces the effective field acting on a given particle by the other moving particles. This effective field cannot destroy the Fermi surface owing to its topological stability, but it can locally shift the position of the Fermi surface. Therefore a collective motion of the particles is seen by an individual quasiparticle as a dynamical mode of the Fermi surface. These bosonic oscillative modes are known as the different harmonics of the zero sound. An example is shown in the upper part of Fig.2.

Note that the Fermi hypersurface exists for any spatial dimension. In the 2+1 dimension the Fermi hypersurface is a line – the vortex line in the 2+1 momentum space as is shown in Fig. 1(a).

### 2.1.3. *Non-Landau Fermi liquids.*

In the 1+1 dimension, the Green's function looses its pole but nevertheless the Fermi surface is still there [9, 10]. However the Landau Fermi liquid transforms to another states within the same topological class with given $N_1$. An example is provided by the Luttinger liquid. Close to the Fermi surface the Green's function for the Luttinger liquid can be approximated as (see [11, 9, 12])

$$G(z,p) \sim (ip_0 - v_1\tilde{p})^{\frac{g-1}{2}} (ip_0 + v_1\tilde{p})^{\frac{g}{2}} (ip_0 - v_2\tilde{p})^{\frac{g-1}{2}} (ip_0 + v_2\tilde{p})^{\frac{g}{2}} \tag{6}$$

where $v_1$ and $v_2$ correspond to Fermi velocities of spinons and holons and $\tilde{p} = p - p_F$. The above equation is not exact but reproduces the momentum space topology of the Green's function in Luttinger Fermi liquid. If $g \neq 0$ and $v_1 \neq v_2$, the singularity in the $(\tilde{p}, z = ip_0)$ momentum space occurs on the Fermi surface, i.e. at $(p_0 = 0,\ \tilde{p} = 0)$. The momentum space topological invariant in Eq.(4) remains the same $N_1 = 1$, as for the conventional Landau

Fermi-liquid. The difference from Landau Fermi liquid occurs only at real frequency $z$: The quasiparticle pole is absent and one has the branch cut singularities instead of the mass shell, so that the quasiparticles are not well defined. The population of the particles has no jump on the Fermi surface, but has a power-law singularity in the derivative [10].

Another example of the non-Landau Fermi liquid is the Fermi liquid with exponential behavior of the residue [13]. It also has the Fermi surface with the same topological invariant, but the singularity at the Fermi surface is exponentially weak.

2.1.4. *Superconducting transition: from pole to zero*
In this section we do not consider the bifurcations which lead to the overall reconstruction of the particle spectrum, which occur, say, during the transition to superconducting state. We assume that the temperature, though is small compared to the "Planck" scale $E_{\rm P} = \mu$, is still above the transition to superconducting state: $T_c < T \ll E_{\rm P}$.

But even in the superconducting region $T < T_c$ the winding number for the conventional Green's function is preserved: the invariant in Eq.(4) is again $N_1 = 1$. Only instead of pole at $p = p_F$ in the Green's function of the normal state, one has zero at $p = p_F$ in the superconducting state:

$$G = -\frac{ip_0 + M(p)}{p_0^2 + M^2(p) + \Delta^2} \;,\; M(\mathbf{p}) = \frac{p^2}{2m} - \mu \approx v_F(p - p_F) \;. \tag{7}$$

Here $\Delta$ is the gap in the quasiparticle spectrum. This demonstrates the stability of the singularity in the Green's function, though the Fermi surface as the manifold of zeroes in the energy spectrum disappears in superconductors.

## 2.2. FULLY GAPPED SYSTEMS

Although the systems we have discussed contain fermionic and bosonic quantum fields, this is not the relativistic quantum field theory which we need for the simulation of quantum vacuum: There is no Lorentz invariance and the oscillations of the Fermi surface do not resemble the gauge field even remotely. The situation is somewhat better for superfluids and superconductors with fully gapped spectra. Examples, which provide useful analogies with Dirac fermions and spontaneously broken symmetry in quantum fields, are conventional superconductors [14] and superfluid $^3$He-B [15]. In $^3$He-B the Hamiltonian of free Bogoliubov quasiparticles is the $4\times4$ matrix:

$$\mathcal{H} = \begin{pmatrix} M(\mathbf{p}) & c\vec{\sigma}\cdot\mathbf{p} \\ c\vec{\sigma}\cdot\mathbf{p} & -M(\mathbf{p}) \end{pmatrix} = \tau_3 M(\mathbf{p}) + c\tau_1\vec{\sigma}\cdot\mathbf{p} \;, \tag{8}$$

$$c = \frac{\Delta}{p_F} \quad , \quad \mathcal{H}^2 = E^2 = M^2(\mathbf{p}) + \Delta^2 \,, \tag{9}$$

where the Pauli $2 \times 2$ matrices $\vec{\sigma}$ describe the conventional spin of fermions and $2 \times 2$ matrices $\vec{\tau}$ describe the Bogoliubov isospin in the particle-hole space. The Bogoliubov Hamiltonian asymptotically approaches the Dirac one in the low $p$ limit. This is however is not the low energy limit, which for the typical $^3$He-B parameters occurs for $p$ close to $p_F$. The low $p$ and low $E$ limits coincide at large negative $\mu$, which is not the case in $^3$He-B. Nevertheless the $^3$He-B also serves as a working model system for simulations of many phenomena in particle physics and cosmology including experimental verification [16] of the Kibble mechanism describing formation of cosmic strings in the early Universe [17].

Note that even for the fully gapped systems there can exist the momentum-space topological invariants, which characterize the vacuum states. Typically this occurs in 2+1 systems, e.g. in the 2D electron system exhibiting quantum Hall effect [18]; in thin film of $^3$He-A (see [19] and Sec.9 of Ref.[20]); and in the 2D superconductors with broken time reversal symmetry [21]. These invariants are responsible for quantization of physical parameters, such as Hall conductance and spin Hall conductance. The quantum (Lifshitz) transition between the states with different values of momentum-space topological invariant occurs through the intermediate gapless regime (momentum-space instanton)[20].

### 2.3. FERMI POINT

#### 2.3.1. *Systems with Fermi points: $^3$He-A and Standard Model*

Now we proceed to the universality class which most fully exhibits the fundamental properties needed for a realization of the relativistic quantum fields, analogous to those in particle physics and gravity.

It is the class of systems, whose representatives are superfluid $^3$He-A and the vacuum of relativistic left-handed and right-handed chiral fermions (another example of this class in condensed matter has been discussed for gapless semiconductors [22]). This universality class is characterized by points in the momentum space where the (quasi)particle energy is zero. In particle physics the energy spectrum $E(\mathbf{p}) = cp$ is characteristic of the massless left neutrino (or any other chiral lepton or quark in the Standard Model above the elctroweak transiton) with $c$ being the speed of light. The energy of a neutrino is zero at point $\mathbf{p} = 0$ in the 3D momentum space. The Hamiltonian for the chiral neutrino – the massless spin-1/2 particle – is a $2 \times 2$ matrix

$$\mathcal{H} = \pm c\vec{\sigma} \cdot \mathbf{p} \tag{10}$$

which is expressed in terms of the Pauli spin matrices $\vec{\sigma}$. The sign + is for a right-handed particle and − for a left-handed one: the spin of the particle is oriented along or opposite to its momentum, respectively.

The Bogoliubov matrix for the $^3$He-A fermions is

$$\mathcal{H} = \tau_3 M(\mathbf{p}) + c(\vec{\sigma}\cdot\hat{d})(\tau_1 \mathbf{e}^{(1)}\cdot\mathbf{p} - \tau_2 \mathbf{e}^{(2)}\cdot\mathbf{p}) , \quad (11)$$

where $\mathbf{e}^{(1)}$ and $\mathbf{e}^{(2)}$ are real vectors, which in equilibrium or ground state are unit orthogonal vectors. The energy of the fermions

$$E^2_{\mathbf{p}} = M^2(\mathbf{p}) + c^2(\mathbf{p}\times\hat{\mathbf{l}})^2 , \ \hat{\mathbf{l}} \equiv \mathbf{e}^{(3)} = \mathbf{e}^{(1)}\times\mathbf{e}^{(2)} , \quad (12)$$

is zero at 2 points, at $\mathbf{p} = \pm p_F \hat{\mathbf{l}}$.

### 2.3.2. *Topological invariant for Fermi point*

Let us show that point zeroes, Fermi points, in the spectrum of the chiral fermions and in the spectrum of $^3$He-A fermions are topologically stable and described by the same momentum space topological invariant. Let us again consider the propagator of the particle $\mathcal{G} = (ip_0 - \mathcal{H})^{-1}$ on the imaginary frequency axis, $z = ip_0$. One can see that this propagator still has a singularity, which is now not on the surface but at the points in the 4D momentum space: $(p_0 = 0, \mathbf{p} = 0)$ and $(p_0 = 0, \pm p_F\hat{\mathbf{l}})$ for relativistic chiral fermions and the $^3$He-A fermions respectively. An integer topological invariant which supports the stability of Fermi points is expressed in terms of the propagator in the following way [20]:

$$N_3 = \frac{1}{24\pi^2} e_{\mu\nu\lambda\gamma}\ \mathbf{tr} \int_\sigma dS^\gamma\ \mathcal{G}\partial_{p_\mu}\mathcal{G}^{-1}\mathcal{G}\partial_{p_\nu}\mathcal{G}^{-1}\mathcal{G}\partial_{p_\lambda}\mathcal{G}^{-1} , \quad (13)$$

where $\sigma$ is the 3-dimensional surface embracing the Fermi point in the 4-momentum space.

For the chiral fermions in Eq.(10) this invariant is $N_3 = \pm 1$, where the sign is determined by the chirality of the fermion. The meaning of this momentum space topological invariant can be easily visualized. Let us consider the spin of the particle $\mathbf{s}(\mathbf{p})$ as a function of the particle momentum $\mathbf{p}$ in the 3D-space $\mathbf{p} = (p_x, p_y.p_z)$. For right-handed particles $\mathbf{s}(\mathbf{p}) = \mathbf{p}/2p$, while for left-handed ones $\mathbf{s}(\mathbf{p}) = -\mathbf{p}/2p$. In both cases the spin distribution in the momentum space looks like a hedgehog (see Fig. 1(b)), whose spines are represented by spins: spines point outward for the right-handed particle and inward for the left-handed one. In the 3D-space the hedgehog is topologically stable.

### 2.3.3. *Spin from isospin.*

For the $^3$He-A fermions in Eq.(11) the momentum space topological invariant $N_3 = \mp 2$, where the sign is determined by the position of the node,

$\mathbf{p} = \pm p_F \hat{\mathbf{l}}$. The topological invariant is twice larger than that for neutrino because of the double degeneracy of the Fermi point: there are two projections of the conventional spin of the $^3$He atom. For each projection of spin one has $N_3 = \mp 1$.

Note that the Bogoliubov spin $\vec{\tau}$ in $^3$He-A plays the same role as the conventional spin $\vec{\sigma}$ of chiral fermions in Eq.(10). On the other hand the conventional spin of the $^3$He atom is responsible for the degeneracy, but not for chirality, and thus plays the part of the isospin (see also Sec.3.8). This means that the origin of the spin responsible for the chirality of the (quasi)particle is fully determined by the matrix structure of the Fermi point. In this sense there is in principle no difference between spin and isospin: changing the matrix structure one can convert isospin to spin without changing of the topological charge of the Fermi point.

## 2.4. FERMI LINE

### 2.4.1. *Superconductivity in cuprates.*

The high-temperature superconductors in cuprates most probably contain zeroes in their quasiparticle energy spectrum. The ARPES experiments [23] show that these are four lines in the 3D momentum space where the quasiparticle energy is zero (or, equivalently, there are four point zeroes in the 2D momentum space relevant for the 2D $CuO_2$ planes). The high-T superconductors thus belong to class of systems with Fermi lines: the dimension $D$ of the manifold of zeroes is 1, which is intermediate between a Fermi surface with $D = 2$ and a Fermi point with $D = 0$. The system belongs to the same class as 2+1 QFT with massless relativistic fermions. The energy spectrum of quasiparticles near each of the 4 gap nodes can be written as

$$\mathcal{H} = \tau_1 c^x (p_x - p_x^{(0)}) + \tau_3 c^y (p_y - p_y^{(0)}) \ . \tag{14}$$

The "speeds" of light $c^x$ and $c^y$ are the "fundamental" characteristics determined by the microscopic physics of the cuprates.

### 2.4.2. *Scaling law near zeroes.*

As in the two universality classes of fermionic systems with the gapless quasiparticles discussed above, all low-energy (infrared) properties of cuprate superconductors are determined by zeroes. In particular, the density of the fermionic states is determined by the dimension of the zeroes:

$$N(E) = \sum_{\mathbf{p}} \delta(E - E(\mathbf{p})) \sim E^{2-D} \ . \tag{15}$$

Many low-temperature properties of these superconductors are obtained from a simple scaling arguments. For example, an external magnetic field

$B$ has dimension of $E^2$ and thus of $T^2$. At finite $B$, the density of states is nonzero even at $E = 0$. Substituting $B \sim E^2$ to Eq.(15) one obtains $N(0, B) \sim B^{(2-D)/2}$ and the following scaling law for the heat capacity:

$$C(T, B) = B^{(2-D)/2} T \ f(\frac{B}{T^2}) \ , \tag{16}$$

where $f$ is some function with the known asymptotes (see [24]). An experimental indication of such scaling with $D = 1$ was reported for $YBa_2Cu_3O_7$ in Ref. [25].

2.4.3. *Topological instability of Fermi line.*
Lines of zeroes do not belong to any universality class: there is no momentum space invariant, which can support the topological stability of the Fermi line. One can imagine the singular line in the momentum space from which the spines (now it is the vector $\vec{\tau}$) point outward (see Fig. 1(c)), but this singularity can be elmininated by the escape of the $\vec{\tau}$-vector to a third dimension. This can be accomplished by an operation similar to the folding of an umbrella (see Fig. 1(d)).

The existence of the Fermi lines can be prescribed, however, by the symmetry of the ground state. There are many nontrivial classes of superconductors, whose symmetry supports the existence of nodal lines insymmetric positions in the momentum space [26]. But the symmetry violating perturbations, such as impurities, an external magnetic field, etc., destroy the lines of zeroes [8]. Depending on the type of perturbation, the Fermi line evolves into different universality classes. Impurities, for example, can: (i) produce the gap in the fermionic spectrum [27] (see Fig.1(d)), which corresponds to appearance of mass for the 2+1 relativistic fermions; (ii) lead to the finite density of states [28], thus transforming the system to universality class of Fermi surfaces; (iii) produce zeroes of fractional dimension, which means that the exponent in the density of states $N(E) \propto E^{2-D}$ is non-integral [29] and thus corresponds to a fractional $D$ of the manifold of zeroes; and (iv) lead to localization [30]. Note that the univeraslity classes of the inhomogeneous systems are essentially richer. An open question is: Can the quantum fluctuations destroy the Fermi line, in particular, can they change the effective dimension of zeroes?

## 3. Properties of system with Fermi points

### 3.1. RELATIVISTIC MASSLESS CHIRAL FERMIONS

Close to the Fermi point $p_\mu^{(0)}$ in the 4D space one can expand the propagator in terms of the deviations from this Fermi point, $p_\mu - p_\mu^{(0)}$. If the Fermi point is not degenerate, the general form of the propagator is

$$\mathcal{G} = \tau^a e_a^\mu (p_\mu - p_\mu^{(0)}) \ . \tag{17}$$

Here we returned back from the imaginary frequency axis to the real energy, so that $z = E = -p_0$ instead of $z = ip_0$; and $\tau^a = (1, \vec{\tau})$. The quasiparticle spectrum $E(\mathbf{p})$ is given by the poles of the propagator:

$$g^{\mu\nu}(p_\mu - p_\mu^{(0)})(p_\nu - p_\nu^{(0)}) = 0 \ , \ g^{\mu\nu} = \eta^{ab} e_a^\mu e_b^\nu \ . \tag{18}$$

Thus in the vicinity of the Fermi point the massless quasiparticles are described by the Lorentzian metric $g^{\mu\nu}$. It is most important that this is the general form of the energy spectrum in the vicinity of any Fermi point, even if the underlying Fermi system is not Lorentz invariant; superfluid $^3$He-A is an example. The fermionic spectrum necessarily becomes Lorentz invariant near the Fermi point. If one applies this reasoning to our quantum vacuum, one may conclude that possibly the observed Lorentz invariance of the physical laws is not fundamental but a low-energy property resulting from the topologically stable Fermi points of the fermionic vacuum.

### 3.2. COLLECTIVE MODES – ELECTROMAGNETIC AND GRAVITATIONAL FIELDS

Let us consider the collective modes in such a system. The effective fields acting on a given particle due to interactions with other moving particles cannot destroy the Fermi point. That is why, under the inhomogeneous perturbation of the fermionic vacuum the general form of Eqs.(17-18) is preserved. However the perturbations lead to a local shift in the position of the Fermi point $p_\mu^{(0)}$ in momentum space and to a local variation of the vierbein $e_a^\mu$ (which in particular includes slopes of the energy spectrum (see Fig. 2). This means that the low-frequency collective modes in such Fermi liquids are the propagating collective oscillations of the positions of the Fermi point and of the slopes at the Fermi point. The former is felt by the right- or the left-handed quasiparticles as the dynamical gauge (electromagnetic) field, because the main effect of the electromagnetic field $A_\mu = (A_0, \mathbf{A})$ is just the dynamical change in the position of zero in the energy spectrum: in the simplest case $(E - eA_0)^2 = c^2(\mathbf{p} - e\mathbf{A})^2$.

The collective modes related to a local change of the vierbein $e_a^\mu$ correspond to the dynamical gravitational field. The quasiparticles feel the inverse tensor $g_{\mu\nu}$ as the metric of the effective space in which they move along the geodesic curves

$$ds^2 = g_{\mu\nu} dx^\mu dx^\nu \tag{19}$$

Therefore, the collective modes related to the slopes play the part of the gravity field (see Fig. 2).

Thus near the Fermi points the quasiparticles are the chiral massless fermions moving in the effective dynamical electromagnetic and gravitational fields.

### 3.3. GAUGE INVARIANCE AND GENERAL COVARIANCE

Another important property which results from the above equation is that the fermionic propagator in Eq.(17) is gauge invariant and even obeys the general covariance near the Fermi point. For example, the local phase transformation of the wave function of the fermion, $\Psi \to \Psi e^{ie\alpha(\mathbf{r},t)}$ can be compensated by the shift of the "electromagnetic" field $A_\mu \to A_\mu + \partial_\mu \alpha$. These attributes of the electromagnetic $(A_\mu)$ and gravitational $(g^{\mu\nu})$ fields also arise spontaneously as the low-energy phenomena.

Now let us discuss the dynamics of collective bosonic modes in this universality class, $A_\mu$ and $g^{\mu\nu}$. Since these are the effective fields their equations of motion do not necessarily obey gauge invariance and general covariance. However, in some special cases such symmetries can arise in the low energy corner. What are the conditions for that?

The effective action for the collective modes is obtained by integrating over the vacuum fluctuations of the fermionic field. This principle was used by Sakharov and by Zeldovich to obtain an effective gravity [31] and effective electrodynamics [32] respectively, both arising from fluctuations of the fermionic vacuum. If the main contribution to the effective action comes from the vacuum fermions whose momenta $\mathbf{p}$ are concentrated near the Fermi point, *i.e.* where the fermionic spectrum is linear and thus obeys the "Lorentz invariance", the result of the integration is necessarily invariant under gauge transformation, $A_\mu \to A_\mu + \partial_\mu \alpha$, and has a covariant form. The obtained effective action then gives the Maxwell equations for $A_\mu$ and the Einstein equations for $g_{\mu\nu}$, so that the propagating bosonic collective modes do represent the gauge bosons and gravitons.

Thus two requirements must be fulfilled – (i) the fermionic system has a Fermi point, and (ii) the main physics is concentrated near this Fermi point. In this case the system acquires at low energy all the properties of the modern quantum field theory: chiral fermions, quantum gauge fields, and gravity. All these ingredients are actually low-energy (infra-red) phenomena.

There is another important symmetry obeyed by massless relativistic Weyl fermions, the conformal invariance – the invariance under transformation $g_{\mu\nu} \to a(\mathbf{r},t) g_{\mu\nu}$. In the extreme limit when the vacuum fermions are dominatingly relativistic, the effective action for gravity must be conformly invariant. Such gravity, the so-called Weyl gravity, is a viable rival to Einstein gravity in modern cosmology[33, 34]: The Weyl gravity (i) can explain the galactic rotation curves without dark matter; (ii) it reproduces the Schwarzschild solution at small distances; (iii) it can solve the cosmological constant problem, since the cosmological constant is forbidden if the conformal invariance is strongly obeyed; etc. (see [35]).

### 3.4. $^3$HE-A: GAUGE INVARIANCE BUT NO GENERAL COVARIANCE

Let us consider what happens in a practical realization of systems with Fermi points in condensed matter – in $^3$He-A. Close to the gap nodes, *i.e.* at energies $E \ll \Delta$, where $\Delta$ is the maximal value of the gap in $^3$He-A which plays the part of the Planck energy, the quasiparticles obey the relativistic equation

$$g^{\mu\nu}(p_\mu - eA_\mu)(p_\nu - eA_\nu) = 0 \tag{20}$$

Here $e = \pm$ is the "electric charge" and simultaneously the chirality of the quasiparticles. Let us consider the simplest situation, when the $^3$He-A is in its vacuum manifold, which is characterized by two unit mutually orthogonal vectors $\mathbf{e}^{(1)}$ and $\mathbf{e}^{(2)}$, and first neglect the spin degeneracy of the Fermi point. When we consider the low-energy collective modes, these vectors are slowly changing in space-time. In this situation the effective metric and effective electromagnetic field are given by:

$$\mathbf{A} = p_F\hat{\mathbf{l}} \ , \ A_0 = p_F\mathbf{v}_s \cdot \hat{\mathbf{l}} \ , \tag{21}$$

$$g^{ik} = v_F^2(\delta^{ik} - \hat{l}^i\hat{l}^k) + c^2\hat{l}^i\hat{l}^k - v_s^i v_s^k \ ,$$
$$g^{00} = -1 \ , \ g^{0i} = v_s^i \ , \tag{22}$$

where $\mathbf{v}_s$ is the superfluid velocity given by $v_{si} = (\hbar/2m)\mathbf{e}^{(1)}\nabla_i\mathbf{e}^{(2)}$.

¿From the above equations it follows that the fields, which act on the "relativistic" quasiparticles as electromagnetic and gravitational fields, have very strange behavior. For example, the same texture of the $\hat{\mathbf{l}}$-vector is felt by quasiparticles as the effective magnetic field $\mathbf{B} = p_F\vec{\nabla} \times \hat{\mathbf{l}}$ according to Eq.(21) and simultaneously it enters the metric according to Eq.(22). Such field certainly cannot be described by the Maxwell and Einstein equations together.

Actually the gravitational and electromagnetic variables coincide in $^3$He-A only when we consider the vacuum manifold: Outside of this manifold they split. $^3$He-A, as any other fermionic system with Fermi point, has enough number of collective modes to provide the analogs for the independent gravitational and electromagnetic fields. But some of these modes correspond to deviation from the vacuum manifold and thus are massive in $^3$He-A. For example, the gravitational waves correspond to the modes, which are different from the oscillations of the $\hat{\mathbf{l}}$-vector. As distinct from photons (orbital waves – propagating oscillations of the $\hat{\mathbf{l}}$-vector) the graviton is massive [20].

All these troubles occur because in $^3$He-A the main contribution to the effective action for bosonic fields come from the vacuum fermions at the"Planck" energy scale, $E \sim \Delta$. These fermions are far from the Fermi points and thus their spectrum is nonlinear. That is why in general the effective action is not symmetric in $^3$He-A.

### 3.5. ZERO CHARGE EFFECT AND MAXWELL EQUATIONS

There are, however, exclusions. For example, the effective action for the $\hat{\mathbf{l}}$-field contains the term with the logarithmically divergent factor $\ln(\Delta/\omega)$ [20]. It comes from the zero charge effect – the polarization of the vacuum of the massless fermions, for whom the $\hat{\mathbf{l}}$-field represents the electromagnetic field. Due to its logarithmic divergence this term is dominanting at low frequency $\omega$: the lower the frequency the larger is the contribution of those vacuum which are close the Fermi point and thus the more symmetric is the effective action for the $\hat{\mathbf{l}}$-field. As a result, in the very low-energy limit, when the non-logarithmic contributions can be neglected, the effective action for the $A_\mu$ field becomes gauge invariant and even obeys the general covariance:

$$L = \frac{\sqrt{-g}}{24\pi^2} \ln\left(\frac{\Delta^2}{\omega^2}\right) g^{\mu\nu} g^{\alpha\beta} F_{\mu\alpha} F_{\nu\beta} \quad , \tag{23}$$

where $F_{\mu\nu}$ is the strength of the effective electromagnetic field $A_\mu$ from Eq.(21) and $g^{\mu\nu}$ is the effective gravitational field from Eq.(22). In this regime the $A_\mu$ field does obey the Maxwell equations in a curved space.

### 3.6. WHY $^3$HE-A IS NOT PERFECT

On the other hand the "Einstein" action for $g_{\mu\nu}$ is highly contaminated by many noncovariant terms, which come from the integration over the "nonrelativistic" high energy degrees of freedom at "Planck" scale. In this sense the $^3$He-A, with its given physical parameters, is not a perfect model for quantum vacuum. In "perfect" system, the integration must be spontaneously cut-off at energies much below the "Planck" scale, $E \ll \Delta$, say, due to strong quasiparticle relaxation.

The main reason why $^3$He-A is not a good substance, is that the Fermi point of the left particles, i.e. at $\mathbf{p} = +p_F\hat{\mathbf{l}}$, and the Fermi point of the right particles, i.e. at $\mathbf{p} = -p_F\hat{\mathbf{l}}$, are far from each orher in equilibrium. The"perfect" condensed matter would be where all the Fermi points are at the origin, at $\mathbf{p} = 0$, as it happens in the Standard Model.

On the other hand, if the fermionic system has an equal number of left-handed and right-handed fermions, then when all Fermi points are in the same position, the total momentum space topological invariant $N_3 = 0$. This situation occurs in the planar phase of superfluid $^3$He and in the Standard Model: if the righthanded neutrino is present then each generation contains 8 left-handed and 8 right-handed fermions. However in both cases the existence of the massless chiral fermions is protected by internal symmetry [36].

However, inspite of the absence of general covariance even in the low-energy corner, many different properties of the physical vacuum of the Standard Model, whose direct observation are still far from realization, can be simulated in $^3$He-A. One of them is the chiral anomaly.

### 3.7. CHIRAL ANOMALY

The chiral anomaly is the phenomenon which allows the nucleation of the fermionic charge from the vacuum [37, 38]. Such nucleation results from the spectral flow of the fermionic charge through the Fermi point to high energy. Since the flux in the momentum space is conserved, it can be equally calculated in the infrared or in the ultraviolet limits. In $^3$He-A it is much easier to use the infrared regime, where the fermions obey all the "relativistic" symmetries. As a result one obtains the same anomaly equation, which has been derived by Adler and by Bell and Jackiw for the relativistic systems. The rate of production of quasiparticle number $n = n_R + n_L$ from the vacuum in applied, say, hyperelectric and hypermagnetic fields is

$$\partial_\mu J^\mu = \frac{1}{8\pi^2}\left(\sum_R e_R^2 - \sum_L e_L^2\right) F^{\mu\nu} F^*_{\mu\nu} \ , \tag{24}$$

Here $n_R$ and $n_L$ is the density of the right and left (quasi)particles; $e_R$ and $e_L$ are their (hyper)charges; and $F^*_{\mu\nu}$ is the dual field strength. This equation has been verified in $^3$He-A experiments [39, 40], where the "hypermagnetic" $\mathbf{B} = p_F \vec{\nabla} \times \hat{\mathbf{l}}$ and "hyperelectric" $\mathbf{E} = p_F \partial_t \hat{\mathbf{l}}$ fields have been simulated by the space and time dependent $\hat{\mathbf{l}}$-texture. In particle physics the only evidence of axial anomaly is related to the decay of the neutral pion $\pi^0 \to 2\gamma$, although the anomaly is much used in different cosmological scenaria explaining an excess of matter over antimatter in the Universe (see review [41]).

### 3.8. DEGENERACY OF FERMI POINT AS THE ORIGIN OF THE NON-ABELIAN GAUGE FIELD

In $^3$He-A the Fermi point (say, at the north pole) is doubly degenerate owing to the ordinary spin $\vec{\sigma}$ of the $^3$He atom. This means that in equilibtium the two zeroes, each with the topological invariant $N_3 = -1$, are at the same point in momentum space. The collective motion of fermionic vacuum splits the Fermi points: positions of the two points can oscillate separately. Moreover, since the propagator is now the $4 \times 4$ matrix there must be cross terms. If one forgets about the degrees of freedom related to the vierbein, then the collective variables of the system with the doubly degenerate Fermi point enter the fermionic propagator as effective field $W^\alpha_\mu$

$$\mathcal{G}^{-1} = \tau^a e^\mu_a (p_\mu - eA_\mu - e\sigma_\alpha W^\alpha_\mu) \ , \tag{25}$$

which acts on the chiral quasiparticles as $SU(2)$ gauge field. Thus in this effective field theory the ordinary spin of the $^3$He atoms plays the part of the weak isospin [42, 20]. The effective "weak" field $W^\alpha_\mu$ is also the dynamical field, which in the leading logarithmic order obeys the Maxwell (actually Yang-Mills) equations. The higher symmetry groups can arise as a consequence of higher degeneracy of the Fermi points. For example, the 4-fold degeneracy of the Fermi point (say, 2 left-handed and 2 right-handed fermions) can produce the $SU(4)$ gauge group.

In addition, both in particle physics and in $^3$He-A the collective modes related to the shift of the 4-momentum can be discussed in terms of the "generalized covariant derivative" [43, 44]. In this theory the gauge fields, the Higgs fields, and Yukawa interactions, all are realized as shifts of positions of the degenerate Fermi points.

The vierbein can also oscillate differently for each of the two elementary Fermi points. This increases the number of possible collective modes. This is an interesting problem which must be investigated in detail. If the degenerate Fermi point mechanism has really some connection to the dynamical origin of the non-Abelian gauge fields the number of the fermionic species and the dimension of the symmetry group of the gauge fields must be related. In a naive approach the 16 fermions should lead to an extremely high symmetry group. That is why there should be some factors which can restrict the number of the gauge and other bosons. For example, the fermions can be composites of more elementary particles (preons, rishons, etc.); or there can be some special discrete symmetry between the fermions of the degenerate point, which restricts the number of massless bosonic collective modes. Another source of the reduction of the number of the effective fields has been found by Chadha and Nielsen [5]. They considered the massless electrodynamics with different metric (vierbein) for the left-handed and right-handed fermions. In this model the Lorentz invariance is violated. They found that the two metrics converge to a single one as the energy is lowered. Thus in the low-energy corner the Lorentz invariance becomes better and better, and at the same time the number of independent massless bosonic modes decreases.

Since the connections between QFT in the Standard Model and in $^3$He-A have been extensively reviewed [20, 42, 40], we concentrate here on some problems related to gravitational analogy.

## 4. Black hole in thin $^3$He-A film

### 4.1. GRAVITY BY MOTION OF SUPERFLUIDS. SONIC BLACK HOLE

As we have seen from Eq.(22) the gravitational field can be simulated in $^3$He-A by the motion of the liquid with the superfluid velocity $\mathbf{v}_s$ and by the $\hat{\mathbf{l}}$-texture [45]. The propagation of fermions in the moving quantum liquid obeys the same equation as propagation of relativistic particlesin the gravitational field. The same happens for the propagating sound waves in normal fluids [46, 47] and phonons in superfluid $^4$He. In the simplest case of the radial motion of superfluid $^4$He, the effective metric is expressed in terms of the radial superfluid velocity $v_s(r)$ as

$$ds^2 = -\left(c^2 - v_s^2(r)\right) dt^2 + 2v_s(r) dr dt + dr^2 + r^2 d\Omega^2 \ . \tag{26}$$

where the speed of light $c$ is played by speed of sound (phonon velocity). For $^3$He-A fermions the spherically symmetric metric occurs if $\hat{\mathbf{l}}$-field is radial. Then it follows from Eq.(22)

$$ds^2 = -\left(c^2 - v_s^2(r)\right) dt^2 + 2v_s(r) dr dt + \frac{c^2}{v_F^2} dr^2 + r^2 d\Omega^2 \ . \tag{27}$$

Kinetic energy of superflow plays the part of the gravitational potential: $\Phi = -v_s^2(r)/2$. If one chooses the velocity field to mimic the potential of the point body of mass $M$

$$v_s^2(r) = -2\Phi = \frac{2GM}{r} \equiv c^2 \frac{r_h}{r} \ , \tag{28}$$

one obtains the Painlevé-Gullstrand form of the Schwarzschild geometry (see e.g. ref.[47]). Here $r_h$ denotes the position of the event horizon, where the velocity reaches the "speed of light" $c$. If the fluid moves towards the origin, the low-energy quasipartilcles are trapped behind the horizon, since their speed $c$ with respect to the fluid is less than the velocity $v_s$ of the fluid.

Such a sonic black hole was first suggested by Unruh for ordinary liquid [46]. However since all the known normal liquids are classical, the most interesting quantum effects related to the horizon cannot be simulated in such flow. Moreover in such radial flow inward the liquid is accumulated at the origin, so that this sonic black hole cannot be stationary. In the other scenario a horizon, appears in moving soliton, if the velocity of the soliton exceeds the local "speed of light" [45]. This scenario has the same drawback: in finite system the motion of the soliton cannot be supported for a long time. In a draining bathtub geometry suggested in Ref.[48] the fluid motion can be made constant in time. However the friction of the liquid, which moves through the drain, is the main source of dissipation. The superfluidity of the liquid can help but the "superluminal" motion with respect to the boundaries is usually unstable and superfluidity collapses (see [49]). However there is a scenario, in which the collapse can be avoided. The superfluid motion becomes quasi-stationary and exhibits the event horizon; the life time of the flow state would be determined by the process inherent to the black hole – the Hawking radiation.

### 4.2. SIMULATION OF 2D BLACK HOLE

The stationary black hole can be realized in the following modification of the bathtub geometry of Ref.[48] (see Fig. 3(a)). The superfluid $^3$He-A film is moving towards the center of the disk, where it escapes to the third dimension due to the orifice (hole). If the thickness of the film is constant, the flow velocity increases towards the center as $v_s(r) = a/r$ and at $r = r_h = a/c$ reaches the speed of light (now $r$ denotes the radial coordinate in the cylindrical system). Outside the orifice the motion of the liquid is two dimensional and the effective metric for the low-energy Bogoliubov quasiparticles is

$$ds^2 = -\left(c^2 - v_s^2\right) dt^2 + 2v_s dr dt + dr^2 + r^2 d\phi^2 + \frac{c^2}{v_F^2} dz^2 . \qquad (29)$$

Here we took into account that the $\hat{l}$ vector in the film is fixed along the normal to the film. The "speed of light" for quasiparticles propagating along the film $c \sim 3$ cm/sec is much smaller than the Fermi velocity $v_F$ which corresponds to the "speed of light" for quasiparticles propagating along the normal to the film. Note that $c$ is much smaller than the speed of sound in $^3$He-A, that is why the motion of fluid has no effect on the density of the liquid.

The important element of the construction in Fig. 3 is that the moving superfluid $^3$He-A film is placed on the top of the superfluid $^4$He film. This is to avoid the interaction of the $^3$He-A film with the solid substrate. The superfluid $^4$He film effectively screens the interaction and thus prevents the collapse of the "superluminal" flow of $^3$He-A. Since the interactions with walls are removed, the local observer moving with the superfluid velocity cannot detect the relative motion with respect to the wall.

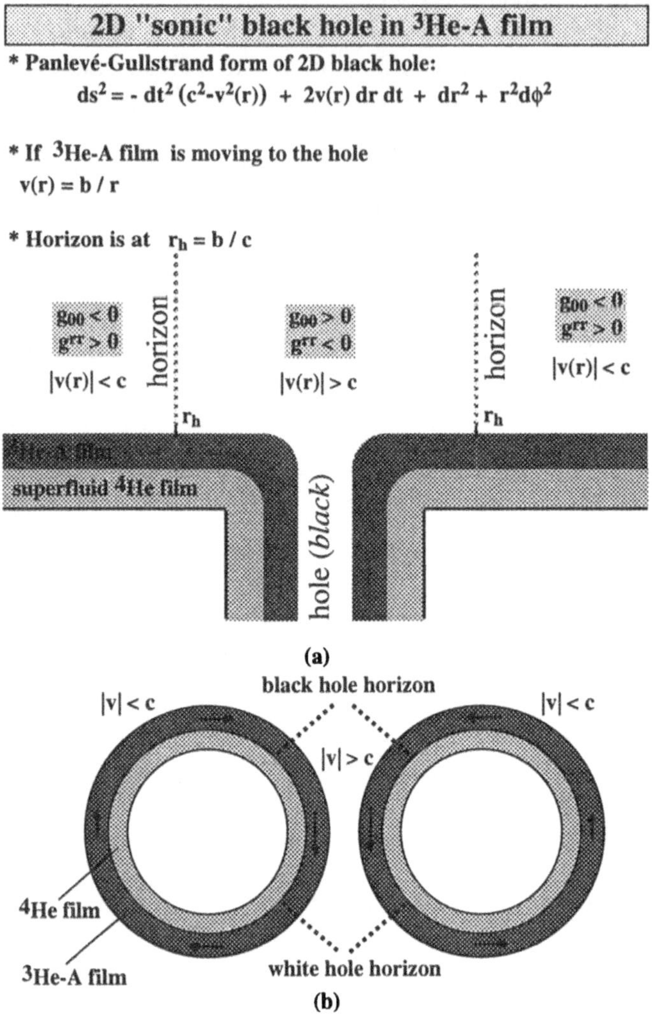


*Figure 3.* Similation of 2D black hole in thin $^3$He-A film. (a) Draining bathtub geometry. (b) $^3$He-A film circulating on the top of the $^4$He film on a torus.

Finally one can close the superflow by introducing the toroidal geometry in Fig. 3(b), so that the superfluid condensate can circulate. In this case in addition to the black hole horizon the white hole horizon appears on the path where the superfluid $^3$He-A flows out from the orifice.

Since the extrinsic mechanism of the friction of $^3$He-A film – the scattering of quasiparticles on the roughness of substrate – is abandoned, we can consider now intrinsic mechanisms of dissipation. The most interesting one is the Hawking radiation related to existence of a horizon.

### 4.3. VACUUM IN COMOVING AND REST FRAMES

Let us consider the simplest case of the 2D motion along the film in the bathtub geometry of Fig. 3(a). This can be easily generalized to the motion in the torus geometry.

There are two important reference frames:

(i) The frame of the observer, who is locally comoving with the superfluid vacuum. In this frame the local superfluid velocity is zero, $\mathbf{v}_s = 0$, so that the energy spectrum of the Bogoliubov fermions in the place of the observer is not Doppler shifted. For the quasiparicles with momentum $\mathbf{p} = p_F\hat{\mathbf{z}} + \mathbf{p}_\perp$ the spectrum is

$$E_{\rm com} = \pm c p_\perp \ . \tag{30}$$

Assuming further a pure 2D motion of the film we omit the index $_\perp$. Note that the speed of light $c$ coincides withthe Landau critical velocity of thesuperfl uid vacuum, $v_{\rm Landau} = \min(|E_{\rm com}(p)|/p) = c$.

The vacuum state as determined by the comoving observer is shown on Fig. 4(a): fermions occupy the negative energy levels in the Dirac sea (the states with the minus sign in Eq.(30)). It is the counterpart of the Minkowski vacuum. This vacuum however cannot be determined globally. Since the velocity of the comoving observer changes with time, the environment is time dependent for the observer, that is why the energy in this reference frame is not conserved.

(ii) The energy is conserved in the laboratory frame (the rest frame): in this frame the system is stationary and the effective metric does not depend on time. The energy in the rest frame is obtained from the local energy in the comoving frame by the Doppler shift:

$$E_{\rm rest} = \pm c p + \mathbf{p}\cdot\mathbf{v}_s(\mathbf{r}) \ . \tag{31}$$

In case of the radial superflow $\mathbf{v}_s(\mathbf{r}) = \hat{\mathbf{r}} v_s(r)$ one has

$$E_{\rm rest} = \pm c p + p_r v_s(r) \ . \tag{32}$$

In this frame the system is not static since there is a superflow and thus the metric contains the mixed component $g_{0i} = v_{si}$.

Figs. 4(b-c) show how the local "Minkowski" vacuum of the comoving frame is seen by the rest observer (note that the velocity is negative, $v_s(r) < 0$). In the absense of horizon, or outside the horizon the local vacuum does not change: The states which are occupied in the Minkowski vacuum have also the negative energy in the rest frame and thus belong to the rest frame vacuum (see Fig. 4(b)).

In the presense of a horizon behind which the velocity of superflow exceeds the Landau critical velocity the situation changes: Behind the horizon the vacuum in the rest frame differs from that in the comoving frame. Let us for simplicity consider the states with zero transverse momentum $p_\phi = 0$. If behind the horizon the system is in the Minkowski vacuum state (i.e. in the ground state as viewed by comoving observer), then quasiparticles on the positive sign branch have a reversed distribution when viewed in the rest frame: the negative energy states are empty, while the positive energy states are occupied (see Fig. 4(c)). Thus the distribution of quasiparticle on this branch as viewed by the rest observer behind horizon corresponds to the negative temperature $T = -0$.

The fermions can tunnel across the horizon from the occupied levels to the empty ones with the same energy. Thus if the system is initially in the Minkowski vacuum in the comoving frame, the tunneling disturbs this vacuum state. Pairs of excitations are created: the quasiparticle, say, is created outside the horizon while its partner – the quasihole – is created inside the horizon. This simulates the Hawking radiation from the black hole.

### 4.4. HAWKING RADIATION

To estimate the tunneling rate in the semiclassical approximation, let us consider the classical trajectories $p_r(r)$ of particles, say, with positive energy, $E_{\rm rest} > 0$, for the simplest case when the transverse momentum $p_\phi$ is zero, Fig. 5. The branch $E_{\rm rest} = (v_s(r) - c)p_r$ describes the incoming particles with $p_r < 0$ which propagate through the horizon to the orifice (or to the singularity at $r = 0$, if the orifice is infinitely small) without any pathology at the horizon. The classical trajectories of these particles are

$$p_r(r) = -\frac{E_{\rm rest}}{c - v_s(r)} < 0 \,. \tag{33}$$

The energies of these particles viewed by the comoving observer are also positive: $E_{\rm com}(r) = -cp_r(r) = E_{\rm rest}(1 - (v_s(r)/c))^{-1} > 0$.

Another branch $E_{\rm rest} = (v_s(r)+c)p_r$ in Fig. 5 contains two disconnected pieces describing particles propagating from the horizon in two opposite directions:

$$r > r_h \;:\; p_r(r) = \frac{E_{\rm rest}}{c + v_s(r)} \,, \quad E_{\rm com}(r) = cp_r(r) > 0 \tag{34}$$

$$r < r_h \;:\; p_r(r) = \frac{E_{\rm rest}}{c + v_s(r)} \,, \quad E_{\rm com}(r) = cp_r(r) < 0 \tag{35}$$

Eq.(34) describes the outgoing particles – the particles propagating from the horizon to the exterior. The energy of these particles is positive in both frames, comoving and rest. The Eq.(35) describes the propagation of

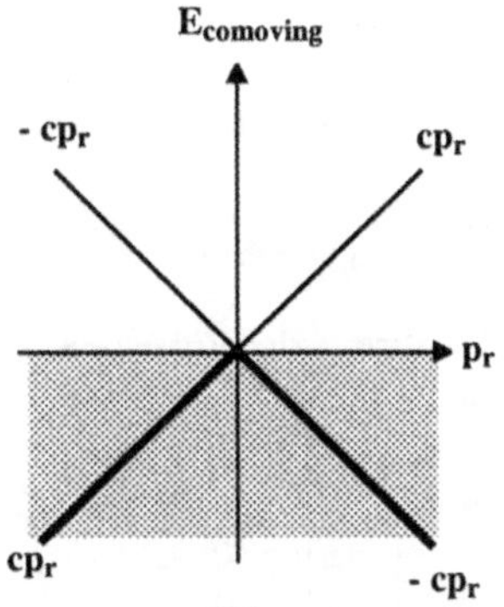


**(a)**
**Vacuum in the comoving frame**

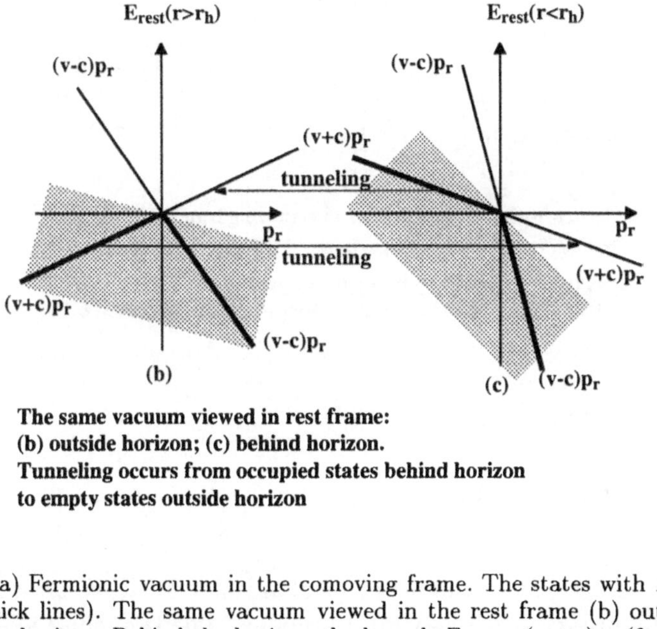


**The same vacuum viewed in rest frame:**
**(b) outside horizon; (c) behind horizon.**
**Tunneling occurs from occupied states behind horizon**
**to empty states outside horizon**

*Figure 4.* (a) Fermionic vacuum in the comoving frame. The states with $E_{\rm com} < 0$ are occupied (thick lines). The same vacuum viewed in the rest frame (b) outside horizon and (c) inside horizon. Behind the horizon the branch $E_{\rm rest} = (v + c)p_r$ (for $p_\phi = 0$) has inverse population as viewed in the rest frame: the states with positive energy $E_{\rm rest} > 0$ are filled, while the states with $E_{\rm rest} < 0$ are empty. The tunneling across horizon from the occupied states to the empty states with the same energy gives rise to the Hawking radiation from the horizon.

particles from the horizon to the orifice (or to the singularity). Though for the rest frame observer the energy of these particles is positive, for the comoving observer they belong to the Minkowski vacuum.

The classical trajectories in Eqs.(34,35) are thus disrupted at the hori-

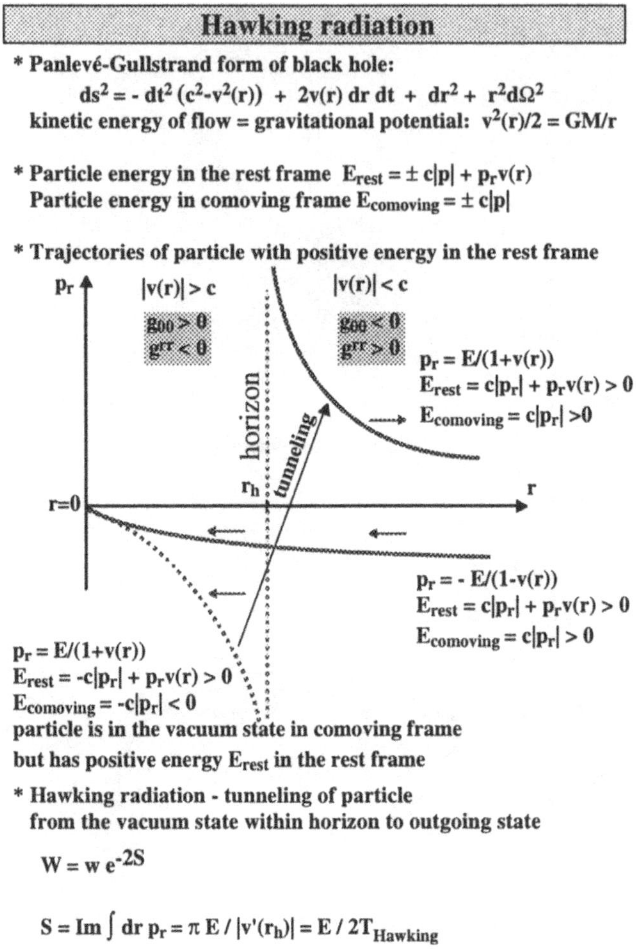


*Figure 5.* Tunneling from Minkowski vacuum within the horizon to the outgoing mode.

zon. There is however a quantum mechanical transition between the two pieces of the branch: the quantum tunneling. The tunneling amplitude can be found in semiclassical approximation by shifting the contour of integration to the complex plane:

$$w \sim \exp(-2S) \ , \tag{36}$$

$$S = \mathbf{Im} \int dr \ p_r(r) = \frac{\pi E_{\rm rest}}{|v_s'(r)|_{r=r_h}} \ . \tag{37}$$

This means that the wave function of the particle in the Minkowski vacuum inside the horizon contains an exponentially small part corresponding to the propagation from the horizon to infinity, and thus describing the radiation from the Minkowski vacuum in the presence of the event horizon. The exponential dependence of the probability on the quasiparticle energy $E_{\rm rest}$ suggests that this radiation looks like thermal one. The corresponding temperature

$$T_{\rm Hawking} = \frac{\hbar |v_s'(r)|_{r=r_h}}{2\pi} \tag{38}$$

corresponds to the Hawking temperature, since the derivative of the superfluid velocity at the horizon corresponds to the surface gravity according to Eq.(29).

The radiation leads to the quantum friction of superfluid, due to which the linear momentum of the flow decreases with time. In the long run the radiation exhausts the superfluid Minkowski vacuum to such an extent, that the superfluidity becomes destroyed somewhere between the horizons, and the phase slip event occurs, after which the number of circulation quanta of superfluid velocity trapped by the torus is reduced. This process will repeatedly continue until the two horizons finally merge.

### 4.5. DISCUSSION

The above construction in Fig. 3(b) allows us (at least in principle) to obtain the event horizon in the quasi-stationary regime, when the main source of nonstationarity is the dissipation coming from the Hawking radiation. As for the practical realization, there are, of course, many technical problems to be solved. On the other hand, if the black hole analog can in principle exist in condensed matter as the quasi-stationary object, its prototype – the real black hole – can also exist, at least, in principle (though it is not so easy to find the scenario of how this object can be obtained from the gravitational collapse of matter [50]).

## 5. Rotating vacuum

### 5.1. UNRUH EFFECT

Another manifestation of the unusual properties of quantum vacuum is the Unruh effect: The body moving in the vacuum with linear acceleration $a$ is believed to radiate the thermal spectrum with the Unruh temperature $T_U = \hbar a/2\pi c$ [51]. The comoving (uniformly accelerated) observer views the vacuum as a thermal bath with $T = T_U$, so that the matter of the body gets heated to $T_U$ (see references in [52]). It is difficult to observe the effect since the linear motion at constant proper acceleration (hyperbolic motion) leads to velocity arbitrarily close to the speed of light. On the other hand one can easily organize the uniform circular motion which has a constant centripetal acceleration (see the latest references in [53, 54, 55, 56]). The latter motion is stationary in the rotating frame, which is thus a convenient frame for study of the radiation and thermalization effects for uniformly rotating body.

## 5.2. ZEL'DOVICH-STAROBINSKY EFFECT

Zel'dovich [57] was the first who predicted that the rotating body (say, dielectric cylinder) amplifies those electromagnetic modes which satisfy the condition

$$\omega - L\Omega < 0 \; . \tag{39}$$

Here $\omega$ is the frequency of the mode, $L$ is its azimuthal quantum number, and $\Omega$ is the angular velocity of the rotating cylinder. This amplification of the incoming radiation is referred to as superradiance [58]. The other aspect of this phenomenon is that due to quantum effects, the cylinder rotating in quantum vacuum spontaneously emits the electromagnetic modes satisfying Eq.(39) [57]. The same occurs for any rotating body, including the rotating black hole [59], if the above condition is satisfied. When discussed in the rotating frame, the radiation occurs because in this frame the vacuum is not well defined at the distance $r > r_e = c/\Omega$ (in the so called ergoregion).

Distinct from the Unruh effect for the linearly accelerated body and the Hawking radiation, the radiation by a rotating body does not look thermal. Also, the rotating observer does not see the Minkowski vacuum as a thermal bath. This means that the matter of the body, though excited by interaction with the quantum fluctuations of the Minkowski vacuum, does not necessarily acquire an intrinsic temperature. Nevertheless, the problems related to the response of the quantum system in its ground state to rotation[53], such as radiation by the object rotating in vacuum [57, 60, 59, 58] and the vacuum instability caused by the existense of ergoregion [61], etc., are important and must be simulated in superfluids.

In superfluids the superfluid ground state plays the part of the quantum vacuum, and the spontaneous emission of quasiparticles leads to the rotational quantum friction [62].

## 5.3. CYLINDRICAL GEOMETRY

Let us consider a cylinder of radius $R$ rotating with angular velocity $\Omega$ in the (infinite) superfluid liquid. There are again two important reference frames.

The first one is the laboratory frame. Distinct from the previous Section, the energy of quasiparticles is not well determined in the laboratory frame: since the rotating body is never perfectly axysymmetric, its rotation always produces time dependent perturbations of the liquid. The vacuum can be determined only at distances far enough from the surface of the cylinder, where the influence of the rotating cylinder on the liquid is small. In this region the superfluid vacuum is in rest with respect to the laboratory frame, $\mathbf{v}_s = 0$. That is why the quasiparticle energy is that as in the comoving frame of previous Section. It is $E_{\rm com} = cp$ if we consider superfluid $^4$He, or the Eq.(30), if we consider a pure 2D motion of $^3$He-A with $\hat{\mathbf{l}} = \hat{\mathbf{z}}$. Such energy spectrum corresponds to the effective Minkowski metric of flat space

$$ds^2 = -c^2dt^2 + r^2d\phi^2 + dr^2 + a^2dz^2 \ . \tag{40}$$

Here $a = 1$ for isotropic $^4$He and $a = c/v_F$ for anisotropic $^3$He-A.

## 5.4. CONICAL TEXTURE WITH NEGATIVE ANGLE DEFICIT

In the 3D case of $^3$He-A there can be another effective metric far from the body. It can be caused by the normal boundary condition on the $\hat{\mathbf{l}}$-vector, which prefers the radial orientation of $\hat{\mathbf{l}}$-vector. In the case of the radial distribution, $\hat{\mathbf{l}} = \hat{\mathbf{r}}$, the effective metric for the quasiparticles moving outside the cylinder follows from Eq.(22):

$$ds^2 = -c^2dt^2 + dz^2 + r^2d\phi^2 + \frac{c^2}{v_F^2}dr^2 \ . \tag{41}$$

Such an effective space is conical: The space outside the cylinder is flat, but the proper length of the circumference of radius $r$ around the cylinder is not equal to $2\pi r$. In the relativistic theories such conical metric can arise outside the local cosmic strings, where there is an angle deficit. In our case the length of the circumference is $2\pi r v_F/c$, which is much larger than $2\pi r$: This effective space exhibits a "negative angle deficit" (for details see Ref.[63]).

## 5.5. ROTATING FRAME

Far outside the rotating body, the system is not disturbed by rotation and thus remains in or close to the Minkowski vacuum (or vacuum in the conical space in the case of radial $\hat{\mathbf{l}}$-vector) as viewed in the laboratory frame. However this does not hold in the region adjacent to the cylinder, where the superfluid velocity field is nonzero and time-dependent due to e.g. imperfections on the surface of the cylinder. Because of the time-dependent perturbations the vacuum in the laboratory frame is not well determined.

The energy is conserved and vacuum can be determined in the frame corotating with the cylinder. In the corotating frame the cylinder is at rest, and thus all the perturbations caused by rotation are stationary. The metric in the corotating frame is simplest far outside the rotating body, where the superfluid velocity in the corotating frame is $\mathbf{v}_s = -\vec{\Omega}\times\mathbf{r}$. Substituting this $\mathbf{v}_s$ into Eq.(22) one obtains the interval $ds^2 = g_{\mu\nu}dx^\mu dx^\nu$, which determines the propagation of quasiparticles in the corotating frame:

$$ds^2 = -(c^2 - \Omega^2 r^2)dt^2 - 2\Omega r^2 d\phi dt + r^2 d\phi^2 + dr^2 + a^2 dz^2 \ . \qquad (42)$$

Here again $a = 1$ for phonons in isotropic $^4$He and $a = c/v_F$ for fermionic quasiparticles in anisotropic $^3$He-A with $\hat{\mathbf{l}} = \hat{\mathbf{z}}$. These metrics correspond to the interval in the rotating frame discussed in relativistic theories.

It is convenient to write the quasiparticle spectrum in this frame in two different approximations. In the classical description one has $E_{\rm corotating} = E_{\rm com}(\mathbf{p}) + \mathbf{p}\cdot\mathbf{v}_s$. In the other description, an azimuthal motion of quasiparticles is quantized in terms of the angular momentum $L = rp_\phi$, while the radial one is still treated in the quasiclassical approximation. In this case the energy spectrum of phonons in the corotating frame is

$$E_{\rm corotating} = cp + \mathbf{p}\cdot\mathbf{v}_s = c\sqrt{\frac{L^2}{r^2} + p_z^2 + p_r^2} - \Omega L \ , \qquad (43)$$

and the energy spectrum of the Bogoliubov fermions

$$E_{\rm corotating}(\mathbf{p}) = \pm\sqrt{\frac{c^2}{r^2}L^2 + c^2 p_r^2 + v_F^2(p_z \mp p_F)^2} - \Omega L \ . \qquad (44)$$

### 5.6. ERGOREGION IN SUPERFLUIDS

The radius $r_e = c/\Omega$, where $g_{00} = 0$ in the Eq.(42), marks the position of the ergoplane. In the ergoregion, i.e. at $r > r_e = c/\Omega$, the energy of quasiparticle in Eq.(44) can be negative for any $L \geq 1$. This means that in the ergoregion the Minkowski vacuum is not the vacuum for the corotating observer. Situation is similar to the horizon discussed in previous Section. The difference is that the superfluid velocity is now along the boundary of the ergoregion, while in the case of the horizon the flow was across the boundary.

¿From the point of view of the corotating observer, the Minkowski vacuum is unstable towards the filling of the negative energy states in the ergoregion, which means that radiation of the quasiparticles from the rotating cylinder to the ergoregion is energetically favorable. However such radiation can only occur if there is an interaction between the superfluid Minkowski vacuum and the rotating object, which produces the nonzero matrix element. Let us estimate this matrix element.

Let us consider the slow rotations $\Omega R \ll c$. In this case the linear velocity of the cylinder at the surface of the cylinder, $\Omega R$, is much smaller than the Landau critical velocity for nucleation of quasiparticles, $v_{Landau} = c$. This means that quasiparticles cannot be nucleated near the surface of cylinder. On the other hand, the ergoregion, where $|v_s| = \Omega r > c$ and thus quasiparticles can be nucleated, is far from the cylinder, $r_e \gg R$. That is why the interaction with the cylinder, which produces the matrix element for the radiation, is very small. In this situation the most effective mechanism of the quasiparticle radiation is the tunneling of quasiparticles from the liquid ajacent to the surface of the rotating body to the ergoregion.

## 5.7. ROTATING DETECTOR

The part of the fermionic system, which is rigidly connected to the rotating body, plays the part of rotating detector. In superfluids the rotating body can be effectively substituted by the rigidly rotating cluster of quantized vortices (see Fig. 6a). Such clusters, that experience the solid-body rotation, are experimentally investigated in $^3$He-A and $^3$He-B (see e.g. [64]). The vortex cluster rotating in the infinite superfluid liquid at $T = 0$ is the quasiequilibrium state, similar to that of the black hole, whose rotation can be slowed down only by radiation. This quasiequilibrium state represents the vacuum state in the rotating frame. It is quasiequilibrium because this vacuum is not determined globally: far from the cluster it is substituted by the analogue of Minkowski vacuum.

Fig. 6(b) shows the distribution of superfluid velocity $v_s$ in the laboratory frame. Within the vortex cluster, i.e. at $r < R$, the superfluid velocity, being averaged over the vortices in the cluster, follows the velocity of the solid body rotation of the cluster: i.e. $< \mathbf{v}_s >= \vec{\Omega} \times \mathbf{r}$ in the laboratory frame and thus $< \mathbf{v}_s >= 0$ in the frame corotating with the cluster. Outside the cluster the superfluid velocity decays as $N\kappa/2\pi r$, where $N$ is number of vortices in the cluster and $\kappa$ is superfluid circulation around individual vortex.

Since $< \mathbf{v}_s >= 0$ in the frame of rotating detector, one can expect that the spectrum of quasiparticles, which live within the cluster, has no Doppler shift. In other words this spectrum has no $-\Omega L$ shift of the energy levels, as distinct from quasiparticle spectrum far from the body, Eqs.(43,44), which is Doppler shifted when measured in the same corotating frame. This appears to be true inspite of the fact that the equation $< \mathbf{v}_s >= 0$ does not imply that $\mathbf{v}_s = 0$ locally: superfluid velocity essentially depends on the position in the vortex lattice. The topology prescribes that there are rotation independent zero energy states in the vortex core – the fermion zero modes (see left part of Fig. 6(c)). The energy spectrum of fermion zero modes in the vortex core, has been first calculated by Caroli, de Gennes and Matricon for the Abrikosov vortex in $s$-wave superconductors [65]; on

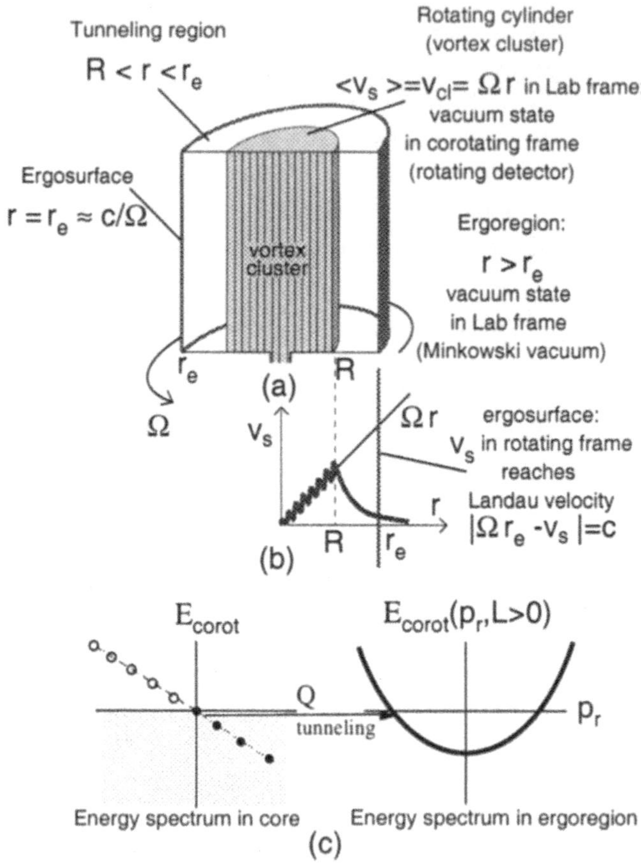


*Figure 6.* (a) Rotating body – rigidly rotating cluster of vortices. Within the cluster, at $r < R$, the average superfluid velocity equals the velocity $v_{cl}$ of the solid body rotation of the cluster. Superfluid state within the cluster plays the part of rotating detector, which is sopposed to be initially in the vacuum state as viewed in the corotating frame. Far outside of the cluster, where velocity of the superfluid is zero in the Laboratory frame, the system is assumed to be in Minkowski vacuum. (b) Distribution of the superfluid velocity in the laboratory frame. At $r = r_e \approx c/\Omega$ the superfluid velocity in the corotating frame reaches the "speed of light" – the Landau critical velocity $v_{Landau} = c$. At $r > r_e$ there is an ergoregion in the corotating frame, where the quasiparticle negative energy states are empty. (c) Tunneling of quasiparticles from the vacuum state of the rotating detector to the ergoregion. This produces radiation from the rotating body (from vortex cluster) and excitation of the rotating detector (excitations of fermion zero modes in the cores of vortices)

the relation between the number of fermion zero modes and the winding number of the vortex see [66].

## 5.8. RADIATION INTO THE ERGOREGION

The radiation of Bogoliubov quasiparticles can be considered as the process in which the particle tunnels from the zero energy state in the detector, $E_{\text{corotating}} = 0$, to the scattering state at the ergoplane, where its energy is also $E_{\text{corotating}} = 0$, but due to the $\Omega L$ shift(Fig. 6c). In the quasiclassical approximation the tunneling probability is $e^{-2S}$, where at $p_z = \pm p_F$ and $\Omega R \ll c$:

$$S = \mathrm{Im} \int dr\; p_r(r; E=0) = L \int_R^{r_e} dr \sqrt{\frac{1}{r^2} - \frac{1}{r_e^2}} \approx L \ln \frac{r_e}{R} \,. \tag{45}$$

The particles with all positive momenta $L > 0$ are radiated, but the radiation probability is smaller the higher $L$:

$$w \propto e^{-2S} = \left(\frac{R}{r_e}\right)^{2L} = \left(\frac{\Omega R}{c}\right)^{2L} = \left(\frac{\omega R}{cL}\right)^{2L} ,\; \Omega R \ll c\,. \tag{46}$$

Here $\omega = \Omega L$ is energy (frequency) of the radiated quasiparticles in the laboratory frame.

If $c$ is substituted by the speed of light, Eq.(46) is proportional to the superradiant amplification of the electromagnetic waves by rotating dielectric cylinder derived by Zel'dovich [58, 60].

Since each radiated fermion carries the angular momentum $L$, the vortex cluster rotating in superfluid at $T = 0$ is loosing its angular momentum and thus experiences the quantum rotational friction. The radation also leads to excitation of the detector matter.

## 5.9. DISCUSSION

The rotational quantum friction experienced by the body rotating in superfluid vacuum, is caused by the spontaneous quantum emission of the quasiparticles from the rotating object to the "Minkowski" vacuum into the ergoregion. The emission is not thermal and depends on the details of the interaction of the radiation with the rotating body. In the quasiclassical approximation it is mainly determined by the tunneling exponent, which can be approximately characterized by the effective temperature $T_{\text{eff}} \sim \hbar\Omega(2/\ln(c/\Omega R))$. The quantum friction can be observed only if the effective temperature is not very small compared to the temperature of the bulk superfluid, $T_{\text{eff}} \geq T$. For the body rotating with $\Omega = 10^3$rad/s, $T$ must be below $10^{-8}$K. However, the higher rotation velocity can be obtained for clusters. The pair of identical vortices rotate around their center of mass with $\Omega = \kappa/4\pi R^2$, where $R$ is the radius of the circular orbit. If the radius

$R$ is of order of superfluid coherence length, the effective temperature can reach $10^{-4}$K.

The process discussed in this Section occurs only if the ergoregion is present, i.e. the radius of the ergosurface is less then the radius of container, $r_e(\Omega) = c/\Omega < R_{\rm ext}$. On the instability of the ergoregion in quantum relativistic vacuum towards emission see also in Ref.[61].

If $r_e(\Omega) > R_{\rm ext}$ and the ergoregion is absent, the interaction between the coaxial cylinders via the vacuum fluctuations becomes the main mechanism for dissipation. Such quantum friction is similar to dynamic Casimir forces between the walls moving laterally in the quantum vacuum (see Review [67]). As in [67] the nonideality of the cylinders, i.e. violation of the rotational symmetry by the body, is the necessary condition for quantum friction.

## 6. Discussion

In the above examples of the nontrivial space, the effective gravitational field acts as a fixed external field. The dynamics of this field has not been discussed here. In most cases the effective gravity field does not obey the Einstein equations. This is the main drawback of superfluid $^3$He-A: Since the Fermi points in $^3$He-A are too far apart from each other, the dynamical equations for the gravitational field are not "relativistic". However many properties of the quantum vacuum in curved space, which are determined by the geometry, do not depend on the dynamical origin of the geometry. For example, it is well known that the Hawking radiation is a purely kinematic effect and occurs in any geometry, if it exhibits an event horizon [47]. That is why the $^3$He-A superfluid vacuum is a right object for simulation of many aspects of physics of quantum vacuum in a curved space. In particular it would be interesting to know how the breakdown of the Lorentz invariance at higher energy influences the Hawking radiation. The entropy of the black hole can be also investigated using the above model, since the microstates within the horizon are well determined and (at least in principle) are completely known in the whole energy range including the "transPlanckian" region. All these are different aspects of the problem of stability of the vacuum in strong gravitational and other fields. The superfluids provide us with many examples of the nonlinear behavior of the superfluid vacuum. Probably they tell us something on the quantum physical vacuum – the ether.

## References

1. Hu, B.L. (1988) Cosmology as 'Condensed Matter' Physics, *3rd Asia-Pacific Conf. Proceedings, Physics*, 1, 301-314.

2. Wilczek, F. (1998) The Future of Particle Physics as a Natural Science, *Int. J. Mod. Phys.*, **A 13**, 863-886.
3. Jegerlehner, F. (1999) The "Ether-world" and Elementary Particles, *Proceedings of 31st International Ahrenshoop Symposium on the Theory of Elementary Particles, Buckow, Germany, 1997*, 386–392.
4. Jackiw, R. (1998) Field theory: Why have some physisists abandon it? *Proc. Natl. Acad. Sci. USA*, **95**, 12776–12778.
5. Chadha, S. and Nielsen, H.B. (1983) Lorentz invariance as a Low-Energy Phenomenon, *Nucl. Phys.* **B 217**, 125–144.
6. Hu, B.L. (1996) General Relativity as Geometrohydrodynamics, Expanded version of an invited talk at 2nd International Sakharov Conference on Physics, Moscow, 20 - 23 May 1996, e-Print Archive: gr-qc/9607070.
7. Volovik, G.E. and Mineev, V.P. (1982) Current in Superfluid Fermi liquids and the Vortex Core Structure, *Sov. Phys. JETP* **56**, 579-586.
8. Grinevich, P.G. and Volovik, G.E. (1988) Topology of Gap Nodes in Superfluid $^3$He, *J. Low Temp. Phys.* **72**, 371-380.
9. Volovik, G.E. (1991) A New Class of Normal Fermi Liquids, *JETP Lett.***53**, 222-225.
10. Blagoev, K.B., Bedell, K.S. (1996) Luttinger Theorem in One Dimensional Metals, e-Print Archive: cond-mat/9611240.
11. Wen, X.G. (1990) Metallic non-Fermi-Liquid Fixed Point in Two and Higher Dimensions, *Phys. Rev.* **B 42**, 6623-6630.
12. Schulz, H.J., Cuniberti, G., and Pieri, P., Fermi liquids and Luttinger liquids, e-Print Archive: cond-mat/9807366.
13. Yakovenko, V.M. (1993) Metals in a High Magnetic Field: A Universality Class of Marginal Fermi liquid, *Phys. Rev.* **B 47**, 8851-8857.
14. Nambu, Y. and Jona-Lasinio, G. (1961) Dynamical model of elementary particles based on an analogy with superconductivity. I. *Phys. Rev.* **122**, 345-358; (1961) Dynamical model of elementary particles based on an analogy with superconductivity. II. **124**, 246-254.
15. Alford, M., Rajagopal, K., Wilczek, F. (1998) QCD at Finite Baryon Density: Nucleon Droplets and Color Superconductivity, *Phys. Lett.* **B 422**, 247-256; Wilczek, F. (1998) From Notes to Chords in QCD *Nucl. Phys.* **A 642**, 1-13.
16. Ruutu, V.M.H., Eltsov, V.B., Gill, A.J., Kibble, T.W.B., Krusius, M., Makhlin, Yu.G., Placais, B., Volovik, G.E. and Wen Xu, (1996) Vortex formation in neutron-irradiated superfluid $^3$He as an analogue of cosmological defect formation, *Nature* **382**, 334-336.
17. Kibble, T.W.B. (1976) Topology of cosmic domains and strings, *J. Phys.* **A 9**, 1387-1398.
18. Ishikawa, K. and T. Matsuyama, T. (1987) A microscopic theory of the quantum Hall effect *Nucl. Phys* **B 280**, 523-548.
19. Volovik, G.E. and Yakovenko, V.M. (1989) Fractional charge, spin and statistics of solitons in superfluid $^3$He film *J. Phys.: Cond. Matter* **1**, 5263-5274.
20. Volovik, G.E. (1992) *Exotic properties of superfluid $^3$He*, World Scientific, Singapore.
21. G. E. Volovik (1997) On edge states in superconductor with time inversion symmetry breaking, *JETP Lett.* **66** , 522-527.
22. Abrikosov, A.A. (1998) Quantum Magnetoresistance *Phys. Rev.* **B 58**, 2788-2794.
23. Ding, H., Norman, M.R., Campuzano, J.C., Randeria, M., Bellman, A.F., Yokoya, T., Takahashi, T., Mochiku, T., Kadowaki, K. (1996) Angle-resolved photoemission spectroscopy study of the superconducting gap anisotropy in $Bi_2Sr_2CaCu_2O_{8+x}$ *Phys. Rev.*, **B 54**, R9678-R9681.
24. Volovik, G.E. (1997) Fermionic entropy of the vortex state in *d*-wave superconductors, *JETP Lett.* **65** , 491-496.
25. Revaz, B., Genoud, J.-Y., Junod, A., Meumaier, K., Erb, A. and Walker, E. (1998)

*d*-Wave Scaling Relations in the Mixed-State Specific Heat of $YBa_2Cu_3O_7$ *Phys. Rev. Lett.*, **80**, 3364-3367.
26. Volovik, G.E. and Gor'kov, L.P. (1985) Superconductivity classes in the heavy fermion systems, *Sov. Phys. JETP*, **61**, 843-854.
27. Pokrovsky, S.V. and Pokrovsky, V.L. (1995) Energy gap Induced by Impurity Scattering: New Phase Transition, in Anisotropic Superconductors, *Phys. Rev. Lett.*, **75**, 1150-1153.
28. Sun, Y. & Maki, K. (1995) Transport Properties of *D*-wave Superconductors with Impurities, *Europhys. Lett.* **32**, 355.
29. Nersesyan, A.A., Tsvelik, A.M. and Wenger, F. (1994) Disorder Effects in Two-Dimensional *d*-wave Superconductors, *Phys. Rev. Lett.*, **72**, 2628-2631.
30. Lee, P. (1993) Localized States in a *d*-wave Superconductors, *Phys. Rev. Lett.* **71**, 1887-1890.
31. Sakharov, A.D. (1968) *Sov. Phys. Doklady* **12**, 1040-1041.
32. Zeldovich, Ya.B. (1967) Interpretation of Electrodynamics as a Consequence of Quantum Theory, *JETP Lett.* **6**, 345-347.
33. Mannheim, P.D. (1998) Implications of cosmic repulsion for gravitational theory, *Phys. Rev.* **D 58**, 1-12.
34. Edery, A. and Paranjape, M.B. (1998) Classical tests for Weyl gravity: Deflection of light and time delay, *Phys. Rev.* **D 58**, 1-8.
35. Mannheim, P.D. (1999) Cosmic acceleration and a natural solution to the cosmological constant problem, e-Print Archive: gr-qc/9903005.
36. Volovik G.E. (1999) Momentum-Space Topology of Standard Model, e-Print Archive: hep-ph/9907456.
37. Adler, S. (1969) Axial-vector vertex in spinor electrodynamics, *Phys. Rev.* **177**, 2426-2438.
38. Bell, J.S. and Jackiw, R. (1969) A PCAC Puzzle: $\pi_0 \to \gamma\gamma$ in the $\sigma$ Model, *Nuovo Cim.* **A 60**, 47-61.
39. Bevan, T.D.C., Manninen, A.J., Cook, J.B., Hook, J.R., Hall, H.E., Vachaspati, T., Volovik, G.E. (1997) Momentogenesis by $^3$He vortices: an experimental analogue of primordial baryogenesis, *Nature* **386**, 689-692.
40. Volovik, G.E. (1998) Axial anomaly in 3He-A: Simulation of baryogenesis and generation of primordial magnetic field in Manchester and Helsinki, *Physica* **B 255**, 86-107.
41. Trodden, M. (1999) Electroweak Baryogenesis, *Rev. Mod. Phys.* to be published in October 1999, e-Print Archive: hep-ph/9803479.
42. Volovik, G.E. and Vachaspati, T. (1996) Aspects of $^3$He and the standard electroweak model, *Int. J. Mod. Phys.* **B 10**, 471-521.
43. Martin, C.P., Gracia-Bondia, J.M., Varilly, J.S. (1998) The Standard Model as a Noncommutative Geometry: The Low-Energy Regime, *Phys. Rep.* **294**, 363-406.
44. Sogami, I.S. (1995) Generalized covariant derivative with gauge and Higgs fields in the Standard Model, *Prog. Theor. Phys.* **94**, 117-123; (1996) Minimal $SU(5)$ grand unified theory based on generalized covariant derivatives with gauge and Higgs fields, *Prog. Theor. Phys.* **95**, 637-655.
45. Jacobson, T.A. & Volovik, G.E. (1998) Event horizons and ergoregions in $^3$He *Phys. Rev.* **D 58**, 1-7; (1998) Effective spacetime and Hawking radiation from moving domain wall in thin film of $^3$He-A, *JETP Lett.* **68**, 874-880.
46. Unruh, W.G. (1981) Experimental Black-Hole Evaporation, *Phys. Rev. Lett.* **46**, 1351-1353.
47. Visser, M. (1999) Acoustic Black Holes, e-Print Archive: gr-qc/9901047, to appear in the Proceedings of the 1998 Peniscola Summer School on Particle Physics and Cosmology (Springer-Verlag).
48. M. Visser (1998) Acoustic black holes: horizons, ergospheres, and Hawking radiation, *Class. Quant. Grav.* **15**, 1767-1791.
49. Kopnin, N.B. and Volovik, G.E. (1998) Critical velocity and event horizon in pair-

correlated systems with "relativistic" fermionic quasiparticles, *JETP Lett.* **67**, 140-145.
50. Callan, C., Giddings, S., Harvey, J. and Strominger, A. (1992) Evanescent black holes, *Phys. Rev.* **D 45**, 1005-1009.
51. Unruh, W. G. (1976) Notes on black hole evaporation, *Phys. Rev.* **D 14**, 870-892.
52. Audretsch, J. and Müller, R. (1994) Spontaneous excitation of an accelerated atom: the contributions of vacuum fluctuations and radiation reaction, *Phys. Rev.* **A 50**, 1755-1763.
53. Davies, P.C.W., Dray, T. and Manogue, C.A. (1996) Detecting the rotating quantum vacuum, *Phys. Rev.* **D 53**, 4382-4387.
54. Leinaas, J.M. (1998) Accelerated Electrons and the Unruh Effect, e-Print Archive: hep-th/9804179.
55. Unruh, W.G. (1998) Acceleration Radiation for Orbiting Electrons, *Phys. Rept.* **307**, 163-171.
56. Barber, D., IV. Unruh effect, Spin Polarisation and the Derbenev-Kondratenko Formalism, e-Print Archive: physics/9901043.
57. Zel'dovich, Ya.B. (1971) Generation of Waves by a Rotating Body, *JETP Lett.*, **14**, 180-181.
58. Bekenstein, J.D. and Schiffer, M. (1998) The many faces of superradiance, *Phys.Rev.* **D 58**, 1-12.
59. Starobinskii, A.A. (1973) Amplification of Waves During Reflection from a Rotating "Black Hole", *Sov. Phys. JETP* **37**, 28-32 .
60. Zel'dovich, Ya.B. (1971) Amplification of Cylindrical Electromagnetic Waves Reflected from Rotating Body, *Sov. Phys. JETP* **35**, 1085-1087.
61. Kang, G. (1997) Quantum aspects of ergoregion instability, *Phys. Rev.* **D 55**, 7563.
62. Calogeracos, A. and Volovik, G.E. (1999) Rotational quantum friction in superfluids: Radiation from object rotating in superfluid vacuum, *JETP Lett.* **69**, 281–287.
63. G.E. Volovik (1998) Gravity of Monopole and String and Gravitational Constant in $^3$He-A, *JETP Lett.* **67**, 698-704.
64. Parts, Ü., Ruutu, V.M.H., Koivuniemi, J.H., Bunkov, Yu.M., Dmitriev, V.V. , Fogelström, M., Huebner, M., Kondo, Y., Kopnin, N.B., Korhonen, J.S., Krusius, M., Lounasmaa, O.V. , Soininen, P.I., Volovik, G.E. (1995) Single-vortex nucleation in rotating superfluid $^3$He-B, *Europhys. Lett.* **31**, 449-454.
65. Caroli, C., de Gennes, P.G., Matricon, J. (1964) Bound Fermion States on a Vortex Line in a Type II Superconductor, *Phys. Lett.* **9**, 307-309.
66. G.E. Volovik, (1993) Vortex motion in fermi superfluids and Callan-Harvey effect, *JETP Lett.* **57**, 244-248.
67. M. Kardar and R. Golestanian, (1999) The 'Friction' of Vacuum, and other Fluctuation-Induced Forces, *Reviews of Modern Physics*, **71**, 1233-1245.

# AUTHORS INDEX

# SUBJECT INDEX